AF439562

Maintenance Indices
Meaningful Measures
Of Equipment Performance

(2nd Discipline on World Class Maintenance Management)
Series 10

By Rolly Angeles

Maintenance Indices - Meaningful Measures of Equipment Performance
(2nd Discipline on World Class Maintenance Management)
Copyright © 2022 Rolly Angeles

Original Printing in the Philippines. All rights reserved.

No part of this publication and document may be reproduced, stored in a retrieval system or transmitted in any form or by any means . . . . . electronic, mechanical, photocopying, recording or otherwise . . . . . without the prior written notice from the author of this book.

10 9 8 7 6 5 4 3 2 1

First Edition: Printed in the Philippines by **Central Books**
Head Office: Phoenix Bldg. 927 Quezon Avenue, Quezon City, Philippines 1101

The National Library of Philippine Catalogue

Kindle Asin Amazon   B09RV4NM5Q
Paperback Amazon   979-8410173513
Hardcover Amazon   979-8412423807
Paperback Ingramspark   979-8885260039
Hardcover Ingramspark   979-8885260046

Published by:

Published by Amazon KDP for Kindle Version 2022
Published by Ingram Spark for Hardcover for International Edition 2022

This book was designed and produced by:

RSA Reliability and Maintenance Consultancy Firm
Sta. Rosa, Laguna, Philippines 4026
Website: http://www.rsareliability.com
Email: rollyangeles@rsareliability.com

First Printing: February 2022

Original Concept of World Class Maintenance Management
The 12 Disciplines (Series 1)

Figure A: Original Concept of World Class Maintenance

By: Rolly Angeles

<u>Figures</u>

Note: For simplification purposes, all pictures, graphs, charts, tables, drawings in this book will be referred to as figures, which are consecutively numbered below.

Table of Contents

<u>Acknowledgment</u>

First, I would like to thank all my students who have attended my training on Meaningful Measures of Equipment Performance. I always believe with all my heart that training is a two-way process. I have to admit that I also learned a great deal from the people I taught in the past. I hope that the knowledge gained from reading this book enlightens the reader in selecting the right maintenance indicators for their equipment and assets.

Special thanks to my wonderful family especially my three kids: Marie Vic, Kathleen Kay, and Christian Joseph, my wife, Marites, and my granddaughter Kalie.

I would like to mention my mom Cecilia Angeles who recently passed away this July 2021 for the love, care, values, moral support, and wisdom that this wonderful woman has provided me throughout my life. To me, she is the best mom and I could not ask for more. She also edited my first three books on World Class Maintenance Management, The 12 Disciplines, Maintenance – Roadmap to Reliability and Reliability – A Shared Responsibility for Both Operators and Maintenance.

Lastly, I would like to thank God Almighty for allowing me to complete this book. May he grant me the time and wisdom to complete the remainder of books that still need to be written which I hope to serve as my small way of contributing to the knowledge and wisdom to all maintenance mankind in industries.

Trust in the Lord with all your heart, and lean not on your own understanding, in all your ways, acknowledge Him, and He will make your path straight – Proverbs 3.5

<u>About the Author</u>

Rolly Angeles is a seasoned technical and international reliability and maintenance trainer and consultant. His portfolio of reliability and maintenance training includes maintenance management and reliability courses on Total Productive Maintenance (TPM), Planned Maintenance, Autonomous Maintenance, Lubrication Strategy, Tribology, Oil Contamination Control, Condition-Based, Predictive Maintenance, Reliability-Centered Maintenance (RCM), Root Cause Failure Analysis (RCFA), Planned Maintenance, World Class Maintenance Management (WCM), Proactive Maintenance, Maintenance Indices, KPI and more.

Rolly is a graduate of Mechanical Engineer from Mapua Institute of Technology in the Philippines, batch 1985, and passed the Licensure Board Examination the following year in 1986. With 30 years of solid experience, he had worked in various industries from shipping, woodworking, foundry, cast-iron machining, assembly lines, semiconductor manufacturing, and the mining industry. From 1994 to 2002, Rolly worked as a TPM senior engineer at Amkor Technology Philippines, a Multi-National company engaged in the manufacture of integrated circuit products, and spearheaded Amkor's Planned Maintenance Organization, composed of maintenance managers and engineers. He was responsible for the dramatic reduction of their machine's unplanned breakdowns in their TPM journey as well as RCM implementation on their facilities and utilities Air Handling Units (AHU) and their sub-station equipment. Here is where he had gained hands-on experience and understanding of both TPM and RCM, respectively. His last corporate employment was in 2002, where he worked as a technical training specialist at Lepanto Consolidated Mining Industry. In 2005, Rolly retired early from the industry and decided to establish his own consulting business, **RSA Reliability and Maintenance Consultancy Firm**, where he dedicates his time and passion to work as an independent reliability and maintenance consultant and provides in-house training, consultation, and facilitation to different industries on Maintenance Best Practices. Rolly Angeles can be reached through his email at rollyangeles@rsareliability.com and at his website http://www.rsareliability.com, where he writes his monthly reliability newsletter. Rolly has written the following books in a series that is all about his passion for Reliability and Maintenance.

• Volume 1: World Class Maintenance Management – The 12 Disciplines
• Volume 2: Maintenance – Roadmap to Reliability
• Volume 3: Reliability – A Shared Responsibility for Both Operators and Maintenance
• Volume 4: Cutting – Edge Maintenance Management Strategies
• Volume 5: Problems and Solutions on MRO Spare Parts and Storeroom
• Volume 6: Lubrication Tactics for Industries Made Simple
• Volume 7: Decoding Reliability-Centered Maintenance Process for Manufacturing Industries
• Volume 8: RSA Reliability and Maintenance Newsletter Vault Collection, Subscribers Edition
• Volume 9: Investigating Equipment Failures through Root Cause Failure Analysis
• Volume 10: Maintenance Indices – Meaningful Measures of Equipment Performance

<u>Take Quiz on Meaningful Measures Part 1</u>

1. MTBF stands for;
    a) Mean Time Between Failure
    b) Mean Time Before Failure
    c) Mean Time Backtracking Failure
    d) Mean Time for Breakdown and Failures

2. MTTR stands for;
    a) Mean Time to Respond
    b) Mean Time to Restore
    c) Mean Time to Repair
    d) Mean Time to React

3. In calculating MTTR's repair time, consideration of machine downtime only includes;
    a) Downtime due to repairs on breakdown
    b) Repairs on breakdown and assists
    c) Downtime in the event of short stoppages
    d) All machine downtime including set-up

4. Which of the following is not an equipment loss?
    a) Breakdown Losses
    b) Design Speed Loss
    c) Set-Up and Adjustment
    d) Facility Losses such as no air supply

5. When we speak about MTTF, we refer to;
    a) Repairable components
    b) Non-repairable components
    c) Removable components
    d) Non-removable components

6. When we speak of a failure or breakdown, there is always;
    a) A downtime involved
    b) A component or part involved
    c) An assist or error involved
    d) A saying that the lifespan had been reached

7. OEE stands for;
    a) Overall Equipment Effectiveness
    b) Overall Equipment Efficiency
    c) Operating Equipment Efficiency
    d) Operating Equipment Effectiveness

8. The best way to measure repair time will be;
    a) MTBF
    b) MTTR
    c) MTTS

   d) MTBA

9. MTBF Analysis was developed by;
   a) John Moubray of RCM
   b) Seiichi Nakajima
   c) US Military
   d) Edward Deming

10. Which of the following is a Planned Downtime?
   a) Breakdown Losses
   b) Start-Up Losses
   c) Operator's break time
   d) Set-up losses

11. Which of the following is not a breakdown?
   a) Run to Fail components
   b) Minor Stoppages
   c) Unscheduled replacement of parts
   d) Unscheduled repairs on equipment

12. Weibull data distribution analysis is all about
   a) Being effective at repairing failures
   b) Consequences of Breakdown
   c) Key Performance Indicators
   d) Life Data Analysis and Failure Forecasting

13 The Japanese term for Breakdown is;
   a) Kosho
   b) Chokotei
   c) Efu
   d) Fuguia

14. The best way to measure errors and assists is to perform;
   a) MTBF Snapshot
   b) MTBA Snapshot
   c) MTTR Snapshot
   d) MTTS Snapshot

15. The Japanese term for Minor Stoppages or shortstops in equipment is called;
   a) Chokotei
   b) Kosho
   c) Gemba
   d) Kaizen

16. The Sum of MTTF (Mean Time to Fail) and MTTR or (Mean Time to Repair) is;
   a) Failure Rate
   b) MTBF
   c) Availability
   d) Weibull Plot

17. The best method for depicting Life-Data Analysis will be;
    a) Failure Rate
    b) Lognormal Distribution
    c) Weibull Analysis
    d) Life Cycle Costing

18. Failure of a car's AM radio, air-conditioning unit, tail light in a motor vehicle is considered as what type of breakdown;
    a) Function-loss breakdown
    b) Function- Reduction Breakdown
    c) Primary Function Breakdown
    d) Secondary Function Breakdown

19. For industries without a CMMS or computerized software, the best method to track MTBA will be;
    a) Operator to record all the errors
    b) Prioritize the errors
    c) Perform an MTBA snapshot
    d) Just allow them to occur

20. The best way to improve Set-up time will be to use;
    a) Weibull Analysis
    b) Adopt a CMMS
    c) Improve MTTR
    d) SMED Techniques

21. The best people to improve OEE will be;
    a) Autonomous Maintenance
    b) Quality Maintenance
    c) Focused Improvement
    d) Depends on the Equipment Loss

22. According to the author of this book, the most important measurement for the storeroom people is;
    a) Carrying Cost
    b) Holding Cost
    c) Inventory Accuracy
    d) Number of Line Items

23. According to Shige Shingo's SMED techniques on Set-up, the standard set-up time should be;
    a) 30 minutes
    b) 60 minutes
    c) 45 minutes
    d) 10 minutes

24. TPM's Autonomous Maintenance activities will greatly impact;
    a) Minor Stoppages

b) Breakdown
c) Start-up Failures
d) Design Speed Loss

25. The OEE goal for all industries is to achieve 85% to be considered Word-Class. This statement is;
a) True
b) False
c) 100% always true
d) Not always true

Take Quiz on Meaningful Measures Part 2

1. OEE is designed to measure both primary and secondary functions.
a) True
b) False

2. The MTBF of a component is also the same as its MTTF.
a) True
b) False

3. The reciprocal of MTBF is its Failure Rate.
a) True
b) False

4. According to Shigeo Shingo, the standard Setup time for manufacturing industries should be 45 minutes.
a) True
b) False

5. Trend of both MTTF and MTBF should be the higher the better.
a) True
b) False

6. Assists and errors greater than 6 minutes should not be considered as a breakdown.
a) True
b) False

7. MTBA measurement is highly recommended when there is a high frequency of breakdowns in the equipment.
a) True
b) False

8. For MTTR if the failure keeps on repeating itself, it is good to perform a thorough Root Cause Failure Analysis.
a) True
b) False

9. To perform an MTBA Study on the equipment, every single assist or error encountered by the equipment should be recorded by the operator 24 hours a day.
   a) True
   b) False

10. MTBF can be used to define the frequency of replacement for PM.
   a) True
   b) False

11. According to the author, having reliable equipment is addressing every single possible equipment loss on the equipment and not just breakdowns and failures.
   a) True
   b) False

12. All equipment failures and breakdowns will cause downtime.
   a) True
   b) False

13. Unscheduled repair and replacement of parts are considered as a breakdown.
   a) True
   b) False

14. One of Shigeo Shingo's milestones in his career is his contribution to reducing start-up losses on equipment.
   a) True
   b) False

15. Errors in automated handling equipment where workpiece flow stops operator resets and machine run is called Minor Stoppages.
   a) True
   b) False

16. The best way to track Minor Stoppages and assists is to perform MTBF snapshots on the equipment.
   a) True
   b) False

17. The frequency of Minor Stoppages are much greater than the number of breakdowns encountered
   a) True
   b) False

18. MTBF is highly recommended in determining the interval for Preventive Maintenance Overhauls and Parts Replacement.
   a) True
   b) False

19. The best way to eliminate failures in our equipment is to be efficient in performing repair and improve MTTR or repair time on our equipment.

a) True
b) False

20. The best TPM Pillar that will impact a reduction in Unplanned Breakdown will be the Kobetsu-Kaizen or Focused Improvement Group.
    a) True
    b) False

21. Minor Stoppages usually occur on equipment and machines that have a lot of mechanical parts.
    a) True
    b) False

22. All parts after consistent use will reach a point of wear and tear, hence replacing the part before it fails will inevitably restore back the equipment's reliability and function
    a) True
    b) False

23. The goal of improving design speed loss is to bridge the gap between the current operating speed and the design speed of the equipment. In the majority of cases, improving the design speed will be done through modification and redesign.
    a) True
    b) False

24. Function Reduction Breakdown is when a specific function or part had failed in the equipment, but the equipment is still capable of running
    a) True
    b) False

25. According to James Reasons and Alan Hobbs, most human errors on PM occur during the disassembly process during overhauls.
    a) True
    b) False

Refer to Appendix B for the answers

<u>Preface: Welcome to the World of Measurements</u>

Figure B: Quotes on Measurements

While there are many famous quotes on measuring performance, one of them is from the late Peter Drucker, who quotes, You can't manage something that you cannot control, and you cannot control something you cannot measure. This means that we will never know whether or not we are successful unless measurements are clearly established and tracked. It is quite difficult to recommend the exact maintenance indicators or Key Performance Indicators to the readers on which are probably the best indicators suited for your industry because industries differ in the way they operate and do business. However, there are common indicators that can be used to measure the performance of the maintenance function. What is important is that these KPIs and indicators should be known by all people from the organization, from the top to the bottom line shop floor workers. The problem with some industries is that their indicators are only known up to the level of the supervisory or superintendent. This should be deployed down to the shop floor and rank in file people. Certain key performance indicators and indices may apply to one plant but not exactly to the rest of the other plants. Experience tells me that there is no single or one indicator that will tell you everything. Maintenance should at least have a minimum of 5 indices or measurements. What is important is for your maintenance team to sit down and decide what indicators and measurements matter most that best suit your industry based on the plant's needs and requirements.

There is a wide selection of measurements and indicators for the maintenance function. We can measure the number of breakdowns, downtime, availability, mean time indicators, lubrication costs, set-up time, repair time, maintenance costs, MRO spare parts costs, and so on. Therefore, the industry must understand what measurements and indicators are needed that best suit their plant needs and requirements. It is unlikely for an oil and gas plant to measure OEE since it is difficult to quantify the quality rate of the oil unless we can determine the exact number of bad molecules in every drop of oil. Each industry should focus on measurements and indicators that are likely applicable to their needs. Indices from manufacturing industries will greatly differ in other sectors, such as power plants, distribution plants, and other assembly plants. For plants implementing TPM, the primary measure of their performance will be OEE or Overall Equipment Effectiveness. It is also difficult to provide a benchmark on what industry standard is for MTBF, cost, and so on because, again, industries are diversified. Having a consistent MTBF of 700 hours in a month for manufacturing will be more than satisfactory, but not for a power plant or distribution industry. This means that if we have an MTBF of 700 for all distribution poles, then the 20 hours means that the city is experiencing a black-out as there are 720 days in a month. While TPM books will say that having an 85% OEE is already considered world-class at its level. Perhaps that would be the case for a manufacturing plant, but I cannot say for a power plant, if 85% would be the world-class level for a power plant, then again 15 percent of the time, the city suffers a blackout. What is important in measuring these indicators will be trending over time. This means that if MTBF is being measured monthly and the trend is increasing, then maintenance is doing well.

As mentioned above, maintenance should understand that no single indicator or measurement will tell us the whole story. What is important is not to manipulate these measurements and indicators just to look good to management. The word key highlights those that are the most important to the maintenance organization. Here is a summary of what the chapters in this book include.

Chapter 1: An Introduction to Maintenance Indicators covers the most common reasons for measuring KPIs and indicators on maintenance. Every industry has some form of measures and indices that they tracked and monitored regularly on maintenance. They provide objectives, goals, and targets on these indices. These indices can either be lagging or leading, but what is important is selecting the right indices that will impact our maintenance performance and there is no single KPI or measurement that will tell us the whole story on our performance that is why the maintenance function should be measuring a minimum of 5 or more indicators.

Chapter 2: Different Types of Equipment Losses explains the different equipment losses that different types of industries experience. These equipment losses include breakdown losses, set-up losses, start-up losses, cutting-blade losses, shutdown losses, minor stoppages, design speed loss, quality defects, and rework losses. There are also losses where the industry is experiencing which may not be present in other industries, or maybe common to one equipment but not to all. Therefore what is important is that our KPIs and measures should be based on the equipment losses so we can find ways to improve them. Also covered in this chapter is a brief discussion about the other losses which includes manpower and energy losses.

Chapter 3: Overall Equipment Effectiveness explained how each of these losses revealed in Chapter 2 connects to one measurement called OEE, which is composed of three components, availability, performance efficiency, and quality rate. This means that if the equipment is suffering from either breakdown, set-up, start-up, or cutting blade loss, then the equipment availability will be affected. If the equipment has a lot of minor stoppages and reduced speed, then the performance rate will be affected, while if a lot of rework and rejects on the product is experienced, then the quality rate will be affected. One interesting topic discussed in this chapter is the many variations on calculating availability. Also, explain in this chapter is why achieving an 85% OEE is not always considered world-class. An important message in this chapter is that OEE is simply a measure of the primary function of the equipment. This means that every equipment has more than one function.

Chapter 4: Understanding MTBF, MTTF, Failure Rate, and Reliability: MTBF or Mean Time Between Failure is a very simple measurement but it is not as simple as we think it is because before we can measure this index, we need to be precise and crystal clear on what failures and breakdown are included. A quiz has been added to this chapter to test the reader on what will constitute a failure and not. Another important topic in this chapter is where can we use these indices as well as their limitations on when not to use MTBF. We also clarify the difference between MTBF and MTTF. We point out the reciprocal of MTBF which is the failure rate. Lastly, we also explained reliability and its formula. However, one important clarification on this chapter is about the term reliability which I also include as an indicator. Although its definition is about the probability that no failure will occur throughout a prescribed operating period, and its formula is based on the failure rate and time, having equipment with no failure does not really indicate the equipment is reliable since failure is just a part of a bigger problem called equipment losses as explained in Chapter 2. A deeper explanation of reliability is explained by a good friend of mine R. Keith Mobley which is extracted from the forward message of my book on Cutting-Edge Maintenance Management Strategies.

Chapter 5: MTTR, MTTS, and MTBA Explained: This chapter explains other Mean Time Indicators such as Mean Time to Repair, Mean Time to Set Up, and Mean Time Between Assists. A survey question and responses have been provided regarding MTTR which provides different opinions among maintenance on whether to include or not in the non-availability of spares in the MTTR calculation. Another highlight of this chapter is how do we track minor stoppages through MTBA which includes an actual case study. Lastly, we cover another Mean Indicator which is about MTTS which is used for non-dedicated equipment that can produce different products and how to reduce the set-up or conversion time using the techniques of Shigeo Shingo's SMED or Single Minute Exchange of Dies.

Chapter 6: Preventive Maintenance Indicators: This chapter examines the different indicators that can be used to measure our efforts on Preventive Maintenance such as PM Compliance, PM Effectiveness, Ratio of Preventive Maintenance versus Breakdown Maintenance, Maintenance Cost, Percentage of Maintenance Cost to RAV (Replacement Asset Value), Maintenance Backlog and Wrench Time. Also detailed in this chapter are some weaknesses of these measurements. Some useful measurements and indicators for Predictive Maintenance are likewise explained.

Chapter 7: MRO Spare Parts and Storeroom Indicators: Managing spare parts simply means how fast we can respond in acquiring the right part during the time when maintenance and operations needed them most. This means that a good MRO Spare Parts Management system ensures the right parts get to the right place at the right time. The goal of any MRO spare parts or materials management is to create a balance on minimizing the cost of spares inventory as well as providing all materials required to keep the plant operating. Although the goals may sound contradicting, it is not if we know and understands what to stock or not to stock in the storeroom. This chapter specifies the different indices and measurements for the storeroom and explains what the author thinks is the most important indicator for the storeroom.

Chapter 8: Indicators for Autonomous Maintenance Operators details the importance of operators in the reliability and maintenance journey and explains the reasons why. This chapter spells out the indicators and measurements operators that will be used when implementing the TPM Pillar of Autonomous Maintenance. Likewise explained in this chapter are some unique and simple indices to measure the effectiveness of conducting the Basic Equipment Condition which definitely contributes to the overall performance of the equipment. The irony of this chapter is that these simple measurements such as tracking abnormalities detected and corrected, detecting loose or missing bolts are the rare indicators that can be used as a measurement for the secondary functions of the equipment since most measurements and indicators are focused on its primary function.

Chapter 9: Having a Crystal Clear Direction on Maintenance divulges the importance of having a strategic planning to review the previous performance and plan the next year's performance. What seems to be the obstacles and deploy the strategies needed to overcome these hurdles. There are many so-called Giant Companies in the Fortune 500 in the past that no longer exist today and we do not want our industry to be one of them. Also explained in this chapter are the different people and their functions in maintenance, and the entire maintenance organization's strength depends upon its weakest link. Finally, we end this chapter by informing the reader that there is a way of knowing if your industry is in the right direction in achieving a World Class Maintenance level, but what is important is that we measure performance along the way.

Chapter 10: Strategies to Improve these Maintenance Indicators: Although there are many strategies to adapt and improve these indicators, we should not be too overwhelmed with so many improvements in the plant. This chapter covers different approaches and methods to improve some of these KPIs such as Preventive Maintenance Indices, Breakdown Rate, Machine Downtime, OEE, and Maintenance Cost

Chapter 11: FAQ, and TIPS, on Maintenance Indicators is a collection of Frequently Asked Questions on these Meaningful Measures of Performance. Some of these questions have been raised during my training on this subject matter. This chapter also provides some helpful tips on these maintenance indicators that can benefit industries and provide more accurate measurements.

Chapter 12: Conclusion on Meaningful Measures reflects on the wisdom that maintenance should be treated as an investment and that the people are the key to improving the plant's performance and indicators especially during this point of writing this book as the whole world are in a crisis due to this pandemic. This book is not just about the formulas but what do we include in them.

Chapter 1

An Introduction to Maintenance Indices

> *We can only improve if we measure what is important to us. What cannot be measured cannot be improve and manage. Remember, if we do not measure performance, then we are just another person with an opinion, and in the real world of doing maintenance, opinions don't last a lifetime.*

1.1: What Does It Take for Maintenance to Get There?

Every industry wants to achieve a level of World Class Maintenance, yet many are struggling with how it can be achieved or if this is really possible. I would say that it is if industries are dead serious about it. Borrowing the concept from the book of Stephen Covey, Principle-Centered Leadership, a 4 Step approach can be applied to answer this question.

1) Determine where we are?
2) Where do we want to go?
3) How will we get there?
4) How will we know we have arrived?

1st: Determine where we are? The first step will define the current industry's situation which we can use as a baseline and benchmark on where the entire organization's maintenance structure and workforce currently is. Make a selection of people and start with a survey or you can select a cross-selection of members and have them conduct an assessment to determine where your industry currently stands from the worst to the best. Knowing where we are is the starting point for any continuous improvement initiative and drive. Another important element in this start-up phase is to decide on the indicators and measurements for their equipment, assets, and people. This is like when a person wants to trim down or reduce their weight, the first thing to do is to measure themselves on a weighing scale. This will serve as the benchmark and the goal in mind is to reduce his weight over a given time.

2nd: Where do we want to go? Once maintenance people know where they currently stand, there should be a clear direction on where they want to go. This will be the direction in which the whole maintenance herd is going. It's important to know the destination and let the people know that this is where our entire maintenance workforce is headed forth in the future, which is similar to having a Vision and Mission. Maintenance should aim for a high or ideal vision so that the people will continuously improve and set a timeframe for achieving it. There should only be one direction and this should be made clear to all maintenance people at all levels of the organization.

3rd: How are we going to get there? Once we have a clear path on where maintenance wants to go, the next step is to develop a path or roadmap, which are the things to be done and accomplished. These will be the activities and steps to be done to implement each of the 12 disciplines on maintenance. Maintenance must understand that implementing these Disciplines is a long-term and not a short-term initiative. Generate a Master Plan for each of these Disciplines and have them reviewed regularly. These steps will determine how we can bridge the gap between where we are, and what we want to achieve on each of the 12 Disciplines on World Class Maintenance which I wrote in my first book in 2009.

Maintenance PM Master Plan

NO.		MASTER PLAN ACTIVITY		INTRO 2021 Q3	Q4	2022 Q1	Q2	Q3	Q4	2023 Q1	Q2	Q3	Q4	2024 Q1	Q2	Q3	Q4	STABILIZE 2025 Q1	Q2	Q3	Q4
1	ZERO BREAKSDOWN ACTIVITIES	PM 7 STEP JOURNEY																			
		Step's 0 : Preparatory Stage	PLAN			Machine are categorized as Rank A, B															
		- Machine Ranking	ACTUAL																		
		Step's 1-3	PLAN		Attain ZERO Breakdown for all Rank A and Rank B Machines																
		- Initial Cleaning, Restore, Standards	ACTUAL																		
		Step 4 - Corrective Maintenance	PLAN				Apply P-M Analysis on Recurring Breakdowns and Feedback to IFCA														
		- Countermeasure for Design Weakness	ACTUAL																		
		Step 5 - Preventive Maintenance	PLAN					Final Inspection Standards - Time Based Maintenance													
		- Periodic - Preventive Maintenance	ACTUAL																		
		Step 6 - Predictive Maintenance	PLAN					Utilize Condition Based Maintenance Instruments &													
		- Overall Audit and Diagnosis	ACTUAL																		
		Step 7	PLAN																		
		- Machine Ultimate Utilization	ACTUAL																		
2	MTCE CONTROL SYSTEM	MRO SPARE PARTS CONTROL	PLAN				Review and Improve Spare Parts Control and Utilization														
			ACTUAL																		
3		MAINTENANCE COST AND BUDGET CONTROL	PLAN			Review Maintenance Cost Control and Utilization															
			ACTUAL																		
4		MAINTENANCE INFORMATION MANAGEMENT & CONTROL SYSTEM	PLAN					Plan for CMMS and Automation													
			ACTUAL																		
5		MAINTENANCE WORK PLANNING AND MANAGEMENT	PLAN					Review PM System													
			ACTUAL																		
6	SUPPORT ACT	GUIDANCE AND SUPPORT FOR JISHU HOZEN	PLAN				PM Guidance and Support for Jishu Hozen Activities														
			ACTUAL																		
7		MAINTENANCE SKILLS ENHANCEMENT	PLAN																		
			ACTUAL																		
7		EVALUATION OF THE PLANNED MAINTENANCE ACTIVITIES	PLAN																		
			ACTUAL																		

Figure 1.1: Maintenance Master Plan

4th: How will we know we have arrived? The first and the last step is to define and measure these indices that not only tell us that we have arrived but whether or not we are headed forth in the right direction. Measurements should be defined at the beginning of implementation and must be monitored and reviewed regularly. The success or failure of all our efforts depends entirely on these measurements. This means that the first and the last step have something to do with these KPIs and measurements. If the industry has achieved its goals and objectives, then we set new targets, if not, we stop for a while and reflect on how we can clear these obstacles and challenge them.

1.2: The Right Reasons Why We Need to Measure Performance

These Maintenance performance indicators are measurements that are tracked and monitored regularly to indicate if an organization is on the right track to hit its goals and objectives. These measures and indicators play a vital role not only in industries but also in our day-to-day lives. We all measure something important. Our kids' performance is reflected in the grades they obtain from school. Their grades indicate what subjects they excel in and those that definitely needed to be improved. When we buy food in the marketplace, we always speak of weight in kilo depending on the amount of money and budget we have. If we want to go on a diet or simply gain weight, the first thing we need to do is to measure our weight through a weighing scale in kilograms or pounds. When we travel to other places or countries, the amount of fare we pay depends upon the distance we want to travel and the comfort we would like to have. Speaking of sports such as sprinting, we measure the time, it takes for an athlete to complete the race. Usain Bolt ran the fastest ever 100-meter sprint at 9.58 seconds at the Olympics in 2009, and this record I believe will stay for several years to come. Likewise, in maintenance, we measure both our people and equipment to indicate if we have been performing well or falling below our goals and targets. These measurements reflect how we do things around the plant, and this takes some form of Key Performance Indicators. The word key indicates that among the many measurements available, these are the ones we have selected and will be used for our equipment, assets, and people. For people, our managers appraise us and tell us our strengths and things to improve. These measurements provide us some sort of benchmark or starting point in our drive to improve both human and asset performance. As the saying goes, if we can measure it, then we can manage it. Hence, before thinking about measuring our performance, let us first understand what we want to manage. What is important is to understand what it is that we want to measure so that we can determine where we are currently heading forth in the future if we want to be ahead of our competitors. These measurements and indices will reflect the success or failure of our maintenance-driven initiative. For industries, this should be done at the very beginning.

However, the most important reason for measuring these KPIs is to improve our industry's performance. By knowing our performance, we ask ourselves, what exactly should we do differently to improve and reach our goals, targets, and objective? Industries today challenge themselves by setting up goals and doing their best to improve them. Their goals cannot remain the same all the time, nor can they have the same goals repeatedly if they want to remain competitive and ahead of their competitors. Moreover, industries should be improving faster than their competition. To do this, they should challenge the goals and targets they set from time to time and continuously improve them. Here are the reasons why we need to measure our performance.

To Benchmark: We need to measure the performance of our equipment and assets to initially determine where we currently stand at the moment. We can just sustain that equipment and assets that are performing well and focus our attention on those that are not performing well at the moment. This will serve as the benchmark for setting our goals for a given time.

To Set Goals and Targets: Once industries decided on what they want to measure, the next thing to do is to set goals and targets on what they want to achieve. Challenging these goals will allow the organization to move forward and aim at something important to them. Once the goals are hit, then they should be changed, while if the performance of the organization is below their desired goals, then they should start to analyze if the goals they set are possible to achieve or not. Goal setting must be done before the current year ends so that the people have a clear expectation of the upcoming year to come.

To Challenge: We need to measure these maintenance indicators and KPIs to determine if our equipment and assets are performing well or not. These measurements can tell us exactly whether our equipment is functioning based on our expectations or not. The objective is to determine if our activities and efforts are on the right track. This means that if we are on target, we can challenge ourselves to increase the goal, while if we are way below on target, then we need to slow down and reflect on what is affecting our performance and how we can overcome them.

To Manage and Control: These measurements can tell us if we are in control of the situation or not. These measurements, unless manipulated, can tell us a lot about the way we do things here in our industry. As the late Peter Drucker said we can only control the situation if we are measuring it right. Every department in the organization has its own respective KPI's. What is important is that all KPI's should be aligned to the overall corporate vision, mission, goals, and objectives of the industry. That is what I have indicated in my previous books; for example, if the supply chain or warehouse controls the MRO spare parts of your industry, then maintenance cannot manage it at all because we can only manage what we can control.

To Budget: By measuring performance, we can determine which strategies and activities deserve additional budget for maintenance. Measuring performance can tell us which ones are within the budget and which are exceeding the budget. By measuring these Key Performance Indicators, we can allocate budget wisely on the things that maintenance needed them most. The maintenance budget should not only be limited to repairs and maintaining the equipment, but a budget should be well in place for maintenance training, improvements, modifications, and instruments they need to perform their jobs correctly, such as the acquisition of Predictive Maintenance instruments and certification of the users as well.

To Make Decisions: Developing performance measures allows an organization to determine its mission, goals, targets, and desired results. These indicators and measurements will also identify methods of measuring how well these results were achieved or not. These measurements can aid decision-makers in making sound decisions based on the outcome of their performance. What is important for every manager is to look at the results or lagging indicators and understand the process that led to the result, creating better communication with their subordinates. As the saying goes, a good manager will always look at the results, but the better manager will not only look at the results but at the process of how the results were achieved.

To Learn: Performance measurements contain information that can be used to evaluate and learn. It is acceptable to have a low score in our measures for as long as we learn and do something about it. The objective of measuring these maintenance indicators is to provide management with real numbers to manage and not to blame people. If your competitor's cycle time is 24 hours, and your industry is 48 hours, we need to learn why they can deliver faster than us. What are they doing differently? If the PM Compliance is high, yet emergency repair is also high, then something is definitely wrong with the way we maintain and perform Preventive Maintenance on our equipment and assets.

To Improve: Finally, the most important reason for measuring these maintenance indicators and KPIs is to improve the organization's performance. By knowing our performance measurement, we can ask ourselves, what exactly should we do differently to improve our performance? Industries today challenge themselves by setting goals and improving them. Industries cannot have the same goals repeatedly if they want to remain competitive and stay in business. Moreover, they should be improving at a faster rate than their competition. They do this by challenging the goals they set and continuously improving them. When I was working in Sharp Philippines, we at the Television and VHS department bought a 14 inch Samsung unit and brainstorm how come this unit sells half the price compared to our own 14-inch unit? What are they doing differently? It was a real challenge to improve and lower the cost of our own unit without sacrificing the reliability and quality of the product.

These Key Performance Indicators must not only be known by the management and decision-makers but also to all levels of the organization down to the shop floor people. All employees must understand if we are below or right on target. These goals and targets must be visible not only on the laptops but also on bulletin board updates.

Each function of the organization will have its own indices to measure. What is important is that first, whatever maintenance indicator is set, it should not be in conflict with other departments, or functions. Second, whatever indicators, measures, or indices are being measured should be aligned with the overall corporate goals and objectives. Quality people will be interested in the yield and number of defects, while Safety will be monitoring the number of accidents, and near-misses.

1.3: What Should We Measure in Maintenance?

Measuring Performance and KPIs should be decided at the very beginning of the plant's operations, which means that whatever indicators and measures must be discussed and agreed upon by key people in the organization. There will be several measurements and indicators that will be done for the whole summation of all equipment, machines, and assets. Measurements can also be done on non-dedicated machines. Non-dedicated machines are those machines in manufacturing that are designed to manufacture different products. For example, in manufacturing, some machines are used to manufacture not just one but different products, hence, one measurement of importance will be the set-up or conversion time. For process plants, one important thing to measure is the bottleneck equipment that constrained the process. This will usually be the equipment or machine with the lowest UPH or units per hour. In figure 1.2, machine 3 will be considered the bottleneck equipment since it is the equipment with the lowest UPH (units per hour). This means that the inventory will be constrained on machine 3. If these five machines represent one process and one measurement being tracked is the OEE or Overall Equipment Effectiveness, then the machine with the lowest OEE will represent the OEE of the whole process.

Every organization has several functions from the Human Resources, Finance, Maintenance, Safety, Quality, Warehouse, Purchasing, MRO Storeroom and so on. Each of these organizational functions has its own distinct measures and indices they track, measure and review if they are within their goals or not. The most important thing is that their goals, indices, KPIs should not be in conflict with other functions. The goal of the storeroom may be to reduce inventory, they can do this by limiting their purchases and they can be successful. On the other side of the story, the maintenance suffers since a lot of stock-out is experienced from the storeroom.

Deciding on what maintenance measurements and KPIs will depend on the losses suffered by the equipment. Different industries experience different losses in their equipment, while others can experience a combination of different equipment losses at the same time. If a piece of equipment is suffering from a lot of breakdowns, then the measurement suited for this will be MTBF, Availability, Number of Breakdowns, Machine Downtime in hours, Ratio of PM versus CM in hours, and other related measurements.

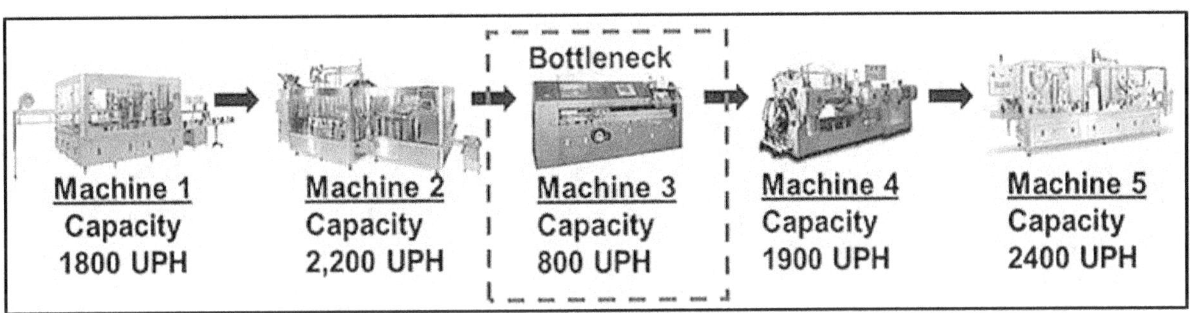

Figure 1.2: Bottleneck or Constrained Equipment

Besides measuring the performance of equipment, we can also measure the performance of our maintenance people. Some of these indicators in measuring our people include:

• Overtime Hours
• Mean Time to Repair (MTTR)
• Mean Time to Set-up (MTTS)
• Mean Time to Assists (MTTA)
• Total Training Hours
• Total Number of Suggestions Implemented
• Total Number of Improvements Generated
• Wrench Time
• Maintenance Backlog

One question that is often asked during my training on Meaningful Measures of Equipment Performance is what should be the benchmark for this particular KPI, let us say MTBF for industries. This is one question that probably does not have a direct answer since assets and equipment in industries are simply not identical. If there is only one type of industry on this planet, then that would be an easy question. Second, it depends on how we measure MTBF, for example in power plants most of their assets, equipment, and components are interconnected, hence, measuring MTBF can be done on a system based, while for a manufacturing plant whose equipment and machines are independent of each other, MTBF can be measured per machine. An MTBF of 700 hours for one month, assuming 30 days period would be difficult to achieve for a manufacturing plant, but this value would be unacceptable for a distribution plant.

OVERALL PM COST SAVINGS ON IMPROVEMENTS

Sa Planned Maintenance, Isang Misyon, Isang Direksyon pa rin.

DEPT.	PM Comm	AREA	MACHINES		HEAD COUNT	TOTAL PM IMPROVEMENT's COSTS					RANK	REMARKS
			Types	Total		1998	1999	2000	2001	TOTAL		
PLCC	RPAND	FOL	29	269	62	29,795.50	0.00	0.00	0.00	29,795.50	16	
	BBERN	EOL	26	117	43	89,191.72	19,853.00	0.00	0.00	109,044.72	8	
SOT	EBUST	FOL	3	18	9	0.00	0.00	0.00	0.00	0.00		No Cost Report
	CCRUZ	EOL	4	4	6	0.00	0.00	63.00	3.00	66.00	19	
PSOP	JJAVI	FOL	14	100	28	51,547.50	-210.00	-222.00	1,391.60	52,507.10	11	
	MAPAD	EOL	14	29	17	25,780.00	3,750.00	64,715.00	3,300.00	97,545.00	10	
MQFP	DGALA	FOL	6	210	39	364,389.20	-3,718.50	6,298.00	736.00	367,704.70	6	
	DBALA	EOL	6	59	31	511,512.00	51,105.00	-555.00	15,694.00	577,756.00	4	
SSOP	OPARP	FOL	9	150	29	20.00	9,273.00	20.00	0.00	9,313.00	17	
	GAVYB	EOL	14	33	23	124,260.00	10,712.00	381,150.00	173,201.00	689,323.00	3	
TQFP	RONAO	FOL	9	141	25	3,358.00	495.00	936.00	240.00	5,029.00		
	ROMES	EOL	15	32	20	0.00	32,130.00	130.00	17,055.00	49,315.00	12	
TSSOP	RFABR	FOL	9	218	36	171,590.00	0.00	3,583.00	-184.00	174,989.00	7	
	CGONZ	EOL	17	52	32	0.00	32,130.00	130.00	17,055.00	49,315.00	12	
SOIC	RSUAR	FOL	20	409	71	0.00	0.00	5,990.00	543,765.00	549,755.00	5	
	JIMBOY	EOL	22	82	31	0.00	0.00	0.00	0.00	0.00		No Cost Report
PDIP	JBAYB	FOL	14	149	45	0.00	495.00	1,130.00	0.00	1,625.00	18	
	ZBALO	EOL	19	77	32	232,125.84	1,377,182.86	131,148.00	21,840.00	1,762,296.70	1	
Hermetics	TLIM / KOKO	SSEL / Cerdip	17	92	34	28,420.00	26,183.00	-24,340.00	0.00	30,263.00	15	
	ASOLO	Analog	14	36	19	0.00	0.00	35,308.00	9,590.00	44,898.00	14	
CLF	JERRC	Plating	8	21	30	27,300.00	23,978.00	26,845.00	21,970.00	100,093.00	9	
Facilities	AJUSAY	---	30	199	17	575,963.00	440,185.00	70,588.00	0.00	1,086,736.00	2	
TOTAL			319	2497	679	2,235,252.76	2,023,543.36	702,917.00	825,656.60	5,787,369.72		
	CUMULATIVE					2,235,253	4,258,796	4,961,713	5,787,370			

Figure 1.3: Example of Intangible Measurement (Number of Improvements)

1.4: Tangible and Intangible Measurements

Tangible measurements are those measurements that are easy to quantify and have a direct impact on the bottom-line results and performance of the plant. They can also be referred to as quantitative measurements. While an intangible measurement can also be quantified but will be difficult to determine its direct impact on the results or lagging indicator. Intangible measurements can also be termed as qualitative measurements. Although they are not converted easily to monetary values, they are still an important part of industries' success. They can serve as a driving force in motivating the people to deliver and perform better. The main difference is that maintenance Key Performance Indicators will be more measurable, while for those implementing Autonomous Maintenance for operators' measurements, their indicators will be more intangible or qualitative measurements, but they can still be measured and quantified. Samples for intangible measurements may include the following.

• Number of suggestions of operator versus those implemented
• Number of questions generated and answered
• Number of kaizen improvements generated with success
• Number of abnormalities detected and corrected
• Number of One Point Lessons generated
• Number of training hours generated
• Number of Teams that completed Step 1, Step 2, Step 3 of Autonomous Maintenance
• Number of loose and missing bolts detected and corrected
• Number of Hard to Clean areas detected and corrected
• Number of leaks detected and corrected

1.5: Leading and Lagging Indicators

Lagging Indicators are indicators and measurements that indicate the results. Examples of this will include profit, sales, revenue, output, repair time, yield, cycle time, maintenance costs, total downtime, labor costs, profit, revenue earned, the total cost of goods sold for the week, overall inventory cost, and so on. They are the bottom line results as management always looks at these figures all the time. The leading indicators are those indicators that contribute to the lagging indicators or results. The result or lagging indicator is a compilation of different processes that contributed to the result. These leading indicators will influence the final outcome of the result. The bottom line is that what gets measured gets done. We do not want to measure so many things, but rather, we want to focus on those that are important and relevant to the organization, especially if measurements are tracked manually.

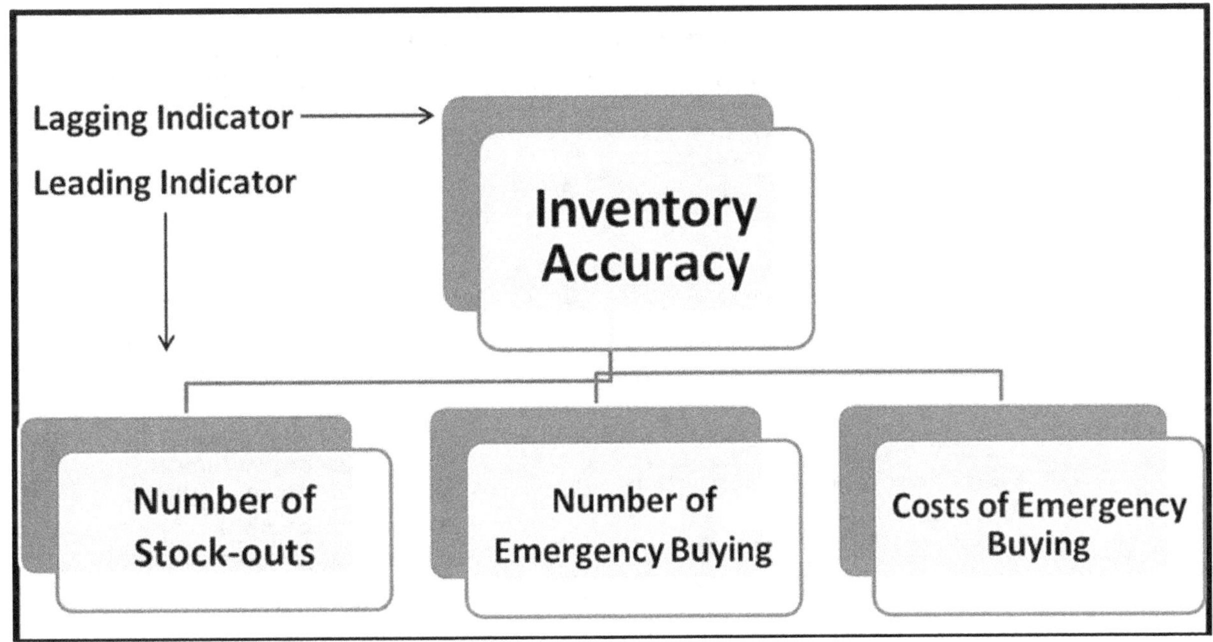

Figure 1.4: Leading and Lagging Indicators

In Figure 1.4, this indicates that the lagging indicator is the inventory accuracy. This is the result of a few stock-outs and less emergency buying. This is what most storeroom managers are looking for as they called this the bottom line. This means that the inventory accuracy was high because the number of stock-outs and emergency buying is low. This means that the MRO storeroom is capable of providing the parts needed by maintenance when they needed it most. The number of stock-outs, number of emergency buying, and the costs of emergency buying are low, which greatly contributes to the high inventory accuracy. If we can state this in a very simple way, the lagging indicator was the result, and having stock out cases and costs of emergency buying is the process that eventually led to the results. What is important for maintenance is not to confuse what is a leading and lagging indicator. If this is the primary measurement that you track, then let this be a lagging indicator. At the same time, if this is a secondary measurement, then this is a leading indicator. This will also depend on who is looking at the measurement. The production lagging indicator might be the total number of products produced, while for the Quality people, the yield will be their lagging indicator.

While many managers always look at the results (lagging indicator), in contrast, the better manager will look at the process (leading indicator) on how the results were achieved. This allows better communication between the manager and their subordinates. Also, there are instances that a lagging indicator may become a leading indicator and a leading indicator may turn out to be a lagging indicator depending on the primary measurements and who is actually reading the report. The CEO will be interested in the revenue of the plant which is a lagging indicator. The revenue was high since downtime was minimal. For the CEO downtime will be one of the leading indicators that contribute to high revenue, while for the maintenance manager, the downtime can be their lagging indicator.

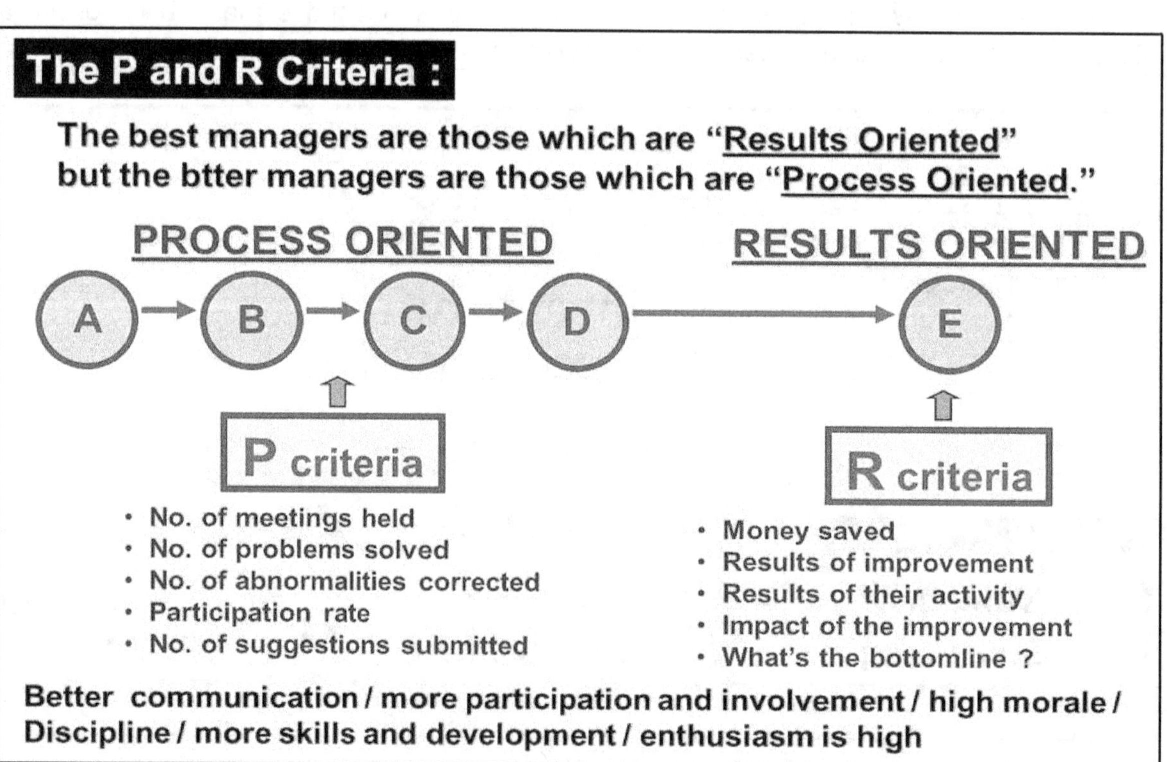

Figure 1.5: The P and R Criteria
1.6: Selecting the Right KPI's for Maintenance

I recalled in July 2013, I was in Botswana, Africa as they hired me to train their people on TPM, Planned Maintenance, Autonomous Maintenance, and Lubrication. During the first day of training, they served roast beef during lunchtime. I was given my meal and sat on one of the tables together with some students. I went to the kitchen and requested a spoon returned to my table and started eating. As I eat, the students were watching me, then perhaps out of curiosity, one of the students asked me why do I use a spoon. I simply told them that this is how we eat in my country and they laughed at me. I told them that you were liberated from the British a long time ago but yet you cannot get rid of some habits you picked up from them. You see in my country, sometimes we eat with our hands if we want to as long as we wash our hands before and after eating. After that, there was silence as we continue to eat. The following day, to my surprise, they started eating with a knife, fork, and spoon. Similarly, when I am in Malaysia eating dinner with my friend CK, I used a spoon and fork, while he used a chopstick. Just like these measures and indices, there might be indices and measurements that will be important to one industry but can be deemed irrelevant to another just like how we want to eat our food either with a fork and knife, chopstick, or just by using our hands.

As stated previously, it is quite difficult to recommend the right measurements and indices suited for your industry because industries differ in the way they operate and do business. Some industries produce products such as manufacturing, while other industries provide services such as power plants. Certain key performance indicators and indices may apply to one plant but not exactly to the rest of the other plants. Experience tells me that there is no one single and perfect KPI that can effectively measure a plant's performance alone; you need to have a variety of these indicators or a minimum of at least 5 KPIs. It is also not recommended to have so many indices on maintenance as this will just confuse the people on what they need to prioritize. As industries differ, what is important is for your people to determine the losses your equipment is suffering and determine what KPIs are important and worth measuring regularly. For powerplants, Oil and Gas plants, the losses their equipment can suffer will be failures, breakdowns, and start-up, while for semiconductor industries, the losses their equipment can suffer can be quality defects, breakdowns, minor stoppages, design speed loss, cutting tool loss, and changeover.

Figure 1.6: Myself at Botswana Africa

Manufacturing industries that produce products may measure their OEE or Overall Equipment Effectiveness, while it would be unlikely for a power plant to use these measurements. OEE is calculated by multiplying the Availability times the Performance Efficiency times the Quality Rate since the Quality Rate will refer to the number of rejected products compared to what the industry produced. The OEE indices are designed for industries that produce a product since this is where the third component, Quality Rate is used. Perhaps the power plant can use Availability and Performance Efficiency and disregard the Quality Rate if there is no way of measuring the Quality. I have a client in the Philippines which is a Geothermal Power Plant in which they can measure the purity of the steam they delivered which is usually at 99% plus, in this case, the three components of OEE can be used. Another thing of importance is that some books may claim that OEE should be 85%. I would not always agree with this being the standard for industries. For a powerplant, an OEE of 85% might indicate that 15% of the time, the city is experiencing a blackout which is simply unacceptable.

There are also common measurements for industries that they can measure. Maintenance can measure the number of breakdowns, downtime, availability, Mean Time Indicators, OEE, repair time, maintenance costs, MRO Spare Parts costs, PM Compliance, PM Effectiveness, and many more. Therefore, the industry must understand what measurements and indices are needed that best suit their plant needs and requirements. Each industry should focus on measurements and indices that apply to them. Indices from manufacturing industries may differ in other industry sectors, such as oil and gas, power plants where their equipment is linked together to form a system. For plants initiating TPM, the primary measure of their performance would be OEE or Overall Equipment Effectiveness.

Common Maintenance KPI and Indices

Tangible Measurements

• PM Compliance in Percentage	• OEE
• Breakdown Occurrence (Unplanned)	• Failure Rate
• PM Effectiveness (PM vs Emergency)	• Total Maintenance Costs
• Repair and Maintenance Costs (Php/USD)	• MRO Spare Parts Costs
• Downtime in hours (both Planned and Unplanned)	• % Work Orders Completed on Time
• Reliability in Percentage	• MTBA in minutes
• Availability in Percentage	• MTBF in hours
• Total Lubrication Costs	• MTTR in hours
• Contract and OEM Costs	• Work Order Completion

Figure 1.7: Common Maintenance KPIs for Industries

1.7: Measurement of Time

There are several indicators whose measurements are based on time such as Availability, Utilization, MTBF, MTTF, MTTS, MTBA, and others. This means that if we have a given day composed of 24 hours, how well did our equipment or machine operates on that specific period minus the losses experienced which are in the form of downtime.

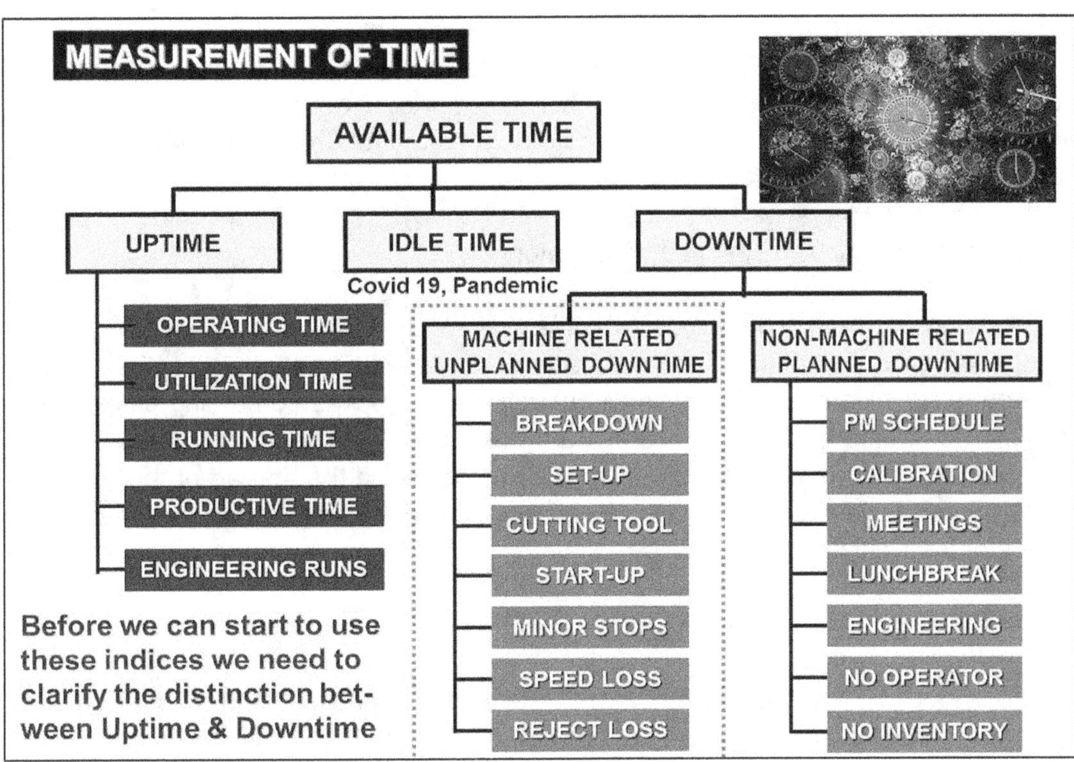

Figure 1.8: Measurement of Time

• **Available Time** is given as the total time in a given period, 24 hours for one day, 168 hours for a week, 720 hours for 30 days, 744 hours for 31 days, or 8760 hours in 1 year.

• **Downtime** is given as the period when a piece of equipment or asset is not operating. This is also when the equipment is unavailable for use. The unavailability of the equipment may result from equipment failure or malfunction, a shutdown or scheduled Preventive Maintenance, or simply when an operator shuts the equipment off to take his lunch break. Downtime does not only refer to a malfunction of the equipment but also a stoppage of the equipment for other reasons such as no work or volume to be produced with the equipment. Downtime can be classified as those which are Machine Related and Non-Machine Related. TPM also termed this as Planned and Unplanned Downtime.

• **Uptime** is the time in which the equipment and assets are operating. It is also the time during which the equipment and asset are working without failure. Other terms to designate uptime is operating time, productive time, utilization time, running time, or time the equipment is utilized.

- **Machine-Related Downtime** is the time the equipment is not operating due to machine-related losses. In manufacturing, failure or breakdown is not the only cause of machine-related downtime. Other causes of machine-related downtime include minor stoppages, change in cutting tools, quality-related problems on the product, set-up and conversion, or changing equipment to produce one product to another product, startup losses, and speed reduction losses. All of these indicate downtime related to the machine. TPM books refer to these as unplanned downtime.

- **Non-Machine Related Downtime** is the time the equipment is not operating due to non-machine-related factors. TPM refers to this as Planned Downtime. Factors reflecting non-machine-related downtime include operators' break time, meetings where equipment is stopped, planned shutdowns, Preventive Maintenance overhauls, and replacements, planned scheduled shutdowns, scheduled outages, and holidays when the machine or equipment is not required to work in the absence of operators.

Most industries will have their own available time for a particular calendar year in which they will be operating. Other industries may or may not be operating on non-working holidays such as Christmas, New Year, and holidays. Although last 2020 was something different since this was the start of the **covid 19** pandemic, where the government itself proclaim a lock-down and ordered several industries to close down temporarily to avoid the spread of the virus. I believe that this type of downtime cannot be considered planned or unplanned so I indicate another dimension of time which I called Idle Time.

1.8: How Do We Track These KPIs?

With today's CMMS, EAM and automation software, these KPIs and measurements can be calculated for the whole summation of all assets, equipment, and machines in the plant. It can also be by process, a group of equipment, or in special cases for non-dedicated machines. An example of a dedicated machine is if we want to track the MTTS or Mean Time to Set-up for non-dedicated machines as we want to know the average conversion or set-up time and the goal is to reduce the time to set-up or convert. Non-dedicated machines are those machines that can produce different products. This means that if 2 or more products will be produced in a given shift using the same machine, a downtime called set-up or changeover will occur to convert the machine to process a new product.

More than a couple of decades ago, during my employment time, we do everything manually where we developed a form for the operator to fill up before their shift ends and to record all the downtime they experience on their machine. The Production Supervisor or the leadman will collect all the forms at the end of their shift manually, and give them to the encoder. These forms and log sheets will be recorded one by one by an encoder on a computer at the end of the shift. The same process will be done for all shifts since we operate 24 hours a day. Although this is not foolproof since the operator can encode the wrong code unintentionally, this is the best that we can think of, since digitalization, IIoT, was not yet around a couple of decades ago, and CMMS features were limited compared to today's software.

But today, it will be much easier and faster with the evolution and upgrades from the CMMS and EAM (Enterprise Management Software), depending on what the company is using. KPIs can be obtained with just a few buttons on the computer. The only issue, in this case, is for old equipment and machines in which the data needs to be encoded manually. What is important is that we provide a code for these losses the equipment is suffering so that we can have a better understanding of what the downtime constitutes. If a piece of equipment is suffering from many minor stoppages, then we need to have a code for all the possible minor stoppages that our equipment is experiencing and orient the operator regarding how to encode the correct code on the system together with the total downtime. This goes for the different types of equipment losses explained in Chapter 2 so we can have accurate indices. The majority of these losses except for design speed loss will constitute a downtime, hence, my point is we need to classify and provide a code for each of these losses so that the cause of the downtime will not be mixed especially for equipment and assets that are directly linked with the system. This means that if we want to get the total downtime last week, and assuming we experienced 45 hours of downtime for Machine 1, we need to have a category on how many hours for breakdown loss, for set-up and conversion loss, for minor stoppages loss, start-up losses and so on. For old equipment suffering from the same loss, in which it cannot communicate with the main system, either this will be done manually or hiring someone to modify and install a PLC so that it can communicate with the main system. Each of the equipment losses experience should have its own codes. The key, in this case, is to educate the operators regarding the different equipment losses so that they can encode the correct code on the system.

What is important is whatever means of extracting the data whether the system is automated or done manually, whoever is populating the data should be consistent so we can have a close to accurate measurement and indices.

1.9: How Many Indices or KPIs Do We Need to Measure?

Although there will be some common and unique measurements depending on the industry, the maintenance function needs to measure a minimum of 5 KPIs, in other cases, perhaps more. The reason is simple, there is no single KPI or measurement that will tell the whole story of whether the maintenance organization is performing well or not in maintenance. A geothermal Power Plant in the Philippines hired me for consultation purposes to help them develop their standard procedures in maintenance. One day as I was in the office, the Vice President called me up and told me that he was confused about the indices maintenance is reporting. I asked the VP what seemed to be the confusion. He said that, isn't it Rolly that if your MTBF is high, then the equipment should be performing well. I replied and said that supposed to be that would be correct. His question was if our MTBF is high for the past month, then how come our maintenance cost is also high compared to other months? I looked up the data and told the VP that your MTBF is high because you are calculating it on a system level and not on a component level. This means that since you have a lot of redundancies and standby, if something failed, then the system will not be affected since you will just switch to standby. If you will ask the maintenance to also calculate the MTBF on a component level, then you will see why your maintenance cost is high since there are a lot of breakdowns causing the maintenance to revert to the standby component.

There is no one KPI I know of that will exactly tell everything and the whole story. The overall inventory cost may be reduced which is a good thing for the storeroom management but the fact lies that a lot of stock-out and emergency buying is happening since the storeroom failed to replenish some items needed by maintenance. Both the Storeroom and Purchasing people are celebrating thinking that they have done a good job in reducing the inventory cost, while on the other side of the building, the maintenance is suffering from long downtime due to no availability of parts.

Or perhaps another classic case is for a manufacturing plant having an OEE of 85% thinking that their equipment already has achieved a world-class level as most books and consultants will declare. The data for their OEE includes Availability at 99%, Performance Efficiency at 95%, and Quality Rate at 90% where the OEE is at 84.64%. This means that for every 1,000,000 products produced by this plant 100,000 products are rejected, scrapped, or reworked. More of this is in Chapter 3 of this book.

1.10: Set Goals and Target for Maintenance

Once these indices on maintenance are laid down and finalized, the next step is to set goals and targets for both the short-term and long-term. Maintenance needs to challenge these goals and targets and deploy the correct maintenance strategies. These goals and targets will be deployed from the top down to the bottom line people. Everyone in the maintenance organization should be aware of the goals and targets they want to improve. This should be a consolidated effort by the whole maintenance organization. The role of management is to let the people know that this is where we want to go, and this is what we want to achieve in maintenance. These goals and targets should not only be confined to management and their supervisors but to all levels of the maintenance organization down to the shop floor people.

PM YEARLY GOALS

• Breakdown Rate :   Reduced by 60%
• Mean Time Between Failure   50 % Increase
• Mean Time to Repair   Less than 0.5 hours
• Repairs and Maintenance Costs 40 % lower
• Availability   99 %
• Machine Downtime   70 % lower
• Number of Accidents   Zero
• Number of Improvements per year 1000
• Number of breakdowns   1 or zero
• No. of MP Design Generated/Team 5

Figure 1.10: Yearly Goals for the Maintenance Function

To start with, I recommend that key people from operations and maintenance set up a budget for their yearly strategic planning, which can be done inside or outside of the plant for 1 or 2 days. Strategic planning should be held before the year ends or at the very start of the year. In the strategic planning session, the maintenance leader will review the highlights of the previous year on what they have accomplished and missed out on and provide a clear direction and strategy for both operations and maintenance. One of the key topics that must be covered is to set goals and define strategies to achieve their goals. Identify all obstacles and pitfalls in achieving a World Class Maintenance Management Structure. If there are industries that can achieve this, then challenge your industry to do the same. What is important is that maintenance should gear themselves towards achieving a proactive maintenance level. Set a clear direction on what maintenance wants to accomplish next year and have a specific time frame of completion for the major activities involved in the undertaking. Create a review team to review and check the progress of its implementation consistently. What is important is that whatever things are agreed upon during the strategic planning will be done based on its schedule, which means that we walk the talk. Prepare a deployment plan on how to disseminate whatever is agreed upon. Goals and measurements are clearly defined and should be understood by all the people involved.

The next level of planning should be scheduled for supervisors and shop floor people in the plant. It is important to clearly state the reason for the change and do it right the first time; otherwise, any initiative for change will result in treating it as another program of the month, just like ice cream. People are needed to implement these strategies, and they need to understand that without a proper workplace culture, it is not possible to achieve these goals. The worst is that we might fail completely even before starting the initiative. To effect the change, people must believe that these new behaviors are good for both the company and the people themselves. Improvement programs fail because they are poorly planned, developed, and implemented. Most programs look good on paper and PowerPoint slides but not in reality. People view these programs as the latest in a series of fads that come and go. When a program takes too long to implement, management will abandon it, and the program fades away making them go back to their old habits of doing things. What's on these people's minds is that these programs are fads and will just fade away just like the other programs in the past.

These measurements discussed in this book are very important since they will spell the success or failure of any drive for maintenance to improve their equipment and assets. These indices that will be tracked should be agreed upon at the very beginning. They should be monitored and reviewed regularly. What we are after is the trend of the graph. These indices will tell us if we are moving forward in the right direction to achieve our vision and mission.

Chapter **2**

Different Types of Equipment Losses Explained

> *Have the right reasons to measure equipment performance. Set goals on what you want to achieve on these indices and aim your people to challenge them. Determine what reliability and maintenance strategies should be implemented and executed to ensure that goals will be achieved. Once the goals are achieved, set new goals, and challenge them.*

2.1: Equipment Losses Explained

An ideal industry would operate 100% of the time at 100% capacity with 100% good quality, however, in real-life situations, that is not true. The difference between an ideal and actual state is due to the losses experienced in our equipment and assets since we encounter these losses on a day-to-day basis. First and foremost is to understand the losses in your equipment and learn why these losses occur in the first place, so that we can think of ways to prevent, predict, prolong, reduce, control, manage, modify, or anticipate these losses at all. Industries vary from one another and the type of equipment they used. Equipment and assets in some industries may be system based such as oil and gas plants while others are process-based. Still, others will be machine-based which can be independent of each other. Therefore, these equipment losses also vary from one industry to another depending on the type of industry. This means that these losses may or may not be present for all industries. According to TPM (Total Productive Maintenance), there are 8 equipment losses and 6 are considered the big ones. These losses may or may not include downtime caused by outside factors such as loss of power, energy losses, facilities losses, and manpower losses. It is important to determine what type of equipment losses are encountered most of the time and who should be responsible for addressing them. It is highly recommended that each of these equipment losses must be treated separately, as each type of these losses needs a different approach in improving them. Depending on the type of losses experienced on the equipment, each of these losses must not be solely addressed by the maintenance people alone but by a cross selection of people familiar with the problem in the plant.

Reliability-Centered Maintenance or RCM for short will deal with the failures, but not all equipment losses, especially for manufacturing industries. Failures and breakdowns are just a sub-set of these losses. These equipment losses are much broader in scope since equipment failures are just a part of the losses. This is just one whole pizza pie which is cut evenly into 8 pieces. Failures and breakdowns are just one slice of the pizza. Addressing all the possible losses on the equipment or asset will require different strategies and methodologies way beyond just improving the way we maintain the equipment. The key is to identify the losses your equipment is experiencing and adopt the most feasible method to address them. Hence, before we discuss the different KPIs and measurements for maintenance, we need to know the losses our equipment is experiencing so we can select the correct and appropriate KPI to measure, but first, let us define each of these losses accordingly.

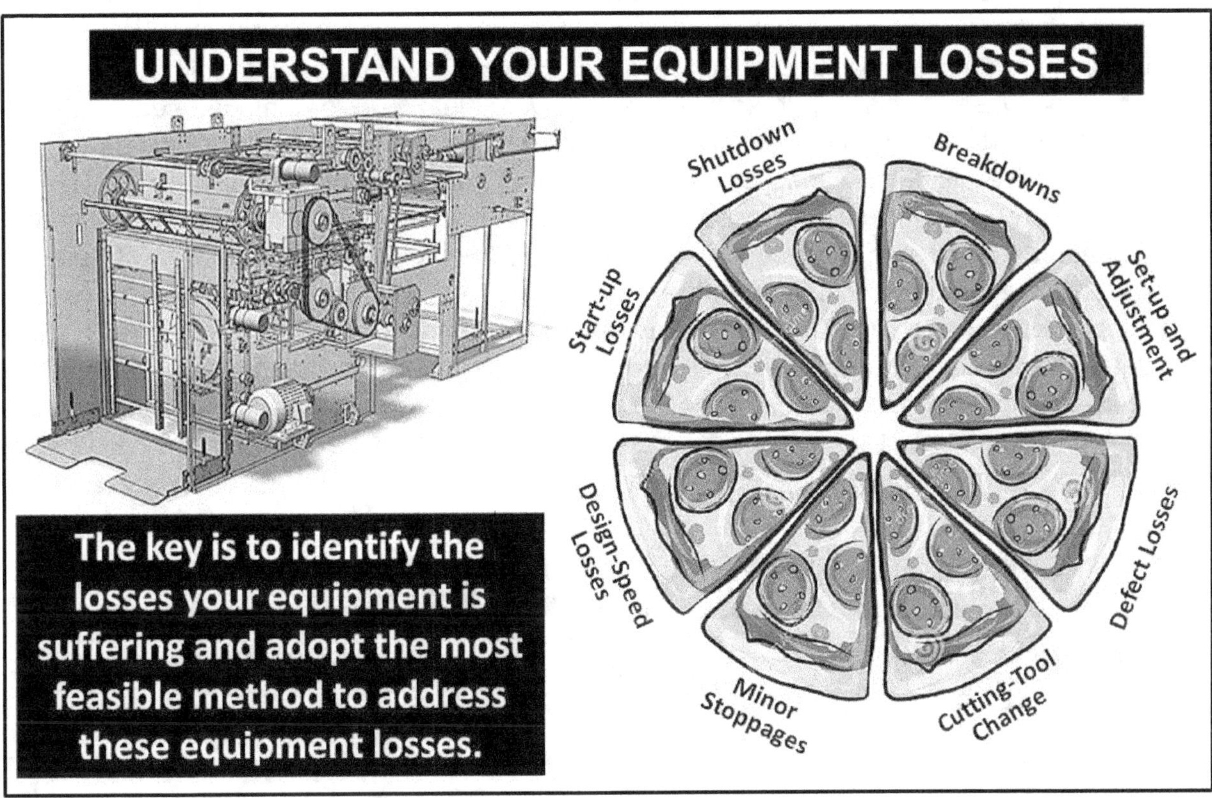

Figure 2.1: The Key is to Understand your Equipment Losses

2.2: Breakdown Loss

Also called equipment failure. This is a loss when a machine stops due to the loss in its specified functions. According to **Wikipedia**, failure, in general, refers to the state or condition of not meeting a desirable or intended objective. It may be viewed as the opposite of success. **Dictionary.com** defines failure as the inability of a system or system component to perform a required function within specified limits. A failure may be produced when a fault is encountered. [1]While from the book of John Moubrey on RCM2, failure can be defined as the inability of any asset to do what its users want it to do. This definition does not distinguish clearly between the failed state or functional failure and the events which caused the failed state or failure modes. Failures can also be classified as the failure of the primary or secondary functions of the equipment.

---

[1] Moubrey, John, **_Reliability-Centred Maintenance II_**, Butterworth-Heinemann Page 46

A breakdown is a lost time due to equipment failure, which is considered a machine downtime loss. The Japanese term for breakdown is **Kosho**. Sometimes refers to as machine-related downtime, unplanned downtime, tooling failures, unplanned replacement of spares and components. Equipment breakdown refers to any event in which equipment cannot accomplish its intended purpose. It may also mean that the equipment stopped working since something failed or is not performing its function. There are two types of breakdowns, **function-reduction breakdown**, which is when the equipment causes other losses in a secondary function even when the equipment can still operate and provide its primary function. Imagine your car running and stating that a car's primary function is to travel from distance A to distance B. It can comply with the primary function, but one of the car's headlights is busted, the side mirror is missing, the hub of the wheel is missing, the car has excessive oil leak due to damaged seal, car's air-conditioning is not working, the wiper is missing, the seat belt is not functioning and so on, but the car can deliver its primary function which is to travel from distance A to distance B. This is what function reduction breakdown is all about, and most equipment suffers from this type of loss. It seems that most of the time, this type of loss is frequently being neglected. These are breakdowns that account for the largest proportion of overall equipment losses. The second type of breakdown is a **function-loss breakdown**, a failure in which the equipment stops completely. These are losses in which production is stopped. This means that for a function reduction breakdown, this will not cause an equipment downtime, while for a function loss breakdown, downtime will be experienced on the equipment. This is also considered as a failure of the primary function of the equipment.

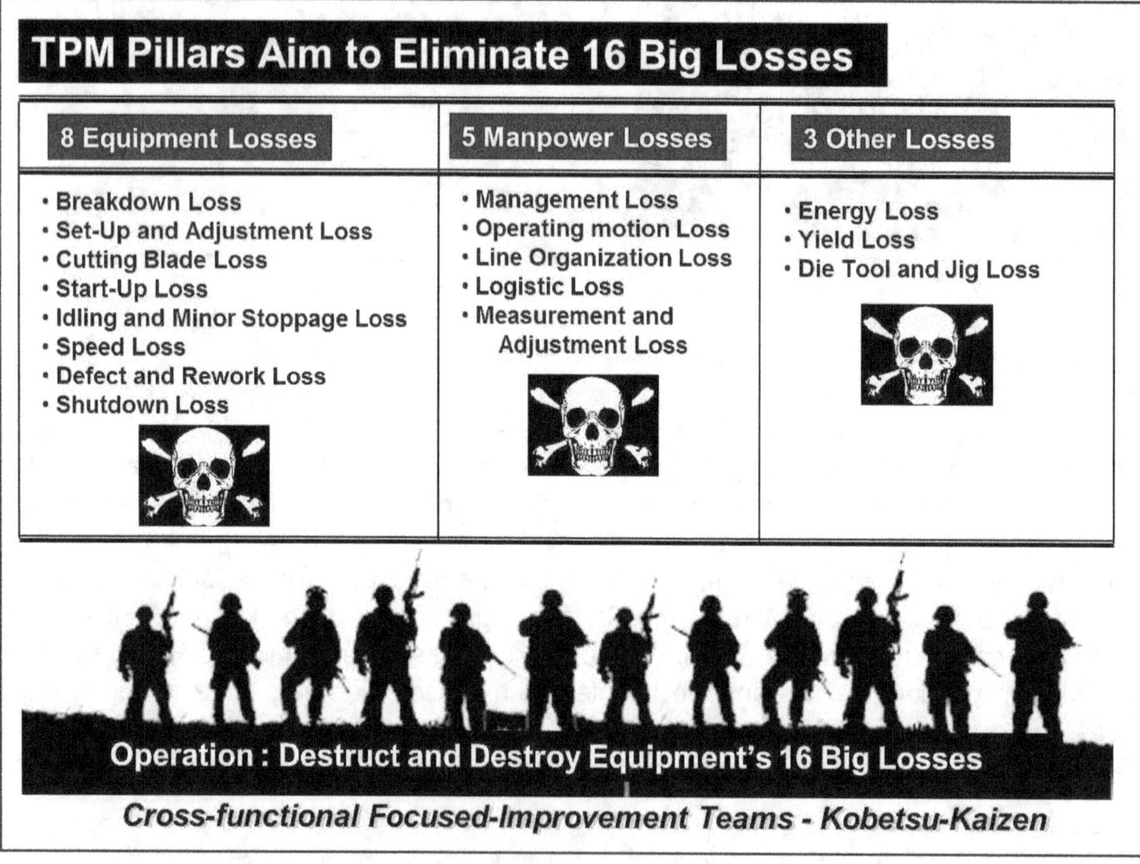

Figure 2.2: TPM 16 Manufacturing Losses

For industries, a function reduction breakdown may include a sensor that had been bypassed, a defective emergency stop button, defective protective devices, leaks in which the equipment is still capable of running and delivering the output needed. Depending on the type of breakdowns, it may or may not halt production. But for a function loss breakdown, machine downtime will be experienced, deliveries are delayed, cycle time is prolonged, quality problems arise, maintenance cost is expected to increase, and a single breakdown can create havoc throughout the plant. Analysis in the breakdown failure can be attributed mostly to human errors, basic neglect of design problems, lack of operator skills, lack of repair skills, and failure to address equipment basic condition. That is why a basic understanding of the failure must be addressed thoroughly. Breakdowns are caused by many factors, and most of the time, slight deteriorations are overlooked, which contributes highly to equipment breakdown. Improvement in equipment performance can be done by simply addressing minor problems such as loose bolts and missing screws; abrasion, debris, and contaminant are addressed. Some of the indices to measure breakdown include:

• Mean Time Between Failure (MTBF)
• Number of Breakdowns
• Maintenance Cost
• Mean Time to Repair (MTTR)
• Total Downtime due to Breakdown in hours
• Overall Repair and Spare Costs
• Availability
• Reliability

Once a breakdown occurred, be certain to learn everything you can and study the causes. The basic responsibility of each maintenance is not to repair breakdowns but to analyze what had caused the breakdown problem and draw measures to prevent the recurrence of the problem. If this is not performed, we will just be wasting your precious time on doing repairs and quick fixes on the same problem again and again. If the same breakdowns recur, then we need to perform a Root Cause Failure Analysis investigation thoroughly.

All breakdowns and failures have one thing in common, it is a result of human error. This means that somewhere in time, a human error was committed either during the design phase, commissioning, installation, how the equipment was operated, or how it was maintained. But we also believed that not all human errors are the cause of a person, as there is always a deeper reason why a person committed the error due to flawed procedures, inadequate training, stress, fatigue, and so on. Repairing the equipment when a failure occurs, will just repeat the problem, maintenance must also understand how to investigate equipment-related problems.

2.3: Conversion and Set-Up Loss

Conversion, change over, or set-up loss is the time required to remove dies, jigs for one product, clean-up, prepare dies and jigs for the next product, reassemble the equipment, adjust the equipment, perform trial runs and make further adjustments until the product of acceptable quality is finally obtained from the equipment. There are machines mostly on manufacturing plants that are not dedicated, and it is used to process more than one type of product. This usually begins when the production of one product is completed and ends when standard quality is attained in producing the next type of product to be processed. Shigeo Shingo's (SMED) Single Minute Exchange of Dies deals in techniques in reducing set-up time and adjustment time without reducing its accuracy. According to him, a good set-up time in manufacturing must fall between 10 minutes and below. This means that if the time to convert from one product to another product will take an hour or more, then there is much room for improvement to shorten the change-over or set-up time. Remember that when we perform changeover or convert the equipment to another product, the machine will be down and the equipment is not producing any revenue. The longer the set-up time, the fewer products the plant is producing. Findings of Shigeo Shingo, why set-up time is prolonged is due to the following factors:

- Preparation of materials, jigs, tools, and fittings   20 %
- Removal and attachment of jigs, tools, and dies   20 %
- Centering and Dimensioning   10 %
- Trial Processing and Adjustments   50 %

Figure 2.3: Racing Pit Crew World Record at 1.82 Seconds

According to Shigeo Shingo, the first step in reducing set-up or change-over is distinguishing the activities performed while the equipment is still running from those performed when it is shut down. This means differentiating the external from internal setup. External set-up is those activities that can be performed while the machine is running, while an internal set-up is those activities that can be performed when the machine had been shut down for conversion. The goal of set-up is to minimize the time to perform it. One of the techniques is to write down all the steps performed in doing the set-up, then analyze the activities one by one and locate those activities that can be converted from internal to external set-up. This can be accomplished using a standard one-touch jig, comparing the shapes of tools and jigs for different products, and considering preparing a standard or universal jig that can be shared by all. Eliminate adjustments during internal set-up time by using intermediate jigs.

Just imagine a Formula 1 racing pit-crew as seen in figure 2.3. Observe how organized each of these crew involved in changing the 4 tires. Each of these pit crew members, which is composed of 20 people, excluding the driver have a crucial role to do. Their actions are well synchronized and consistent. The sequence is very crucial, in which they took a considerable amount of time practicing to change the tires in the shortest possible time. The fastest racing pit crew has set a new record beating the old record of 1.91 seconds to a new world record of 1.82 seconds at the Brazilian Grand Prix last November 18, 2019. If you can watch this in a slow-motion process, you can see how synchronized this team of 20 people is on who will be first, who will be second, and so on. Although we may not have 20 people just like in the racing pit crew to perform the conversion, what I believe is we can still learn from this experience by being precise and synchronizing the set-up and conversion activities. The application of Precision Maintenance can also help reduce human errors during the conversion process. This means that whoever is performing the set-up or changeover whether the person is the most or least experienced must result in the same outcome of the process.

Here are a couple of steps that can be taken to reduce the need for adjustment. First, in many cases, adjustments can be scaled down simply by improving the precision of the equipment, jigs, and tools. The accumulation of imprecise settings creates the need for many adjustments which can be avoided. Second, standardize the procedures. Lack of consistency in the standards for measurement, quantification, and other operation and maintenance procedures is another cause of unnecessary adjustments. Set-up losses cannot be eliminated, but they can be reduced dramatically. Poor set-up and conversion can cause other losses such as breakdown, and quality problems to occur. Procedures of the conversion process of each product must be standardized. One maintenance can set up one piece of equipment differently and take 15 steps, while one may take as much as 20 to 25 steps to complete the conversion. It is important that whoever is performing the set-up procedure has a set of detailed step-by-step standards to follow. Much can be learned by knowing the difference between what are those that are considered as internal and external set-up practices, and most of them only need a modicum of common sense. Remember that the longer the set-up or conversion time, the company is not generating any revenue and profit. Although Shigeo Shingo pointed out that the challenge in manufacturing is that set-up time should fall between 10 minutes and below, this is not a one-time improvement but rather, is a continuous improvement process. If in our initial attempt, we have lowered the conversion time by 5 minutes, we challenge it again. Do we need one more person for the conversion? Would it be possible if we use a universal guidepost so that centering would take lesser time? Have we listed all internal set-up activities that can be converted to external activities? Can operators assist in the conversion process?

PRACTICAL IDEAS TO REDUCE SET-UP

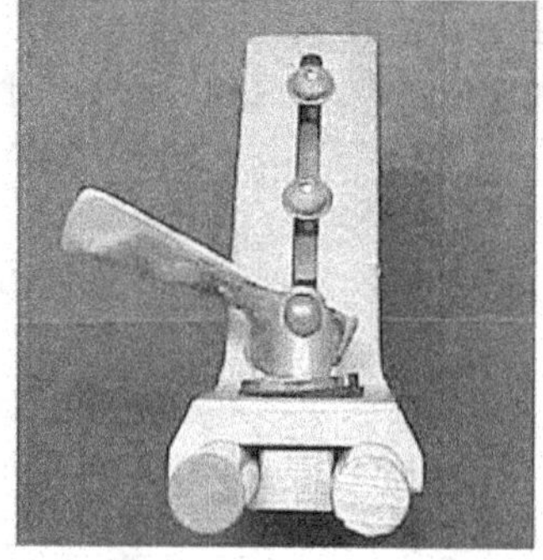

Cam locking clamps provide a simple quick release

Similar to a bike rear wheel quick release

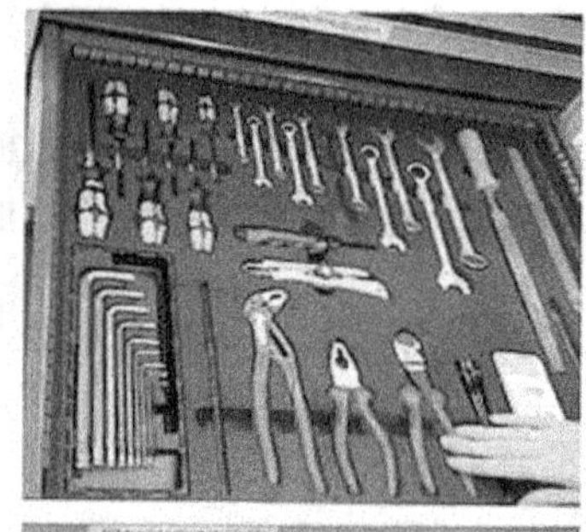

Organized tools to be used for changeover

Well organized change part storage carts are key to both protecting the parts and speeding Changeover process

Use of guide pins for easy adjustments

Wing-nut instead of bolts for easy and convenient release

Figure 2.4: Practical Ideas to Reduce Set-Up Time

SET-UP & ADJUSTMENT

Shigeo Shingo's SMED (Single Min. Exchange of Dies) to improve set-up

SHIGEO SHINGO
A Revolution in Manufacturing: The SMED System

• SHIGEO SHINGO's FINDINGS

• Preparation of materials jigs tools and fittings	20%
• Removal and attachment of jigs, tools and dies	20 %
• Centering and Dimensioning	10 %
• Trial processing and adjustment	50 %

Figure 2.5: Shigeo Shingo's Single Minute Exchange of Dies (SMED)

In 1955, Shigeo Shingo was acquired by Toyota Company as a consultant to improve the manufacturing process. From 1956 to 1958, Shigeo Shingo was responsible for reducing the time for hull assembly of 65,000 tons super-tanker from 4 to 2 months. This established a new world record in shipbuilding and the system spread to every shipyard in Japan. From 1961 to 1964, Shigeo Shingo extended the ideas of quality control to develop the Poka-Yoke, Mistake-Proofing, or Zero Defects Concept and by 1968 till 1970, he developed the SMED (Single Minute Exchange of Dies) to reduce set-up time and changeover for manufacturing. The success of this system was illustrated in 1982 at Toyota when the die punch set up time in the cold-forging process was reduced over three months from one hour and forty minutes to three minutes.

Some Tips on Shortening the Internal Set-Up Time

• Simplify clamping mechanism by using quick-fitting jigs.

• Adapt parallel operations; two people working together can perform a set-up faster.

• Optimize the number of workers and division of labor, especially for large and complicated set-ups. Doing these simple basic steps will greatly reduce your set-up time.

• Understand the difference between internal and external activities.

• List and identify activities that are internal and external.

• Treat external activities as true external activities. For example, if the machine operator also is responsible for getting parts for the next product, have someone else do this while the machine is still running.

• Change as many internal activities as possible to external activities. Obtain the parts, tools, jigs, dies, and other needed items ahead of time.

• Make sure everything needed for the changeover is organized and on hand before set-up and changeover begin. This means that when the last lot of production is being processed, gather all the needed items on the equipment such as tools, dies, jigs and others.

• Set goals and objectives. A 50 percent reduction, or 75 percent reduction in set-up time. Document achievements on set-up.

• Reduce the time it takes to complete internal activities. Use two people to perform the changeover, replace bolts with dowel pins or notches and install quick disconnects.

• Videotape the changeover and review what activities took longer to accomplish and generate improvement plans for reducing the time to perform such activities.

• Use wingnuts instead of screws to remove the cover.

Eliminating Small Losses in Set-Up

• What type of preparations needs to be made in advance?

• What tools must be on hand?

• Are the jigs and tools to be installed in good condition?

• What type of workbench is needed?

• Where should jigs and dies be placed after removal?

• How will they be transported?

• What types of parts are necessary?

• Are guide posts universal?

• How many maintenance people are needed to perform the set-up?

• Will operators assist in the set-up and conversion?

2.3.1: Traditional Setup Approaches

Processing set-up and conversions are the main emphases in most manufacturing plants. First, Shigeo Shingo the author of SMED, Single Minute Exchange of Dies addresses the traditional strategies for improving setup in operations. In traditional manufacturing operations, efficient set-up changes require knowledge, precision, and skill. As a result, set-up and changeover require highly skilled workers or "set-up engineers." While a set-up engineer performs the set-up, the machine operator is idle or performing other insignificant or miscellaneous tasks. Shigeo Shingo points out that this approach on set-up and conversion is a common misconception and is very inefficient. Many companies have set up policies to raise the skill level of workers, while few have tried to implement strategies that lower the skill level required by the set-up itself. Traditionally, manufacturing companies have also increased lot size to hide the effects of longer setup times. Large-scale production seems the easiest and most effective way to minimize set-up time's undesirable effects because the set-up time is minimal compared to each unit's operating time. This large lot size also increases inventory. This means that the machine will continue to run the same products which can delay the processing and delivery of the new product. The inventory itself does not produce added value, so the space it ties up is wasted, and the inventory costs increase. Inventory stock is at the risk of becoming outdated or even damaged. This approach assumes that reductions in set-up time are impossible. The SMED developed by Shigeo Shingo system makes these reduced set-up times a reality and a better alternative to increasing lot size.

2.3.2: The SMED System (Single Minute Exchange of Dies)

Shigeo Shingo found that set-up operations are composed of two different types, internal set-ups, and external set-ups. Internal set-ups are set-ups that can be performed only when a machine is stopped, while external set-ups are set-ups that can be conducted while the machine is in operation. This distinction between internal and external set-ups is one of the driving factors behind the SMED techniques; reduce set-up times by converting internal set-ups to external set-ups. SMED was developed over nineteen years due to closely investigating the theoretical and practical aspects of set-up improvement. It is a scientific approach to set-up time and reduction applied in any factory to any machine. It was implemented first into the Toyota Production System and has helped them to become the leading production system. In traditional setup operations, internal and external setups are mixed up. Some set-ups that could be done externally are performed as internal set-ups, causing machines to remain idle for extended periods. The first and most important stage in implementing SMED is to identify which set-ups are internal and external. The next stage is to convert the internal set-ups to external set-ups. The third and final stage is streamlining all aspects of the set-up operations. Many different kinds of waste occur in traditional set-up operations because internal and external operations are not distinguished properly. For example, in some manufacturing facilities, the machine will be turned off when the finished goods are transported to the warehouse. This will also be the time for the maintenance to get the tools, jigs needed which is a complete waste of valuable time.

The first stage of SMED is to separate internal and external set-ups. This task sounds easier than it actually is. Therefore Shigeo Shingo suggests using various techniques to accomplish this task. Checklists, charts, tables, visual control, functionality checks, identification, and transportation improvement to and from the machines are all suggested to help facilitate this stage. The next stage is to convert the internal set-up to external set-up operations. This stage is different from each process, and Shigeo Shingo provides various examples. Shigeo Shingo introduces function standardization, where he standardized only those parts whose functions are necessary from set-up operations. Efficient function standardization requires analyzing each piece of apparatus's functions, element by element, and replacing the fewest parts possible. When converting internal set-ups to external, standardizing the process as much as possible will reduce the set-up time. Standardizing is accomplished through different jigs and clamps, making everything the same, so set-up changes are minimal. Standardization reduces set-up time substantially, simplifies the organization, eliminates the need to search for appropriate tools, and eliminates the need for adjustments. The final stage in SMED is streamlining all aspects of the set-up operation. Improvements in transportation and storage of all parts, products, and tools can assist in streamlining operations. This stage doesn't necessarily reduce set-up time by itself, but it does aid SMED in creating a continuous flow. Having a standard and universal guiding pin will likewise shorten the time of making adjustments during the conversion time. If several bolts need to be removed during the conversion process, perhaps the use of wingnuts will shorten the time of removing them.

2.3.3: Techniques to Implement SMED

Some work in a manufacturing facility involves work at both the front and the machine's back. A single person working at one of these machines wastes time and movement when they continually walk around the machine. Shigeo Shingo suggests parallel operations involving more than one worker in this situation. The most important issue in parallel operations is safety and the delegation of jobs. Some sort of signal system has to be worked out so that workers in the process know when to do their respective jobs. The concept of parallel operations is also a technique that can be used to streamline the setup process. A functional clamp is an attachment device that is used to hold objects in place with minimal effort. Shigeo Shingo points out that a nut and bolt are used to fasten or tighten a clamp most of the time. If the bolt has fifteen threads on it, it can't really be tightened until the last turn, loosened in the first turn. The other fourteen turns are wasted. It is with this observation that Shingo spent time developing one-turn attachments and one-motion methods such as simple wing nuts which do not require any tools except for the hands.

There are also several techniques used to eliminate wasted adjustments. Adjustments and test run normally account for as much as 50% of the set-up time. Therefore, if there is a decrease in the adjustments, there is a reduction in set-up time. Inaccurate centering, dimensioning, and other procedures in the internal setup necessitate test runs and adjustments. To reduce the time for these adjustments, Shigeo Shingo says that we must improve the stages of internal setup. Minimizing these adjustments can be accomplished by fixing numerical settings, setting centers on machines, using gauges, and using reference planes. Shigeo Shingo also introduces the Least Common Multiple systems as a technique for eliminating adjustments. Finally, after every attempt has been made to improve set-ups, mechanization could reduce set-up times. Mechanization should be considered last to reduce set-ups because it is an inefficient setup operation that will achieve time reductions, but it will do little to remedy the basic faults of a poorly designed setup process. Mechanization can also cost a lot of money to implement. It is more efficient to mechanize set-ups that have already been streamlined. Shigeo Shingo strongly believes that SMED success involves knowing why the system works rather than just knowing how to implement it. Therefore, he provides a plethora of actual examples of the SMED system in operation so that the reader can gain further insight into the concepts and their principles.

2.3.4: Effects of SMED (Single Minute Exchange of Dies)

Although set-up time and conversion can be eliminated by not changing the equipment to another product and continuing processing the same products. However, this will cause a delay in the delivery of the product to the customer which can damage the reputation of the industry. The goal of SMED is to reduce set-up times and conversion of equipment that produce different products, but there are also other effects that SMED has on a production system. One effect that SMED has on a production system is that inventory is minimized. An inventory reduction can lead to more efficient use of plant space. The unusable stock due to model changeovers or mistaken estimates is eliminated. Production is increased because stock handling operations are eliminated. Goods are no longer lost through deterioration, mishandling, or damage. The ability to mix the production of various types of goods leads to further inventory reductions. Some other beneficial effects of SMED are that machine work rates and productive capacity are increased since downtime is reduced due to set-up and conversion. There is an elimination of human errors on set-up and conversion, and the elimination of trial runs lowers the occurrence of quality defects and trial processing on products. Quality and safety are both improved. Standardization reduces the number of tools required, and those that are still needed are organized more functionally for easy retrieval and access. Tool changes are quick and simple. Eliminating the need for highly skilled workers to perform the set-up, therefore lowering the skill level requirements for the process. SMED increases the manufacturing flexibility because of faster time in doing changeovers. Therefore a company can increase its production flexibility because it will respond rapidly to changes in demand. SMED can also have an effect on the attitude of people in the company. SMED welcomes employee involvement towards continuous improvement. Its effects consist of more than just reduced set-up times, conversion, and improved work rates. Industries can gain more from Shigeo Shingo's SMED technique than they realize.

2.3.5: Is Set-up Time a Planned or Unplanned Downtime?

Most manufacturing plants especially semiconductor industries consider setup changeover or conversion as a Planned Downtime. However, if we follow the Total Productive Maintenance process, it is clearly indicated that set-up is considered a machine downtime, therefore, this is unplanned downtime. This means that we will be having a different value if we compute for the machine utilization. Assuming that in one week, the equipment consumed 15 hours of set-up and conversion time, and 20 hours due to breakdown. Utilization will be;

If we consider Set-up as a Planned Downtime,
• Utilization = [(Loading Time – Planned Downtime) – Unplanned Downtime] / Loading Time
• Utilization = [(168 – 15) – 20] / (168 – 15)

• Utilization = 86.92%

If we consider Set-up as a Unplanned Downtime,
• Utilization = [(Loading Time – Planned Downtime) – Unplanned Downtime] / Loading Time
• Utilization = [(168 - 0) – 35] / (168)
• Utilization = 79.17%

Set-up time will always be a machine or unplanned downtime as this is included as one of the equipment losses in manufacturing industries.

2.4: Idling and Minor Stoppages

As industries' equipment becomes more complex and automated, more losses are attributed to Idling and Minor Stoppages. Some industries termed this as assists or errors. First, let us define this type of loss. A minor stoppage occurs when a failure or an error in automatic handling, processing, assembly of parts and workpieces, or an equipment stoppage is due to quality-related abnormality. This type of loss can be experienced mostly in automated processes and includes the following:

• Workpiece flow stops
• Operator resets workpieces correctly
• Operator reactivates process and machine runs

The problem with this type of loss is that the number of occurrences far exceeds the number of breakdowns, and fixing this type of failure requires a shorter time. This type of error is often left unrecorded since the time to fix it can be done by just resetting the buttons, and the operator will spend more time recording and writing down the error. Idling and Minor stoppages sometimes are mistaken for breakdowns, which must not be the case. These types of losses must be separated from breakdowns.

This type of equipment loss can be reduced through the implementation of Autonomous Maintenance through initial cleaning. JIPM experts believe that having equipment free from dirt and dust can reduce minor stoppages from 20 to 60%. While more complicated minor stoppages that cannot be reduced through cleaning must be addressed by cross-functional Kobetsu-Kaizen or Focused Improvement teams composed of process engineers, quality, production, and maintenance people. Some examples of Idling and Minor Stoppages include workpiece jamming, resetting of sensors, an error reading on the monitor of computerized equipment where the machine is stopped temporarily, operator resets, and the machine starts running again. These problems occur more frequently in automated equipment. This means that if the equipment contains many electronic parts, minor stoppages are highly possible to occur. In addressing minor stoppages, ensure the following, observe what is happening. These can be done by carefully observing the equipment until a minor stoppage occurred and planning corrective measures. Correcting slight or minor defects, a small dent in the chute may cause minor stoppages, and frequently, we overlook this factor. Maintaining the cleanliness of the equipment can reduce minor stoppages. Remember that the main distinction between a minor stoppage and a breakdown is that there is nothing that failed on the equipment. The equipment just stopped and must be reset and restarted once again. But there are also cases that a minor stoppage can lead to a breakdown. Industries must be familiar with the distinction, if this will be mixed, then we will be having variations in our indices especially in calculating MTBF and other metrics.

2.5: Design Speed Loss

Design Speed Loss is the loss of production or output caused by the difference between the design speed or theoretical speed and the equipment's actual operating speed. Lack of care at the design stage of the equipment may result in speed reduction. Equipment is operated beyond its operating speed limit, quality defects and breakdowns are encountered. Although Japanese and European maintenance systems had conflicts on this type of loss, Japanese TPM experts still consider this a type of loss. According to them, to reduce speed loss, we must address the following:

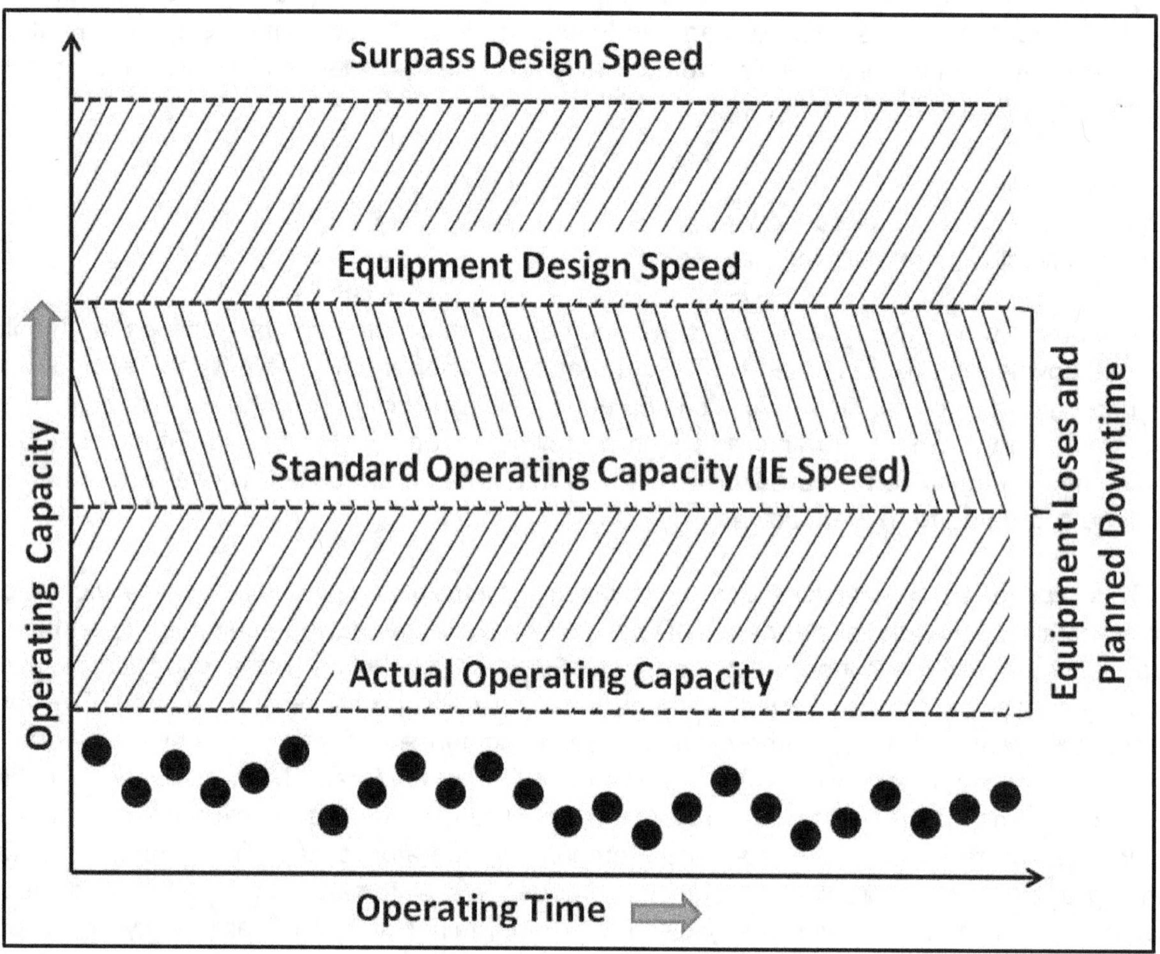

Figure 2.6: Design Speed Loss

• Level 1: Achieve Standard Operating Speed for each product
• Level 2: Increase Standard Operating Speed for each product
• Level 3: Achieve Design Speed
• Level 4: Surpass Design Speed

This type of equipment loss will not experience downtime since the equipment is still running at a reduced speed. This means that the availability of the equipment may be high, however, the Performance Efficiency of the equipment will be lower. The design speed is actually the desired performance that the users want the asset to perform, while the actual operating speed is also termed as the built-in capability in which this is what the asset can perform on day-to-day operations. Equipment may be run at less than the design or ideal speed for various reasons since it can contribute to mechanical problems, quality defects, history of past problems, and sometimes not knowing the optimal speed. On the other hand, deliberately increasing the operating speed actually contributes to quality problems by revealing latent defects in equipment conditions. The goal is to eliminate the gap between the actual speed and the design speed of the equipment. Design speed is reduced due to problems with quality or mechanical trouble, past trouble resulting in short lifespan, equipment being operated with its specification unknown, or increasing the speed will reveal concealed problems either on the product or equipment itself.

Running the equipment on its rated or design speed will produce problems with the quality of the product. Thus, maintenance will reduce the machine's running speed; however, in most cases, production will increase the equipment's speed to deal with productivity issues and provide more serious downtime. Maintenance needs to know what parts are affected when the speed of the equipment is increased. Study the part concerning its design, shape, the strength of materials, environment, and so on. TPM Planned Maintenance Phase 2 will focus on lengthening the equipment lifespan by addressing inherent design weaknesses on the equipment.

2.6: Start-Up Loss

When you want to photocopy something, and the Xerox machine is off, when you turned it on, it will warm up for a few seconds just like the printer we have in our home. Start-Up loss is a type of loss that occurs during an equipment start-up or run-in. Problems arise during starting the equipment. Start-Up Loss means the material loss caused at the initial stage of product launching, namely the loss caused during the start-up of the machine to the stabilized production stage. Its frequency depends on several factors, such as unstable machining conditions, poor maintenance, and operator skills, human errors, and infant mortality failures. If we speak about the 6 Failure Patterns, this refers to D which is also known as the initial break-in curve. This failure starts off with a very low level of failure followed by a sharp rise to a constant level. This pattern accounts for approximately 7% of failures according to the study of Nowlan and Heap from United Airlines.

Start-Up loss can also occur after a poor set-up or conversion, after overhauling equipment subject to a Preventive Maintenance Schedule. It takes some time for a machine to stabilize, or premature failures are experienced right after the PM endorses the equipment back to operators. This can also be considered as pattern F again on the 6 failure pattern in which the curve shows a high incidence of infant mortality failures. This curve usually shows a high initial failure rate followed by a random level of failures. This pattern accounts for 68% of failures. These are failures that occur at the beginning of its life. Others call them start-up failures, commissioning failures that are likely to occur after a major overhaul or a major Preventive

Maintenance had been initiated. The Infant mortality period is a time when the failure rate is dropping but is undesirable because a significant number of failures occur in a short time, causing early customer dissatisfaction and warranty expense. The main cause of infant mortality failures are human errors usually committed during the reassembly process of overhauling. Implementing Precision Maintenance will reduce infant mortality failures. Start-up losses have many causes. Their scope includes the stability of processing conditions, workers' skills, training, machine warming up, the loss incurred by test operations, and other factors.

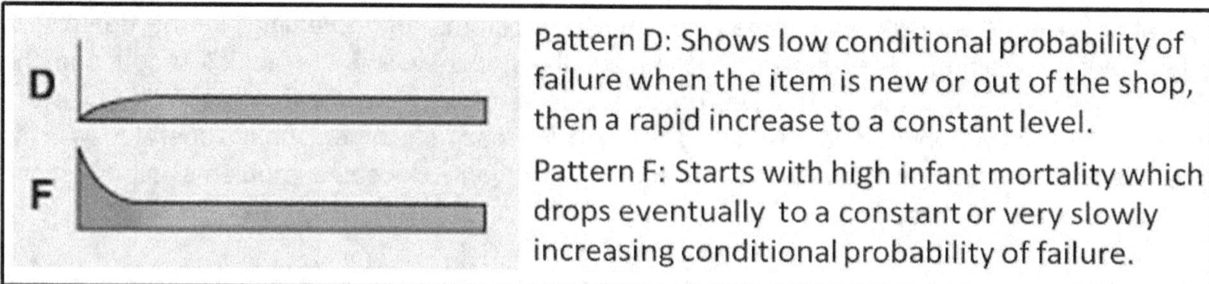

Figure 2.7: Pattern's E and F of the Six Failure Pattern

2.7: Defect and Rework Loss

These are losses caused when defects are found and the product has to be reworked. In general, these defects are likely to be considered a waste that should be disposed of, but even the reworked products need wasted manpower to repair them. Included in this type of loss are products being shipped back due to customer complaints. Products are returned and reworked in the production area.

Quality Defects and rework losses are caused by malfunctioning production equipment. In general, sporadic defects are easily corrected. Chronic defects require a thorough investigation and innovative remedial action. The conditions surrounding and causing the defect must be determined and then effectively controlled. Tools such as P-M Analysis (P stands for Physical, Phenomenon, while M stands for Mechanism and the 4M Man, Machine, Method, and Materials) are suitable in dealing with chronic problems. This tool will aim at reducing the defects to zero. However, P-M Analysis must only be used after extensive use of conventional analytical problem-solving tools, and around 1 to 5 % of the problems still exist; this is where P-M Analysis must be used ideally. Do not use P-M Analysis, when the rate of defects is still high. Sporadic defects are easy to solve, and it is more difficult to solve chronic defects. Although other TPM books include 8 Major Equipment Losses, the other two losses mentioned are shutdown loss, which means stopping the equipment for a periodical or Time-Based Maintenance, and Cutting blade change, which is loss caused by line stoppage for replacing the grinding wheel, cutter, bits which might be broken or worn out after excessive usage.

Quality defects are more difficult to address compared to breakdowns since most of these defects are chronic by nature. The source of quality defects may come not only from the machine but may be due to the environment, human error, or poor quality of raw materials supplied. Chronic defects have multiple causes that make it is difficult to address. A given problem with 10 potential causes 1 to 10 and each time the problem occurs, the cause is different, and always in combination. Most problem-solving tools will measure and focus only on one specific cause, while the problem resurfaced once again. Today problem is caused by 1, 2, 3, and when the second shift comes, the cause is 7, 8, and 10. How do you deal with this type of problem? For Chronic problems, we need to identify all factors that contributed to the loss. We need to investigate thoroughly each factor and eliminate any malfunctions or sub-optimal conditions discovered in the process.

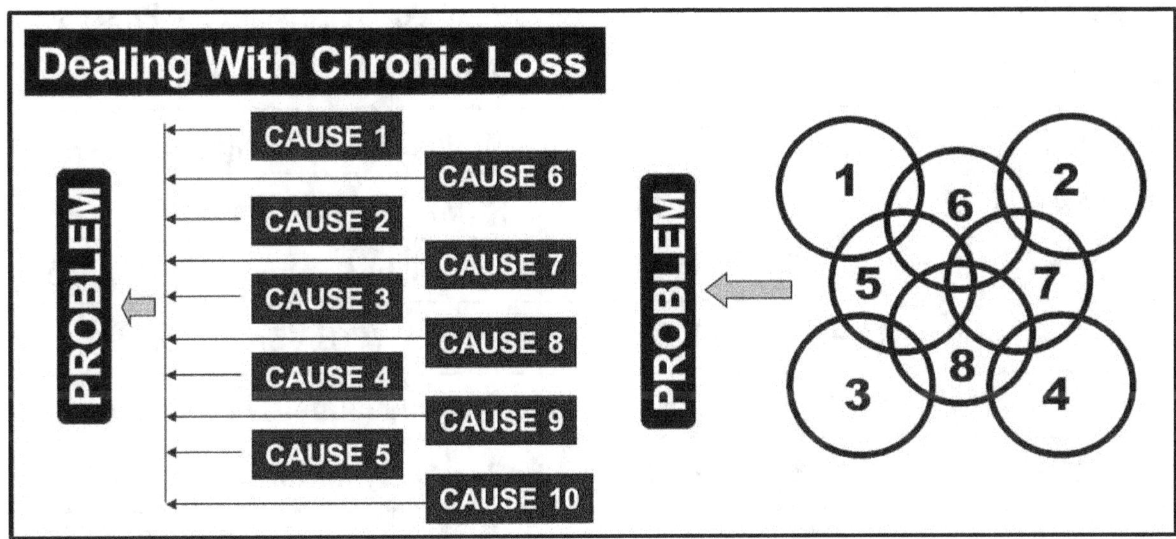

Figure 2.8: Complexity of Chronic Losses

Chronic losses exist since we do not understand the nature of chronic loss as well as using ineffective approaches in dealing with chronic losses. A tool to solve chronic losses is known as P-M Analysis. P stands for Physical and M stands for Mechanism and the 4m (man, machine, method, and materials). Physical analysis physically analyzes chronic losses according to the inherent principles and natural laws that govern them. Its principle asks in precise physical terms what happens when a machine breaks down or produces bad parts and how it happens. This tool clarifies the mechanics of their occurrence and the conditions that must be controlled to prevent them. It will physically analyze chronic problems such as defects and failures according to the machine's operating principles.

2.8: Other Equipment Losses

Shutdown Losses are referred to as line shutdown losses which are caused by stopping the equipment for periodical maintenance and scheduled or planned annual shutdown. Periodic inspections are usually performed during shutdown maintenance, while others may be required by law for compliance. Among all the equipment losses, this will be considered as Planned Downtime.

Cutting Tool Change there are machines where downtime will be caused by line stoppage for replacing the tools, grinder, wheels, blades, dies, punches, cutter which might be broken or have deteriorated to a point that it needs to be replaced as this will eventually contribute to the defects on the product. When a cutting blade is replaced since it has worn out and the replacement was done before the failure, then this is considered as a Planned Downtime, however, if the cutting blade was replaced due to breakage or fracture, then this will be considered as unplanned downtime.   The breakage of the cutting tool can also be considered in this case as a breakdown.

8 EQUIPMENT LOSSES	TPM PILLAR	HOW TO ADDRESS
BREAKDOWN LOSS	PM	Planned Maintenance 4 Phases
SET-UP & CONVERSION	FI	SMED by Shigeo Shingo
CUTTING TOOL CHANGE	PM, FI	Modification, Change in material
START-UP LOSSES	PM	RCM, Precision Maintenance
SHUTDOWN LOSS	PM	Reliability-Centred Maintenance
IDLING & MINOR STOPS	AM, FI, PM	Perform MTBA Snapshots
DESIGN SPEED LOSS	FI	PM Phase, Life Cycle Costing
DEFECT & REWORK LOSS	FI, QM	P-M Analysis for Chronic Defects

Legend: PM – Planned Maintenance, FI – Focused Improvement, AM – Autonomous Maintenance, QM – Quality Maintenance

Figure 2.9: Different Equipment Losses and How to Address Them

2.9: Manpower Losses

Besides the equipment losses, we also have Manpower losses which also contribute to lost productivity and lower production. Although a different approach will be used to reduce these losses. What is important is to identify these manpower losses that are currently present in your industry and perform a time and motion study.   For example, if the operator completed processing the products and will transport the product to the next process or machine, how much time would it take the operator to do that? Would it be possible to streamline the process and move the machine nearer so lesser time will be consumed in transporting the products? These manpower losses include the following:

1. **Management Loss**: This includes the waiting time that is lost caused by management problems and delays such as no available MRO spare parts, lack of manpower resources, lack of insufficient utilities, and work instructions.

2. **Operator Motion Loss:** These are losses generated due to unnecessary or excessive movement by the operator, as a result of poor layout, material transfer, and work organization due to excessive walking, wasted motion, unnecessary reaching or equipment are far apart where the operator will take a longer time to transport the product to the next process in the manufacturing process.

3. **Line Organization Loss:** This loss results from a shortage of operators on the production floor where the operator needs to operate other machines that were originally planned. This

often results in a shorter break time for the operators which can cause the operator to fatigue easily.

4. **Distribution Loss:** This loss occurs due to incorrect or inefficient delivery of raw materials, packaging, or products to and from the factory or the production line. Example: Incorrect delivery of materials from supplier to store, late deliveries, excessive handling of deliveries (double handling)

5. **Measurement and Adjustment Loss:** This loss is caused by the frequent measurement and adjustment to prevent the recurrence of problems. Example: Excessive inspection integrated into the process as a result of poor quality and failure to find the root cause. Operators perform 100% inspection of the products before processing them on the equipment. Adjustment loss is experienced when adjusting equipment back to the standard after routine cleaning and periodic consumable changes.

2.10: Other Losses

1. **Yield Losses:** This is the total loss between the input of raw material and the output of finished goods. Examples: over-pack, giveaway, mass balances

2. **Energy Loss:** In this type of loss, the facilities/utilities are involved in these losses as they supply the power, air, cooling system to most manufacturing equipment. This is the loss in the input energy which cannot be used effectively for processing. Most of these losses happen on the facilities and utility equipment that supports the manufacturing equipment. Examples: Start-up losses, insufficient compressed air for pneumatic machines.

3. **Die, Tool, and Jig Losses:** This is the cost of the physical consumption of the spare parts or the refurbishment of items that are used on the line. Examples: Cost of spares, cost of replacement, and maintenance tooling, dies, and jigs.

2.11: Take Quiz on Equipment Losses

1. Takes 40 minutes to convert, this process to another, and another 30 minutes for adjustment.
   a) Breakdown Loss
   b) Set-up and Adjustment Loss
   c) Minor Stoppages
   d) Design Speed Loss
   e) Defect and Rework Losses
   f) Start-Up Losses

2 Equipment is down due to failure of hydraulic press
   a) Breakdown Loss
   b) Set-up and Adjustment Loss
   c) Minor Stoppages
   d) Design Speed Loss

e) Defect and Rework Losses
f) Start-Up Losses

3. This pump is rated at 300 GPM however it is running only at 220 GPM
a) Breakdown Loss
b) Set-up and Adjustment Loss
c) Minor Stoppages
d) Design Speed Loss
e) Defect and Rework Losses
f) Start-Up Losses

4. After performing PM on this equipment, the operator complains that it ain't running.
a) Breakdown Loss
b) Set-up and Adjustment Loss
c) Minor Stoppages
d) Design Speed Loss
e) Defect and Rework Losses
f) Start-Up Losses

5. A product was stuck on the chute, creates jamming, operator resets the machine and it started running, after a few minutes the same thing happens.
a) Breakdown Loss
b) Set-up and Adjustment Loss
c) Minor Stoppages
d) Design Speed Loss
e) Defect and Rework Losses
f) Start-Up Losses

6. Sporadic equipment failures are part of what type of loss?
a) Breakdown Loss
b) Set-up and Adjustment Loss
c) Minor Stoppages
d) Design Speed Loss
e) Defect and Rework Losses
f) Start-Up Losses

7. Bearing seizure and needs to be replaced immediately.
a) Breakdown Loss
b) Set-up and Adjustment Loss
c) Minor Stoppages
d) Design Speed Loss
e) Defect and Rework Losses
f) Start-Up Losses

8. Customer returns shipment since the product does not conform to their specification.
a) Breakdown Loss
b) Set-up and Adjustment Loss
c) Minor Stoppages

    d) Design Speed Loss
    e) Defect and Rework Losses
    f) Start-Up Losses

9. Three types of products are being operated in this equipment and it takes 1 hour to change one product to another.
    a) Breakdown Loss
    b) Set-up and Adjustment Loss
    c) Minor Stoppages
    d) Design Speed Loss
    e) Defect and Rework Losses
    f) Start-Up Losses

10. This is the only equipment loss that will not experience downtime but can decrease the Performance Efficiency of the machine.
    a) Breakdown Loss
    b) Set-up and Adjustment Loss
    c) Minor Stoppages
    d) Design Speed Loss
    e) Defect and Rework Losses
    f) Start-Up Losses

11. Pattern D of the 6 failure pattern refers to what type of loss?
    a) Breakdown Loss
    b) Set-up and Adjustment Loss
    c) Minor Stoppages
    d) Design Speed Loss
    e) Defect and Rework Losses
    f) Start-Up Losses

12. TPM believes that if the equipment is maintained clean, 30% of what type of loss will be reduced from the equipment.
    a) Breakdown Loss
    b) Set-up and Adjustment Loss
    c) Minor Stoppages
    d) Design Speed Loss
    e) Defect and Rework Losses
    f) Start-Up Losses

13. The design capacity of this equipment is 6000 ups, however, we are only producing 5200 with no breakdowns and adjustments, why?
    a) Breakdown Loss
    b) Set-up and Adjustment Loss
    c) Minor Stoppages
    d) Design Speed Loss
    e) Defect and Rework Losses
    f) Start-Up Losses

14. Maintenance analysis indicates that a low supply of air cause products to clog and jam.
    a) Breakdown Loss
    b) Set-up and Adjustment Loss
    c) Minor Stoppages
    d) Design Speed Loss
    e) Defect and Rework Losses
    f) Start-Up Losses

15. Speed of this equipment was reduced since it keeps on producing scraps and defects.
    a) Breakdown Loss
    b) Set-up and Adjustment Loss
    c) Minor Stoppages
    d) Design Speed Loss
    e) Defect and Rework Losses
    f) Start-Up Losses

Refer to Appendix B for the Answers

Chapter **3**

Overall Equipment Effectiveness Explained

> ***OEE is a good measurement, but not a perfect one, since it is just a measurement of the primary function of the equipment. Remember that equipment have both primary and secondary functions, and there are cases where the failure of a secondary function will be much more dangerous than the failure of the primary function of the equipment***

3.1: Overall Equipment Effectiveness Explained

Chapter 2 explains the different losses equipment can run into. These losses can be measured through an indicator called OEE or Overall Equipment Effectiveness. OEE has been first introduced in the book of Seiichi Nakajima, An Introduction to TPM, Total Productive Maintenance. Every industry has a product or service to render, but more importantly, it is your customer that spells the difference why a company exists and still remains in business. Today, industries adopt the law of the jungle and that is to survive and be left behind. Therefore, to survive in this tough business competition, we need our customers to stay and remain loyal, and to do this, your industry needs to satisfy the following requirements.

1) Those that can produce the lowest possible costs.
2) Those that can produce the highest quality product or service with the best customer satisfaction.
3) Those that can produce the fastest delivery of them all.
4) Those that can make decisions, adapt and innovate to rapid changes in the market trend.

If any of these cannot be provided by the industry, then there will be repercussions, risks, and the possibility of your industry shutting down permanently. In the previous chapter, we discussed the different losses that are possible to be experienced on our equipment and assets. One measurement that will include all these equipment losses is OEE or Overall Equipment Effectiveness. OEE is the primary measure of TPM. Although this is a consolidated KPI, which means that maintenance is not only the sole person responsible for improving OEE as it contains several equipment losses, and Planned Maintenance will only be focusing on breakdowns, which are just one of the major equipment losses. OEE indicates the relative productivity of a piece of equipment compared to its theoretical performance. It will identify the bottleneck equipment critical for the improvement of equipment's productivity.

[2]OEE indicates the relative productivity of a piece of equipment compared to its theoretical performance. It is used to prioritize equipment to improve productivity. OEE contains three components, which are Availability, Performance Rate, and Quality Rate. For those implementing TPM (Total Productive Maintenance), OEE will be the primary measure of performance. It is a function of the complete manufacturing system, process line, or individual piece of equipment. It will measure the effective utilization of capital assets by expressing the impact of equipment related losses which includes the following:

• Machine downtime caused by Breakdowns
• Time required for Set-Up, Conversion, and Adjustments
• Time lost due to inefficient start-up
• Time lost due to tooling
• Time lost due to Minor Stoppages
• Time lost due to Shutdown
• Production lost due to Design Speed Loss
• Production lost due to defective products
• Production lost due to rework and reprocessing

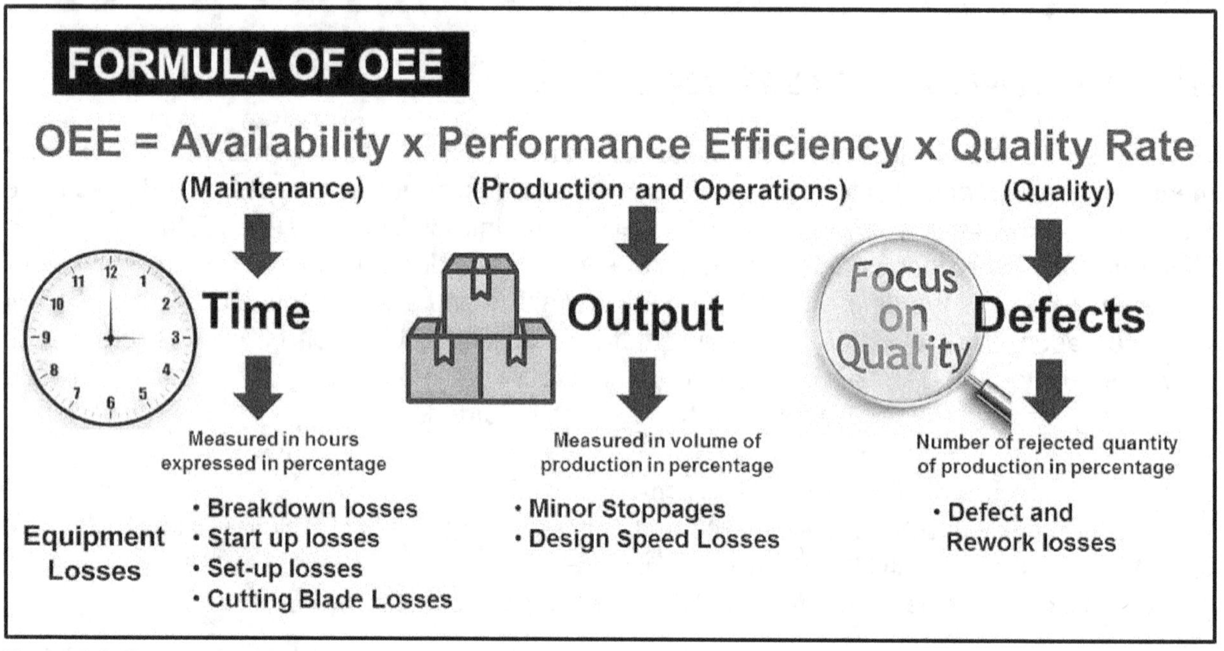

Figure 3.1: Formula for OEE

As we can see in figure 3.1, OEE is composed of three components, Availability, Performance Rate, and Quality Rate. This means that there are three functions involved in this situation, Availability for maintenance, Performance Efficiency for Operations, and Quality Rate for the Quality function. This means that improving the machine's OEE cannot be done solely by just a single function like maintenance which some books claim. Improving OEE is a consolidated effort of these major functions with the help of other functions as well. This means

---

[2] Charles Robinson and Andrew Ginder, *Implementing TPM – The North American Experience,* (Productivity Press Inc, 1995), Page 125

that if the MRO Storeroom cannot provide the parts to maintenance, then downtime caused by the breakdown will be prolonged. If the raw materials supplied are of poor quality, then the number of defective products will increase which can contribute to a low yield. Except for breakdown losses which are addressed by the TPM Planned Maintenance pillar, typically a cross-functional team called Focused Improvement or Kobetsu Kaizen will be deployed to improve the OEE by addressing the losses experienced by the equipment.

3.2: Equipment Downtime Explained

Downtime can be classified as Planned which is also called non-machine related and Unplanned Downtime which is also considered as machine-related downtime. In industries, downtime is a term in which the equipment is not running. There are two reasons why the equipment is not running, first either it was broken or something failed. The second is that the machine had not been utilized. Ok, are you telling me that even if the machine ain't broken, we have downtime. My reply is yes. When the machine is working but has not been used, then we can conclude that the machine is available but not utilized. Downtime is a situation in which the equipment stopped either due to planned or unplanned events in the plant's scheduled operations. Although it might seem unfair and the maintenance will say that why do you consider my car is down if it is just parked in my garage? It is available and I can use it anytime I want. Similarly, for industries, why do you penalize me for having a low Availability if my equipment is running, and available? I have experience feuds between operations, maintenance, and other functions regarding this, especially when the machine is not loaded but available, where the Maintenance Manager defend his crew and question why they should be blamed if the machine is not utilized.

The important thing is that before we discuss any of these KPIs and indices, especially OEE, your industry should have a clear definition and distinction of what will be included as Planned and Unplanned Downtime. Except for Preventive Maintenance, Scheduled Outages, and Shutdown, Planned Downtime is a proportion of time that the equipment is available but not actually utilized. Refer to figure 3.2. In this case, the equipment was actually not in a failed state, but we need to account for this as a planned downtime since it was not used. On the other end, unplanned downtime or machine-related downtime is downtime caused by the different equipment losses except for Design Speed Loss, since there will be no downtime in this case only a reduction in the speed of the equipment which will affect the Performance Efficiency of the machine.

When an operator from a manufacturing plant shuts down the equipment to take his lunch break and consumes 1 hour, this will be included as a Planned Downtime even if the equipment is not in a failed state. Another case is if the delivery of the raw materials needed to produce the product is delayed. Although the machine is available, it was not used. In both these cases, the machine is idle. This is considered a Planned Downtime. Key people from the operations, maintenance, reliability, quality, process, and other functions must have a sit down just like the La Cosa Nostra of Mafia and discuss the lists of planned and unplanned downtime that they experience on the equipment, otherwise, if this will not be done, then we will be having problems and variations in calculating our maintenance indices.

Planned Downtime	Uplanned Downtime
Non-Machine Related Downtime	**Machine Related Downtime**
- Shutdown Loss	- Breakdown Losses
- Preventive Maintenance Activities	- Set-up and Changeover
- Scheduled Outage	- Start-up Losses
- Operator's Breaktime	- Changing Blades, cutters
- No Operator	- Quality Defects where operator
- No Raw Material to Process	stopped the machine
- Christmas and Holidays	- Idling and Minor Stoppages
- Meetings and General Assembly	- All Function Loss Breakdown
- Machine Audits	
- Predictive Maintenance Activities	
- TPM Activities such as Initial Cleaning	
- Operator Cleaning Activities	

Figure 3.2: Planned and Unplanned Downtime

3.3: Availability and Utilization Explained

When you drive your car to go to work, and it takes you one hour to go to your plant and another one hour to go back to your house, then you spend at least two hours driving your car. This means that the car has been utilized only for two hours for the entire day. The remaining 22 hours was either the car was parked inside the plant or in your garage. Assuming that the car has no problems and it runs smoothly, the utilization can be calculated in percentage as follows:

• Utilization = 2 / 24 x 100% = 8.33%, while the Availability = 100%

Since the car is parked inside your garage for around 14 hours and 8 hours in the plant. The total time the car was parked was 22 hours, in which 2 hours was spent driving it to your plant or place of work and back to your house. Since there are no failures encountered during the day, Availability can be determined easily to be 100%. The car is available for 24 hours, in which only 2 hours of it are actually utilized. This sounds simple, but not always in the case for industries, especially manufacturing, since the time it is parked will be considered downtime, whether we agree or disagree, especially for the maintenance function.

Utilization is when a machine is actually used, divided by the time a machine is being loaded and able to work. On the other hand, availability is when a machine is available for work less all the downtime divided by the total available time. Let us provide an example to understand my point of view fully. For example, if the plant operates 330 days in one year, then the 35 days will be considered as the time the machine is idle as it is not included in the plant's operating time. If your plant had shut down during this pandemic, then we cannot declare a downtime but an idle time. But most Industries will only consider both Planned and Unplanned downtime.

DIFFERENT FORMULAS ON AVAILABILITY

According to several books on TPM, the formula for Availability is:

1) Availability = Operating Time / Loading Time
= (Loading time – Downtime) / Loading Time

Introduction to TPM by Seiichi Nakajima page 22
Nakajima is the Father and Founder of TPM

2) Availability = Operating Time / Net Available Time
Where Net Available Time = Total Available Time – Planned Downtime

Implementing TPM The North American Experience
By Charles Robinson and Andrew Ginder, page 120 (Published 1995)

3) Availability = (Calendar Time – Losses) / Calendar Time
What is meant by Calendar Time is the Available Time Industry Operates

TPM in Process Industries
By Tokutaro Suzuki, page 29 (published 1992)

4) Availability = (Required Availability - Downtime) / Required Availability
Required Availability is equal to the Available Time Industry Operates

Total Productive Maintenance
By Terry Wireman page 37 (published 1991)

5) Availability = (Availability – All DT) / Availability Previous JIPM Consultant

---

Figure 3.3: Different Variations in Calculating Availability

In figure 3.3, assuming that operations were 24 hours, 7 days a week. 7 hours were consumed by the operator's lunch break and 20 hours where the machine experienced an Unplanned Downtime due to breakdown during the week. What will be the availability in this case? If we use the formula given in figure 3.3, we will be having some variations in the value for availability.

Given:
• Available Time = 168 hours (7 x 24)
• Planned Downtime = 7 hours due to operator's lunch break during the week
• Unplanned Downtime = 20 hours due to breakdown

From Seiichi Nakajima, Introduction to TPM book

$$\text{Availability} = \frac{\text{Operating Time}}{\text{Loading Time}} = [(168 - 7) - 20)] / (168 - 7) = 87.58\%$$

From Robinson and Ginder, Implementing TPM The North American Experience book

$$\text{Availability} = \frac{\text{Operating Time}}{\text{Net Available Time}} = [(168 - 7) - 20)] / (168 - 7) = 87.58\%$$

From the book of Tokutaro Suzuki, TPM in Process Industries book

$$\text{Availability} = \frac{\text{Calendar Time} - \text{Losses}}{\text{Calendar Time}} = (168 - 27) / (168) = 83.93\%$$

From Terry Wireman, Total Productive Maintenance book

$$\text{Availability} = \frac{\text{Calendar Time} - \text{Losses}}{\text{Calendar Time}} = (168 - 27) / (168) = 83.93\%$$

From my previous JIPM Consultant book

$$\text{Availability} = \frac{\text{Available Time} - \text{All Downtime}}{\text{Available Time}} = (168 - 27) / (168) = 83.93\%$$

Reminiscing my employment days in a large semiconductor industry, I recalled that when I was introducing OEE as one of the indicators that Planned Maintenance need to monitor on our pilot equipment, one manager asked me a question regarding the formula of Availability, which I extracted from the book of Seiichi Nakajima, in which we are going to used.

• Availability = (Operating Time / Loading Time) x 100%

Where Operating Time is equal to the Loading Time minus Unplanned Downtime and Loading is equal to the Available Time minus Planned Downtime.

There was a manager which raised a question, Rolly isn't it that your formula for Availability is actually Utilization? I recalled saying that this is the term that Seiichi Nakajima used and he was the founder of TPM, so let us just comply. But come to think about it, I believed that the manager who raised that question was actually right that it was utilization and not availability. Although we used the formula from the book of Seiichi Nakajima to calculate Availability which we also used for calculating OEE, when we hired a JIPM (Japan Institute of Plant Maintenance), he told us to use his formula which will give the lowest value of Availability. We raised this question to our previous JIPM Consultant on why the founder of TPM used this formula, and the JIPM consultant just looked at me smiled, and said to just use his formula, perhaps out of respect to the original founder of TPM. His reason was that having a low Availability will provide more room for improvement. Since the formula for Availability will be changed, we raised this concern with our CEO and advised us to have 2 Availability, as well as OEE. First, we have the original plant's Availability originally used, and second, is to have an Availability based on the formula given by the JIPM Consultant which will be much lower than our original availability which settles the problem. You see the main difference in the formula between Availability and Utilization is that Availability's denominator will be the available time, while for Utilization, the denominator will be the loading time. Since Nakajima was using Loading Time and not the Available Time, this will indicate the Utilization and not the Availability.

Availability is the proportion of time that a system or equipment is in a functioning condition, whether it will be used or not by operations, while **Utilization** is a proportion of time, the equipment has been used or utilized throughout a given time or period. Availability is a maintenance measurement, while utilization is an operations measurement. The goal of maintenance is to make the equipment available for production use, while the goal of operations is to ensure that the equipment is being utilized. Maintenance should not measure utilization because this is one measurement that they cannot control. If the machine is not utilized, maintenance has no total control over this matter. It is not the job of maintenance to call the vendor and get mad due to the delay of raw materials or drive around the city and look for potential customers so that they can run the machine. I know of some industries that forced their maintenance to track utilization. This is absolutely crazy. Whether the equipment will be used or not, the maintenance role is to make the equipment available. The main difference between Availability and Utilization lies in its denominator. Availability will use Available Time, while Utilization will use Loading Time which is also equal to the Available Time minus the Planned Downtime.

Case 1: A plant produces candies and operates for three shifts covering an 8 hour per shift period. Both availability and utilization are being monitored by this candy plant, and a daily report is forwarded to their computer to determine their daily measurement. During the day, the equipment was operated for 24 hours, of which 3 hours are spent on break time by the operators for their meals. The equipment suffers breakdowns amounting to 6 hours on that day. The equipment is fully loaded. Calculate both the availability and utilization of this equipment.

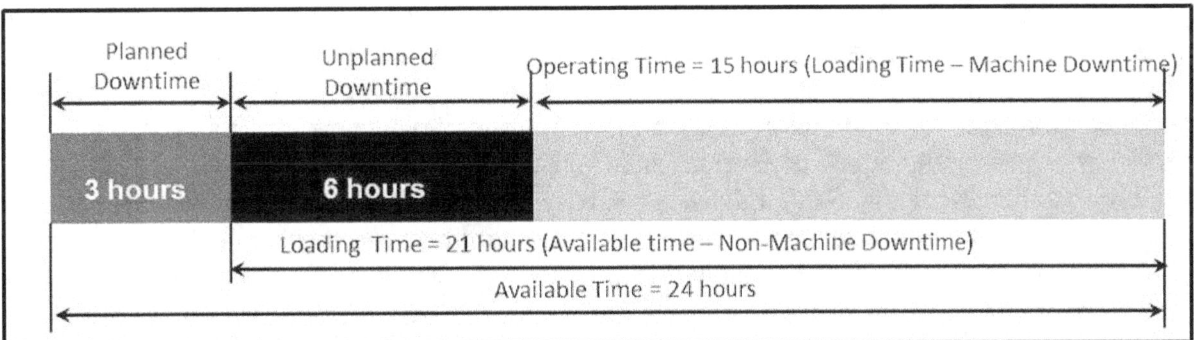

Figure 3.4: Calculating Availability and Utilization

Given:
- Available Time is the total time the machine can operate during a specific given time, assuming that there is no downtime, whether machine or non-machine-related, and is given as 24 hours.
- Planned Downtime also referred to as non-machine-related downtime in the form of lunch breaks and is given as 3 hours.
- Unplanned Downtime is also referred to as machine-related downtime. In this case, the equipment was down due to breakdowns for 6 hours.
- Loading Time is the time the equipment was loaded which is the Available time minus the Planned Downtime at 21 hours
- Operating Time: This is when equipment is operating or is actually loaded, excluding both machine and non-machine-related downtime, and is given as 15 hours.

$$\text{Utilization} = \frac{\text{Loading Time} - \text{Unplanned Downtime}}{\text{Loading Time}} \ (100\%)$$

$$\text{Utilization} = \frac{\text{Operating Time}}{\text{Loading Time}} \ (100\%)$$

$$\text{Utilization} = \frac{(24 - 3) - 6}{21} \ (100\%) = 71.43\%$$

$$\text{Availability} = \frac{\text{Available Time} - \text{All Downtime}}{\text{Available Time}} \ (100\%)$$

$$\text{Availability} = \frac{24 - (3 + 6)}{24} \ (100\%) = 62.5\%$$

On the other hand, Availability is computed as the total available time minus all the downtime encountered on the equipment, whether machine or non-machine related, divided by the available time itself. In the case above, the available time is 24 hours. In TPM books, OEE is calculated as the availability multiplied by the performance rate multiplied by the quality rate expressed in percentage, and availability is computed to be the loading time minus the machine-related downtime divided by the loading time.

The example above is straightforward. The confusion begins when the equipment is not loaded. On the other hand, what if the equipment has not been running for quite a time since the demand is low and the equipment was on standby mode. This is relatively easy if we schedule the equipment for a scheduled Preventive Maintenance since the machine has no load but is relatively difficult to calculate if the equipment is not operating but available. The difficulty is in deciding where the time the equipment is not loaded will be included. Will it be included in the machine-related downtime or in the non-machine-related downtime? For sure, if we include this in the machine-related downtime, both utilization and availability will be low, and maintenance might complain about a low availability and ask us why do we pass the burden to them for things that maintenance has no relative control over the matter? Remarks by maintenance usually indicate that their role is for the equipment to be available, and if the equipment is not used or in standby mode since there is no volume to produce, the blame should not be passed on to the maintenance and definitely, they have a point, and their concern is indeed valid. That is why as said previously, there should be a sit down on what to include or exclude as planned and unplanned downtime. Here is a case where the machine is not loaded for a few hours.

Case 2: During A-Shift, a semiconductor End of Line DTFS machine number SOSY-002 operated under the following conditions. The operator attended a Quality Control meeting for 0.5 hours, the machine encountered a breakdown, and it took the maintenance 2 hours to put the equipment back in operation, and the machine operated for 3.5 hours during the shift since there was no raw material available for 2 hours. Compute for the utilization and availability?

Given:
• Available Time = 24 hours
• Planned Downtime = 0.5 hours

• Unplanned Downtime = 2 hours
• Note that the machine was only loaded for 3.5 hours

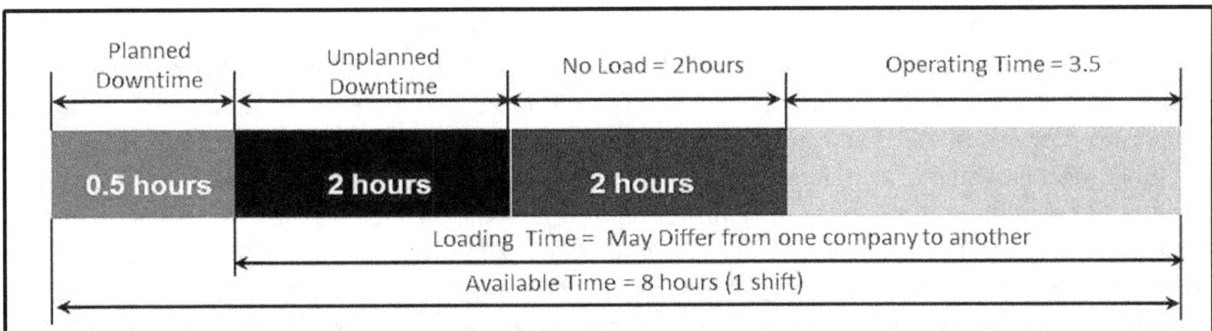

Figure 3.5: Many Ways of Calculating Utilization

Case 1: No load is taken into consideration (Excluding Material Downtime)

$$\text{Utilization} = \frac{\text{Loading Time} - \text{Machine downtime}\,(100)}{\text{Loading Time}} = \frac{(8-0.5) - 2 \times 100\%}{(8-0.5)} = \mathbf{73.33\%}$$

Case 2: No load is included as Planned Downtime

$$\text{Utilization} = \frac{\text{Loading Time} - \text{Machine downtime}\,(100)}{\text{Loading Time}} = \frac{(8-2.5) - 2 \times 100\%}{(8-2.5)} = \mathbf{63.63\%}$$

Case 3: No load is included as Unplanned Downtime

$$\text{Utilization} = \frac{\text{Loading Time} - \text{Machine downtime}\,(100)}{\text{Loading Time}} = \frac{(8-0.5) - 4 \times 100\%}{(8-0.5)} = \mathbf{46.67\%}$$

$$\text{Available Time} = \frac{[8 - (0.5+2+2)]}{8} \times 100\% = \mathbf{43.75\%}$$

For Case 1: Higher utilization percentage at 73.33%, since the two hours the equipment is in standby or idle mode was not included in either the machine-related or non-machine-related downtime. The maintenance point of view would be, why should we penalize ourselves for something we have no control over? Our job is to keep the machine available whenever operations need it, and that is the way it should be. If the equipment is not loaded, then it is not the maintenance fault.

For Case 2: The utilization value is much lower at 63.63%. In this case, since the equipment is in standby or idle time mode, the no-load or standby time is definitely considered and included in the non-machine-related downtime. Industries are aware of their equipment will be utilized or not depending on the volume. Production planning knows when production will be high or low. The only thing Production Planning does not know is that equipment also suffers from breakdown, and its calculation is always based on the capacity of the equipment running at its peak without a single failure multiplied by a factor of nowhere. Industries such as manufacturing and semiconductor have a peak in demand and a fallout season where their products' demand is low. Some will take advantage of this situation to perform some Preventive Maintenance activities on their equipment. If the equipment is not loaded or on a standby mode for Preventive Maintenance schedules, the time spent on Preventive Maintenance would be included in the Planned Downtime or non-machine-related downtime.

For Case 3: The Utilization value will hit its lowest mark at 46.67%. Machine standby or idle time due to no-load will be included in the machine-related downtime.

If we calculate the Availability, it will be given as 43.75%. For plants that operate 24 hours a day such as power plants, Oil, and Gas, this would not be a problem but for manufacturing industries that produce products, the question is what if the raw materials are delayed or there is nothing to process at all. The question is do we include the no materials as part of the Planned or Unplanned Downtime or neither. As noted earlier, our JIPM consultant insists on using this formula in calculating Availability, which means that this will be included in the Overall Machine Downtime. He said that a drop in OEE value due to low market demand will definitely be considered by JIPM and will not serve as a hindrance for plants aiming for TPM JIPM excellence awards, and what is important is that we must show the true value of Availability since this is what they use to call it in this case. The problem with manufacturing plants is when the equipment is not running because of no raw materials or load to manufacture. The machine is not utilized and is on standby mode. The decision is whether to include standby mode as a form of downtime or not. Other plants will include this in the machine-related downtime or Planned Downtime, while others will indicate this in the non-machine-related downtime. Other plants will not consider either placing this no load on the Unplanned or non-machine-related downtime. The difference in indicating this on either side of the downtime will affect having a low or high utilization or availability rate depending on the formula used in the calculation for availability and utilization which will affect the value of OEE.

Figure 3.6: The La Cosa Nostra Sit Down

That is why people in industries need to have a sit-down just like the La Cosa Nostra of Mafia to discuss if no-load should be considered as a Planned Downtime. However, in reality, industries may have different ways of calculating their availability and utilization especially when the plant has a low demand. Both maintenance and operations must decide this situation at the very beginning. Maintenance can control equipment availability, but only operations can control when the equipment can be utilized. That is why, as stated earlier in this chapter, maintenance must have a category and classification on what will be and what will not be included in the Planned and Unplanned Downtime as in the case of figure 3.2. It is not easy to change something in a plant since this is how they derive their equation, and some even go to the depths of their corporate just to amend and approve the formula they are going to use. What I know is that they will use the formula that is in the best interest of their plant and industry.

3.3.1: Which Formula Should We Use on Availability?

As mentioned in figure 3.3, there are many variations and formulas on calculating Availability, the question is, which formula should we use? My advice is that if your plant is implementing TPM under a consultant from JIPM or any other authorized group from Japan, you need to follow them since they will be your lawyers in achieving the TPM Excellence Awards.

Likewise in figure 3.3, items 3 to 5 will yield the same answer, while Nakajima's formula is more about utilization since the denominator for availability should be available time and not the loading time. For simplicity, industries aiming for TPM awards should use this formula for availability.

$$\text{Availability} = \frac{\text{Available Time} - \text{All Downtime}}{\text{Available Time}} \ (100\%)$$

All downtime, in this case, will be the sum of all Planned and Unplanned Downtime on the equipment. Although this will yield a lower value compared when using Seiichi's Nakajima's formula, which also provides more room for improvement on the machine or asset being analyzed.

3.4: Performance Efficiency

Performance Efficiency Rate is sometimes referred to as the Throughput or Equipment Performance Efficiency. This is usually an operation's measurement. It reflects whether the equipment is running at its full capacity or design speed for individual products. Performance Efficiency will measure both the Design Speed Losses and Idling or Minor Stoppages. In calculating the Performance Efficiency, the design speed or the design units per hour should be known. This means that the number of products a machine can produce should be known based on its design capacity. The speed operating rate means the difference between the design speed and actual speed. The net operating rate indicates whether the equipment is operated at a stable speed. It is not used to indicate whether the speed is faster or slower than the standard speed, but only to determine if the equipment is operating at a stable speed continuously.

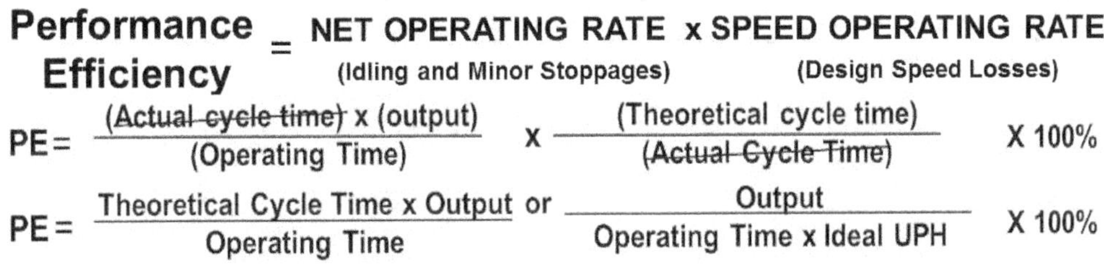

- Net Operating Time is used to identify whether the equipment is operated at the stabilized speed within the unit time.
- Speed Operating Rate is the ratio of the Theoretical Cycle Time divided by the Actual Cycle Time.
- Theoretical Cycle Time also called the Ideal or Design Cycle Time is the Ideal production Rate of the machine and is sometimes referred to as the Design Speed. This is usually given in the equipment manual. The reciprocal of the Designed Cycle Time will be the Designed units per hour (UPH) produced by the equipment or machine.

Case 1: A Semiconductor Back-End PLCC Marking AST 960 – 006 machine running 32 leads. Its Design UPH as specified in the manual is at 500 units per hour, the best output obtained from a single shift was 450 units per hour, IE UPH record was at 430 units per hour. It was able to produce 1800 during A-Shift. Other details include the operator's meeting of 1 hour, Breakdown of 1 hour. What will be the Performance Rate or Efficiency of the machine?

Using the Design Speed

$$\text{Performance Efficiency} = \frac{\text{Output}}{\text{Operating Time x Ideal UPH}} \times 100\%$$

- Performance Efficiency = [(1800) / (7 – 1) (500)]
- Performance Efficiency = 60.00%

Using Highest Output Scored by Operator

- Performance Efficiency = [(1800) / (7 – 1) (450)]
- Performance Efficiency = 66.67%

Using IE Time and Motion Study

- Performance Efficiency = (1800) / (7 – 1) (430)
- Performance Efficiency = 69.76

In calculating the Performance Efficiency, the ideal UPH (units per hour) must be known. However, in case the equipment is old and the Design Speed is not known since the equipment manual was lost or for whatever reasons, we can either base the ideal UPH on the highest output generated by the operator or we can ask an IE (Industrial Engineer) to conduct a time and motion study to determine the UPH for the machine. These cases will only be applied if the design speed of the machine is unknown.

Case 2: A PLCC Marking AST 960 – 008 is running 32 leads for 3 hours producing 1500 units where the Ideal UPH is at 650 units and 28 leads where the Ideal UPH is at 1200 for 3 hours producing 2600 units. It took 1 hour to convert 32 leads to 28 leads, and during the operation, the operator encountered 0.5 hours breakdown and 0.5 hours on PM Schedule, what is the efficiency?

Given:
- 3 hours to process 32 leads at UPH of 650
- 3 hours to process 28 leads at UPH of 1200
- 0.5 hours PM Schedule which is a Planned Downtime
- 1 hour on Set-Up time which will be included as Unplanned Machine Downtime
- 0.5 hours on Breakdown which is also included as Unplanned or Machine Downtime

Getting the Average

- Performance Efficiency = [(1500 + 2600)] / [(8 - 0.5) – (1.5)]
- Performance Efficiency = 73.87%

Computing for the Individual UPH

$$\text{Efficiency} = \frac{1500}{3 \times 650} = 76.92\,\% \qquad \text{Efficiency} = \frac{2600}{3 \times 1200} = 72.22\,\%$$

For 32 lds   For 28 lds

[3]Getting the Performance Rate for machines that runs on a single product is easy, but for non-dedicated machines or those machines that can manufacture different products, it is best to determine the UPH for each product and the number of hours that each of these products run. This is the cleanest method to achieve an accurate OEE. Performance Efficiency can never exceed 100%. If this happens, then check the Ideal Cycle Time or units per hour used. Figure 3.4 was an OEE form we developed during my TPM days, however, it can only be used for dedicated machines. For machines that are non-dedicated or can produce different products, what is needed is to get each of the product's individual UPH to have an accurate measurement on calculating the Performance Efficiency of the equipment.

3.5: Quality Rate

Quality Rate is the product of the following two factors which are the number of items produced and the number of items scrapped, reworked, rejected, and reprocessed once again in the machine during production. Often times referred to as the throughput efficiency or first run capability. Total defects are the total number of rejected, reworked, or scrapped parts produced during a given operating time. In some manufacturing industries, we can refer to the yield of the overall production process which is expressed in percentage.

$$QR = \frac{\text{Number of items produced - items scrapped}}{\text{Number of items produced}} \times 100\%$$

$$QR = \frac{\text{Total parts run - Total Defects}}{\text{Total Parts Run}} \times 100\%$$

Although in manufacturing, there are Quality Inspectors who inspect the products to see to it that they comply with the standards and customer's requirements. The thing is there are cases, where the customer returns the products due to some deviations that have not been captured by these Quality Inspectors. These are considered good products when they left the plant only to be rejected by their customers. Once they are returned, these products will either be scrapped or reworked and reprocessed on the machine. The question is, should we consider this case to be included in the Quality Rate? Technically speaking, Yes, the data on Quality Rate when the product was manufactured should be adjusted and include those shipments that were returned on the calculation of the Quality Rate, so that we can have a realization on what the true OEE figure is. This will result in a lower Quality Rate.

Although some manufacturing firms are not doing this, it is important to note that those products that generate revenue are what are important to industries. If a customer returned a shipment, the industry will not be paid for rejected products and this should be adjusted in the Quality Rate of the product the day it was manufactured. Industries are just fooling themselves with having a high Quality Rate where the fact lies that there are shipments returned from unhappy customers and was not able to generate any revenue and profit.

---

[3] Charles Robinson and Andrew Ginder, *Implementing TPM – The North American Experience*, (Productivity Press Inc, 1995), Page 130

If 2000 products were produced last January 31, 2022, and 20 products were rejected during that date, the Quality Rate was (2000 -20) / 2000 = 99%. However, on February 6, 2022, 500 products were rejected by the customer in which the date it was manufactured was on January 31, 2022, the Quality Rate should be adjusted which will yield a value of (2000 – 520) / 2000 = 74 % on the date it was processed which was on January 31, 2022.

OVERALL EQUIPMENT EFFECTIVENESS WEEKLY MONITORING

AREA : ______________________   MACHINE : ______________   OPERATOR ______________

A	B	C	D	E	F	G	H	I	J	K	L	M	N
DATE	SHIFT	AVAIL TIME	PLANNED DOWNTIME	LOADING TIME	MACHINE DOWTIME	AVAIL (Percent)	OPER TIME	PROD OUTPUT	IDEAL UPH	PERF RATE	REJECT RATE	QUALITY RATE	OEE
		(HRS)	(HRS)	(C - D)	(HRS)	(E-F)/E	(E-F)	(ENCODE)	(ENCODE)	I/(HxJ)	(ENCODE)	(I-L)/I	(G)(K)(M)
30-Nov-21	A	8	0.5	7.5	2	73.33%	6	8000	1500	96.97%	3	99.96%	71.08%
	B	8	0.5	7.5	1	86.67%	7	10000	1500	102.56%	10	99.90%	88.80%
	C	8	0.5	7.5	3	60.00%	5	7000	1500	103.70%	20	99.71%	62.04%
TOTAL		24	1.5	22.5	6	73.33%	17	25000	1500	101.01%	33	99.87%	73.98%
1-Dec-21	A	8	0.5	7.5	1	86.67%	7	5500	1500	56.41%	3	99.95%	48.86%
	B	8	0.5	7.5	4	46.67%	4	6000	1500	114.29%	5	99.92%	53.29%
	C	8	0.5	7.5	3	60.00%	5	4000	1500	59.26%	6	99.85%	35.50%
TOTAL		24	1.5	22.5	8	64.44%	15	15500	1500	71.26%	14	99.91%	45.88%
2-Dec-21	A	8	0.5	7.5	5	33.33%	3	6000	1500	160.00%	7	99.88%	53.27%
	B	8	0.5	7.5	4	46.67%	4	5000	1500	95.24%	6	99.88%	44.39%
	C	8	0.5	7.5	3	60.00%	5	5000	1500	74.07%	7	99.86%	44.38%
TOTAL		24	1.5	22.5	12	46.67%	11	16000	1500	101.59%	20	99.88%	47.35%
3-Dec-21	A	8	0.5	7.5	3	60.00%	5	4000	1500	59.26%	9	99.78%	35.48%
	B	8	0.5	7.5	2	73.33%	6	3500	1500	42.42%	12	99.66%	31.00%
	C	8	0.5	7.5	1.5	80.00%	6	3500	1500	38.89%	15	99.57%	30.98%
TOTAL		24	1.5	22.5	6.5	71.11%	16	11000	1500	45.83%	36	99.67%	32.49%
4-Dec-21	A	8	0.5	7.5	2.5	66.67%	5	6000	1500	80.00%	13	99.78%	53.22%
	B	8	0.5	7.5	3.5	53.33%	4	7000	1500	116.67%	11	99.84%	62.12%
	C	8	0.5	7.5	2	73.33%	6	5500	1500	66.67%	4	99.93%	48.85%
TOTAL		24	1.5	22.5	8	64.44%	15	18500	1500	85.06%	28	99.85%	54.73%
5-Dec-21	A	8	0.5	7.5	4	46.67%	4	5000	1500	95.24%	3	99.94%	44.42%
	B	8	0.5	7.5	3	60.00%	5	6000	1500	88.89%	6	99.90%	53.28%
	C	8	0.5	7.5	5	33.33%	3	7000	1500	186.67%	9	99.87%	62.14%
TOTAL		24	1.5	22.5	12	46.67%	11	18000	1500	114.29%	18	99.90%	53.28%
6-Dec-21	A	8	0.5	7.5	2	73.33%	6	5000	1500	60.61%	10	99.80%	44.36%
	B	8	0.5	7.5	3	60.00%	5	6577	1500	97.44%	2	99.97%	58.44%
	C	8	0.5	7.5	1	86.67%	7	6687	1500	68.58%	8	99.88%	59.37%
TOTAL		24	1.5	22.5	6	73.33%	17	18264	1500	73.79%	20	99.89%	54.06%
WEEKLY		168	10.5	157.5	58.5	62.86%	99	122264	1500	82.33%	169	99.86%	51.68%

(Note : Form recommeded for dedicated equipment's only)

Figure 3.7: Sample OEE Form

3.6: Analyzing OEE

As we have discussed, OEE is composed of three components, which include Utilization (Availability in most TPM books), Performance Efficiency, and Quality Rate. To analyze OEE is to break down each of the three components independently and determine which has the lowest rating. Next is to understand the different losses each component composed as in figure 3.7, and know the losses that your equipment is suffering. After the losses had been determined, deploy the team most suited to deal with these equipment losses. Addressing each of these losses will require a different strategy, hence it is important to know what losses your equipment is suffering. OEE is not only a maintenance KPI since different losses will require different departmental functions. The losses experienced on the equipment are breakdowns, then

Planned Maintenance will handle this problem, however, if Quality defects are the cause of having a low OEE, a Focused Improvement team composing of a cross selection of people from Quality, Process, Production, Maintenance, and others will be deployed. Once the losses are reduced to a minimum, a tool called P-M Analysis will be used by the Quality Team to finally attempt to eliminate the defect permanently on the asset. For the purpose of these examples, *I would prefer to use the term <u>Utilization</u> rather than <u>Availability</u>.*

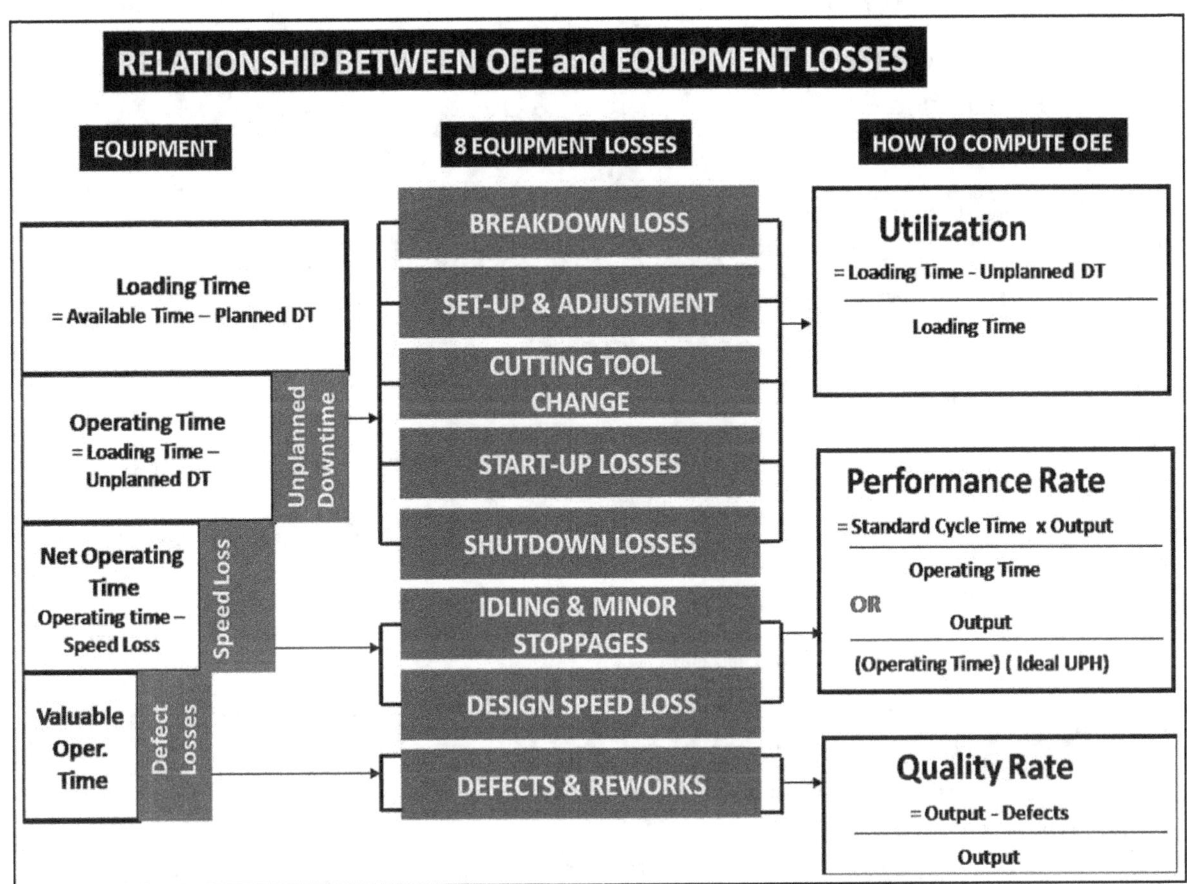

Figure 3.8: Relationship Between OEE and Equipment Losses

Case 1: Machine with Low Utilization: Machine with low utilization may suffer from breakdowns, change-over or set-up, cutting tool change, or start-up losses. The best people to address this type of loss will be the maintenance function. In figure 3.9, the machine is suffering from low utilization due to breakdown losses. The decision, in this case, is to deploy a Planned Maintenance team. The 4 Phases of Planned Maintenance can address these breakdown losses. Phase 1 and 2 will reduce the breakdown losses, while Phase 3 and 4 will create a sustainable process so that the equipment will not revert back to its reactive mode. My second book on Maintenance – Roadmap to Reliability details the activities to be done to reduce these losses. The 4 Phases will include:

• Phase 0: Planned Maintenance Preparatory Stage
• Phase 1: Stabilize MTBF through Restoration
• Phase 2: Lengthen Equipment Lifetime by Addressing Design Weaknesses

• Phase 3: Periodically Restore Deterioration
• Phase 4: Predict Equipment Lifetime

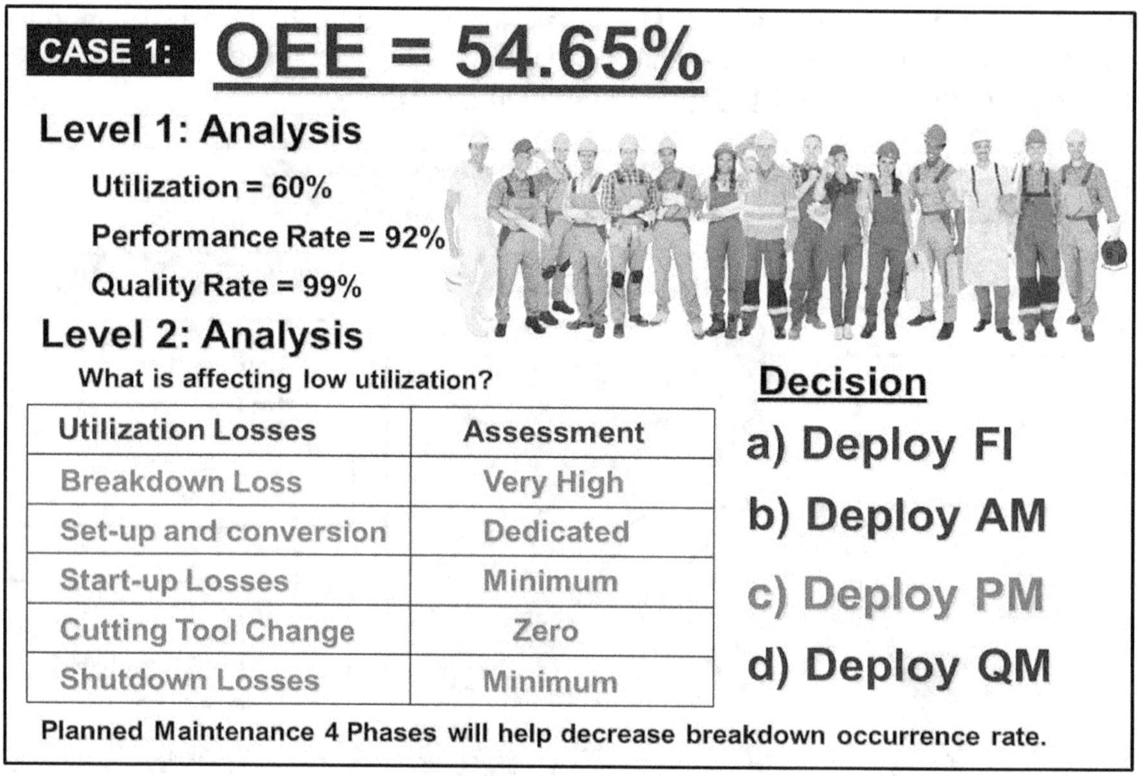

Utilization Losses	Assessment
Breakdown Loss	Very High
Set-up and conversion	Dedicated
Start-up Losses	Minimum
Cutting Tool Change	Zero
Shutdown Losses	Minimum

Figure 3.9: OEE Case 1 Machine with Low Utilization

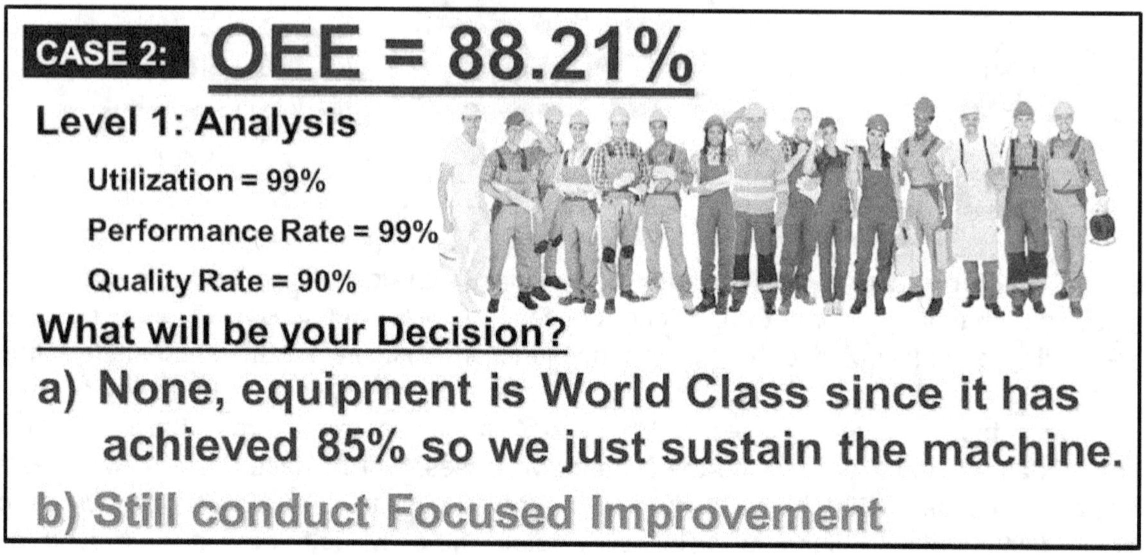

Figure 3.10: OEE Case 2 Machine with Low-Quality Rate

Case 2: Machine with 88.21% OEE: While some books on TPM will indicate that when equipment achieved an OEE of 85% or more, the machine is already considered World Class. This is not actually the case as OEE is not only about achieving 85% but making sure the 3 OEE components are satisfied as in figure 3.10. If we look at this case, the machine is suffering from a lot of defects. This means that for every 1,000,000 products produced, 10% or 100,000 products are considered defective which is simply unacceptable.

Case 3: Machine with 47.02% OEE: Last 2020, when the world was affected by this **Covid 19 Pandemic**, many industries halt their operation as cities were on lock-down to prevent the spread of the virus. Company ABC was one of the industries affected by the lockdown declared by the government since their industry needs to stop operating. The last known OEE as of March 2020 was at 47.42% since most types of equipment were not loaded. In this case, our answer is (e) as in figure 3.11, since nothing can be done about this since this is a government decision and the plant was in idle mode.

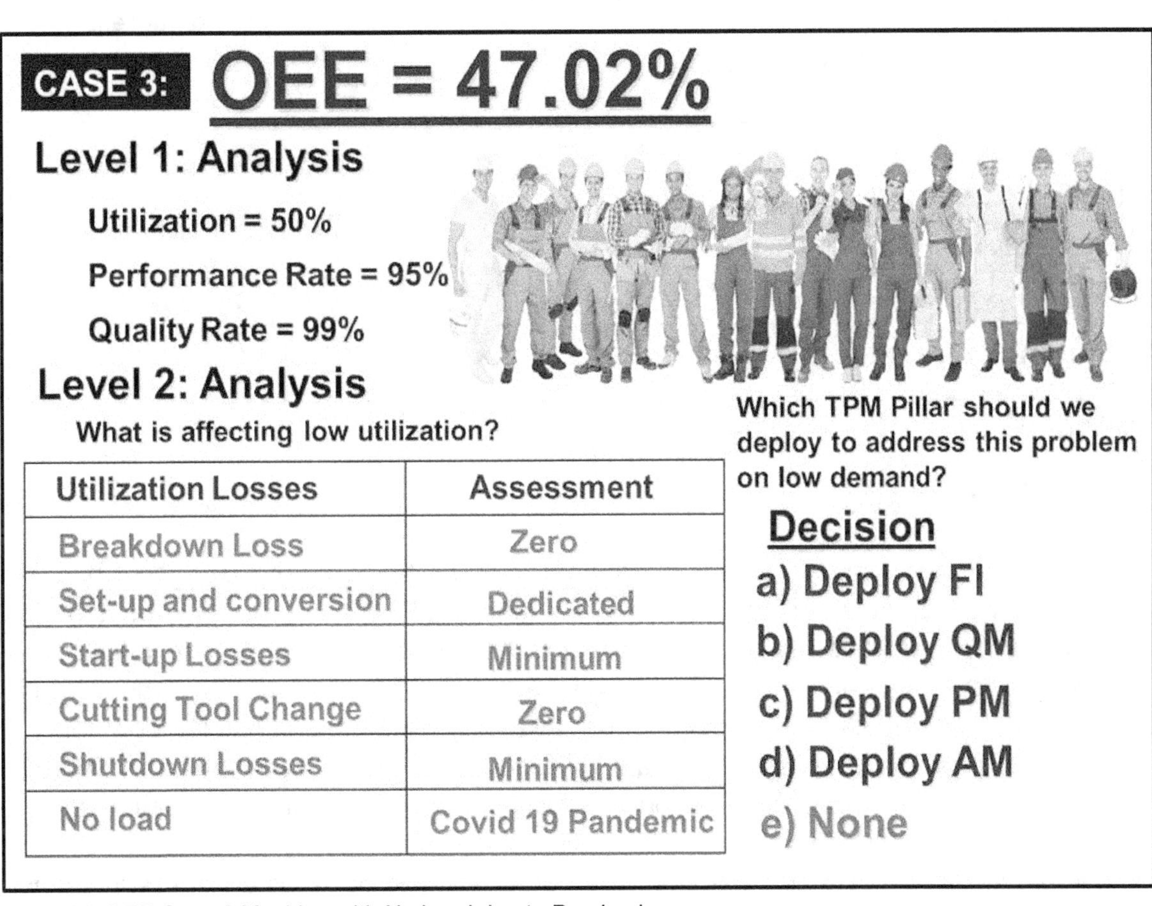

Utilization Losses	Assessment
Breakdown Loss	Zero
Set-up and conversion	Dedicated
Start-up Losses	Minimum
Cutting Tool Change	Zero
Shutdown Losses	Minimum
No load	Covid 19 Pandemic

Figure 3.11: OEE Case 3 Machine with No Load due to Pandemic

Case 4: Machine with 49.00% OEE due to Low-Performance Efficiency Rate: Equipment losses affecting Performance Rate will include Design Speed Loss and Minor Stoppages. In this case, the machine is suffering from minor stoppages or shortstops. In TPM, the best pillar to address this will be Autonomous Maintenance since one of their main focus will be to sustain the cleanliness of the machine. Having a clean machine will definitely reduce minor stoppages. However, for those minor stoppages that still continue to recur, we can deploy a cross-selection of people to compose the Focused Improvement or Planned Maintenance team. Most of the corrective actions needed to address these minor stoppages will include modification or redesign. Low Facilities supply must also be checked as this can contribute to Minor Stoppages such as a low supply of compressed air on pneumatic equipment.

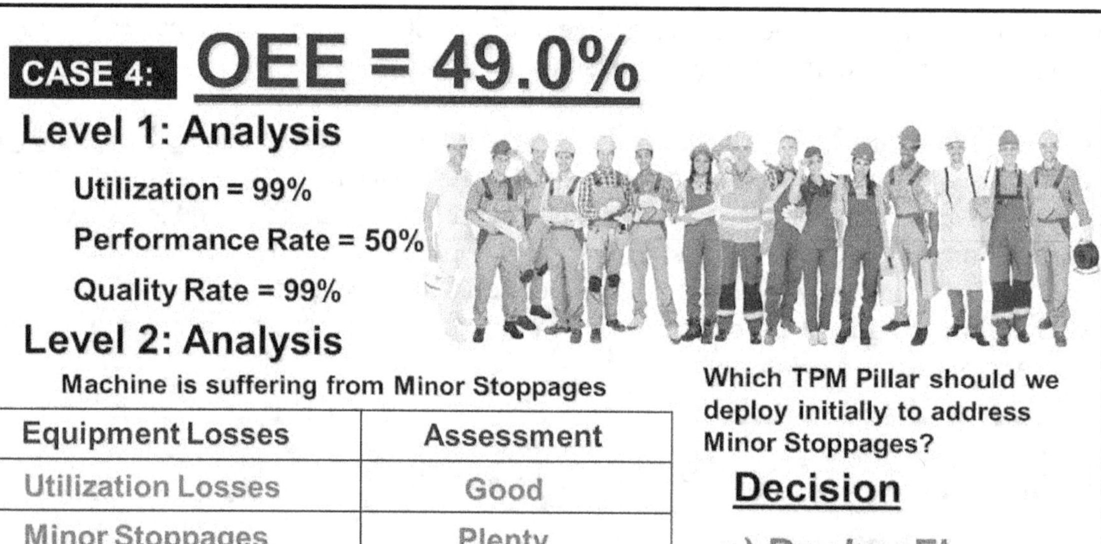

Equipment Losses	Assessment
Utilization Losses	Good
Minor Stoppages	Plenty
Design Speed Loss	Good
Quality Losses	Good
Shutdown Losses	Good
Cutting Blade Losses	Good

Figure 3.12: OEE Case 4 Machine with Low-Performance Rate

3.7: Case Study on OEE

Company ABC is an industry that is engaged in manufacturing automobile parts, household appliances, and other lines of products having a total of 1,500 employees. A-Plant is the main plant of 3 plants they have and is performing the role molding, welding, and assembling of automobile parts. The total number of employees at the A-Plant is about 800. The increase in the orders they received had been dealt with by increasing overtime and holiday work, but it has been requested by the president that planned production should be finished within regular working hours to achieve the cost reduction without sacrificing the required production turnover.

According to Mr. Smith, Plant Manager, the bottleneck and constraint process in Company ABC is the roll-molding line, where continuous cutting, roll-molding, and welding of steel plate are performed. To meet the customer demand for small-lot production of various kinds of products and Just-In-Time deliveries, 3 lines are fully operated with two days and night shifts. Overtime and holiday work is also going on to avoid delay in the delivery. Working hours include 8 hours per shift and one hour is allocated for lunchtime leaving 7 actual working hours. About 20 minutes will be spent on an operator's toolbox meeting before and after work as well as cleaning and inspection. Therefore, the actual loading time is 400 minutes or [(8) (60) - (80)]. Since the standard cycle, time per unit of product calculated from the process amount time of this roll molding line is 0.3 minutes, the theoretical production of 1,333 units of products per shift per line within the regular operating time can be obtained by dividing 400 minutes by 0.3 minutes. However, in reality, the average production record is only 640 units per line within the regular operating time. That is less than half of the theoretical production amount. Even though it is impossible to produce the amount as theoretically calculated, the cost reduction and target production plan could be achieved without working overtime or holidays if a minimum of 1000 units of product can be produced per shift. Mr. Smith is considering a TPM methodology implementation in the plant to achieve this.

According to Mr. Anderson, Chief of Production Engineering Section, the Maintenance of the production equipment is our responsibility. Emphasis is being put on Preventive Maintenance but, regrettably, it is hard to eliminate sporadic and chronic failures. In our plant, we make it a rule to record the line stoppage time and production floor equipment shutdowns of more than 10 minutes due to sporadic failure. When looking into the past record, the sporadic failure in the roll forming line average about 30 minutes per line per shift within the regular operating time. Besides the sporadic downtime, there is the line downtime for set-up and adjustment. For details, would you please ask the Chief of the Production Section? Though the standard cycle time for producing one product is 0.3 minutes as the plant manager mentioned. The actual cycle time measured at the production site was 0.4 minutes. This means that if 400 minutes of loading time would be fully utilized, the production of 1000 units of products per shift per line might be possible. Even if 100 minutes of line downtime due to failure or set-up adjustment has to be considered, 750 units of products could be produced. We made a field investigation into the causes why the production record couldn't reach that level and found that so-called minor or short stoppages were due to such defective line operations as low work supply or work falling from the chute. The frequency of minor stoppages averages 20 times per line per shift.

According to Mr. David, Chief of Production Section, it can be said that the major problem in the roll forming line is the long downtime due to set-up. To meet the production requirement of small-lot production of various kinds of product, about 15 types of products are constantly running on one line every month. They look so similar in shape that it is difficult to identify them. Preferably, a single type of product should be run continuously, but that might cause piling up of inventory for work-in-process for the other products. Therefore, to avoid such a situation, it is inevitable to improve the set-up time to a few minutes, but, regrettably, the current average set-up time within regular working hours is 50 minutes per shift per line.

According to Mr. Charlie, Chief of Quality Assurance Section, it is the personal opinion of the Plant Manager that the roll forming line can theoretically produce more than 1000 product units per shift per line. Of course, it is out of the question that many defective products were produced, even if the amount of production is increased. Fortunately, the quality level of the products our plant produces is on par. The rate of defective products is around 2%, producing 13 defective products out of 640 units which are produced per shift per roll forming line, on average. What will be the answer to the following questions?

1) What is the current OEE of Company ABC?
2) Is it possible to produce more than 1000 products units per shift per roll-forming line at Company ABC?
3) What losses are given in this case study?

4) What is the OEE when 1000 units per shift per line are produced?

5) What is the OEE when 1333 units per shift per line are produced?

Solution on OEE Case Study

• Company ABC industry with 3 plants

• Total Employees = 1500

• A-Plant = roll molding, welding, and assembly

• A-Plant Total Employees = 800

Given:

• A. Actual working hours of shift = 480 minutes (Available Time)

• B. Planned or Non-Machine Downtime = 20 minutes. Shop meeting + 60 min. lunch = 80 minutes

• C. Loading time of shift (A - B) = [(8)(60) - (80)] = 400 minutes

• D. Unplanned or Machine-Related Downtime loss of shift

  = Breakdown = 30 minutes per line per shift

  = Set-up and Conversion = 50 min per line per shift

• E. Operating time of shift (C - D) = Loading Time – Machine Downtime = [(400) - (80)] = 320 minutes

• F. Actual Processing Time (J x G) = [(0.40) x (640)] = 256 minutes

• G. Product units processed by shift = 640 units

• H. Quality products rate = [(640 - 13)] / 640 x 100% = 97.97%

• I. Standard (Theoretical) Cycle time = 0.3 min / unit or 3.33 units / min = 1333.33 units / shift

• J. Actual cycle time = 0.4 min / unit or 2.5 units / min = 1000 units / shift

• K. Availability (E / C) 100% = 320 / 400 = 80%

• L. Speed Operating Rate (I / J) x 100% = 0.3 / 0.4 = 75%

• M. Net Operating Rate (F / E) x 100% = (Actual Cycle Time) (Output) / Operating Time

  = [(0.4) minutes / unit x (640) units] / 320 units = 80%

• N. Performance Rate (L x M) x 100% = [(0.75) x (0.8)] x 100 % = 60%

Computation:

1) What is the current OEE?

• OEE = K x N x H = [(0.80) x (0.60) x (0.9797)] x 100%

• OEE = 47.02%

2) Possibility of producing 1000 units per shift

• Answer: Yes, this is possible if breakdown and set-up losses are both reduced.

3) What losses are given in this case study?

• Answer: Breakdown Loss, Minor Stoppages, Set-Up, Design Speed Loss, Reject and Rework Losses

4) What is the OEE when 1000 units per shift per line are produced?

• OEE when output is 1000 units = 1000 units is possible if set-up and breakdown loss is equal to (400) min / 0.4 min per unit

• Availability = 100% (No breakdown and Set-up loss)

• Performance Efficiency = Speed Operating Rate x Net Operating Rate

• Efficiency = [(0.75) (0.4) x (1000) / (400)] = 75%

• Quality = 97.97 %

• OEE = (0.1 x 0.75 x 0.9797) x 100% = 73.48%

5) What is the OEE when 1333 units per shift per line are produced?
- Actual Cycle Time equal to Ideal Cycle Time
- Availability = 100% (No breakdown and Set-up loss)
- Performance Efficiency = [(100)] x [(0.3) (1333) / (400)] = 100%
- Quality = 97.97 %
- OEE = 97.97%

3.8: Why having an OEE of 85% is Not Always World-Class?

As said previously the primary measure of TPM is OEE or termed as Overall Equipment Effectiveness, not Efficiency. According to TPM books, OEE has three components: Availability multiplied by the Performance Rate multiplied by the Quality Rate and expressed as a percentage. An availability of 95% multiplied by a Performance Rate of 90% and a Quality Rate of 99%. OEE will be 0.95 x .090 x 0.99 = 84.65% will provide a value of 85%. According to TPM books and other consultants, OEE should be at least 85% or more to be considered World Class. OEE indicates the relative productivity of a piece of equipment compared to its theoretical performance. It identifies the bottleneck or constraint equipment and processes critical for the improvement of equipment productivity.

When utilization or availability is low, the equipment may be suffering from either the following equipment losses: breakdown losses, set-up adjustment, conversion, cutting tool change, start-up losses, and shutdown losses. This downtime loss will definitely affect the uptime of the equipment. When the equipment's Performance Rate or Efficiency is low, the equipment may be suffering from either idling, minor stoppages, or design speed loss. Lastly, for equipment that manufactures products, if your equipment suffers from many defects, quality issues, re-runs, reworks, and something of this sort, the equipment's quality rate will be low. What is important is to analyze OEE based on its losses and not only about providing a single rating in percent but rather breaking it down into the three components, which are availability, performance rate, and quality rate, so that we can analyze the weakness of the equipment and determine the corresponding equipment losses the equipment is experiencing. This is how OEE is being analyzed and interpreted, and it is really quite challenging to improve the OEE of the equipment. So, let me start off with the simple ones first so that the reader can understand and recognize what my point of view is all about.

First question: If the equipment availability is 100%, is the equipment said to be at its peak and reliable? The answer is no, the equipment may be 100% available but may have a low utilization rate or may not have been utilized. After all, availability is not a measure of reliability, but it is a measure only when the machine is available. It is just a measure of time. Suppose the equipment is running but is not being utilized; the availability is 100% while the utilization is 0%. In this case, the industry is not generating any revenue and profit.

Second question: At this point, let us assume that if both the equipment availability and utilization are 100%, is the equipment said to be at its peak and reliable? The answer is still no. The equipment may be 100% utilized and 100% available, but the machine may suffer from errors and speed loss. Example: The speed of a machine producing 1,000 units per hour is reduced by the maintenance to 800 units per hour since running at 1,000 units per hour provides many quality problems and defects on the product. A pump have a rated capacity of 1000 gallons per minute, but over time, the impeller will start to erode as it is in direct contact with whatever it is discharging, so over a given period, the rated capacity drops to 920 gallons per minute. Still, the pump or equipment is utilized 100% of the time and is available for use since it is running. Therefore, when the machine speed is reduced, it still runs. Hence, if the machine is loaded, it will give us a high utilization and availability rate, but the efficiency will be low, or the performance rate of the equipment will suffer since both availability and utilization are just a measurement of time.

Last question: When you consider the efficiency and quality rate, then this is where OEE will come in. Now let us proceed with OEE. If the equipment has an OEE of, say, around 90% and beyond, is the equipment peak and reliable? Answer: Not exactly, for two reasons. First, the real objective of OEE is not about achieving 85% or more, but OEE is about satisfying the three components. In the case below, the equipment suffers from many defects, which is simply not acceptable. The equipment utilization here is perfect, so the Performance Rate has a high rate, but the Quality Rate indicates that the equipment is producing many defects in the product; hence, there is no need to celebrate, and we are just fooling around with the data.

OEE = 100% (Avail.) x 99% (Performance. Rate) x 94% (Quality Rate) = 93.06%

Every single piece of equipment has more than one function. For manufacturing industries, the primary function of the equipment will have something to do with producing products to earn some revenue and profit. This means that if we have cases of function reduction breakdown in our equipment or simply some of the secondary functions of the equipment are not capable of working at all, and there are cases where the failure of secondary functions is much more important than the failure of the primary function. The equipment may be producing the desired output, but for some of its secondary functions, such as protective devices which may not be functioning. There is a possibility that the OEE can still be high. Let me provide you the following cases to shed light on this matter:

Case 1: The Case of the Marking Machine: This is actually a true story. In the semiconductor industry I previously worked, after molding the IC (Integrated Circuit) package for encapsulation, the next process would be either to go to the Deflash Station, plating and then marking station. Anyway, if you have seen an IC, there is a mark with numbers, letters, and symbols that, you can see on the package. This is the black thing with legs that you see on electronic boards. This process is done in the Marking Station. Previously, most of the marks used were white ink, but nowadays, most marks are done using a laser. We have several Marking Machines, and during those times, the top-of-the-line Marking Machine was a Dongyang Machine. Although the marking machine never obtained a high OEE than the Molding machine, let me explain the importance of the equipment's secondary functions. The IC package we used to produce was a PLCC package. It was square in shape and of different sizes depending upon the number of leads it has. The more leads, the larger the package. This Dongyang machine marks different packages such as 28 leads, 32 leads, 44 leads, etc. The IC package is still in strip form, which means that it is still not yet cut. It is grouped in sets of 10 or more depending on the package. Marking is done per stroke for every strip or group of 10 IC's. Since the package was square, the mark's orientation is critical. This means that if the orientation of the strips into the machine was reversed, the markings will be reversed too and the customer will reject this. Before the operator can place the strips on the Dongyang machine, the operator must check the strips randomly in the tray to check their orientation on the machine. One day, the production department received a love letter from one of their customers. It was a customer complaint. We call it CC, which stands for customer complaints, and actions that are taken to correct this issue are called CCART or Customer Complain Action Response Team.

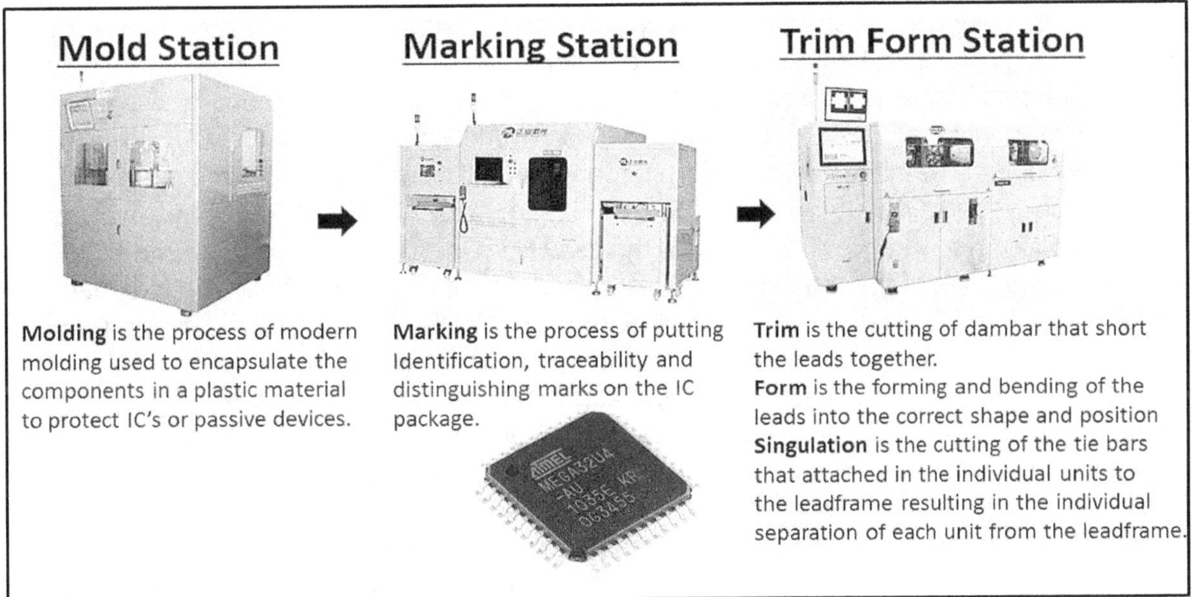

Figure 3.13: Semiconductor End of Line Process

The production manager called the Quality, Process, Supervisor, and Maintenance for a meeting. The maintenance said that he knew the equipment quite well and said that the equipment was equipped with a sensor to detect reversed strips. They all went to the machine and the maintenance locate the sensor which was clogged with dirt. He wiped it with a clean cloth and got a dummy strip, deliberately reversed the orientation, and process it on the machine. This time the machine stopped because it detected a reversed strip. The engineers prepared a containment plan which is to check the strips 100% before processing them on the equipment. Originally, the equipment was designed that if the IC package orientation was not correct during the loading process, the equipment was supposed to stop automatically. But everything was too late since the customer walked away and placed his business elsewhere. It was just only a small piece of a sensor, a failure of a secondary function of the equipment, yet the impact of this failure was a complete loss of business for the company. And that customer that walked away was the second biggest customer of the plant. Although, there was no breakdowns experience, and it was just a small proximity sensor that caused the damage and loss of that customer amounting to millions of dollars in revenue. There are several cases where the criticality of the secondary function poses a much bigger threat than the failure of the primary function of the equipment, and OEE will not capture those failures of the secondary function since OEE is only designed to measure the primary function of the equipment. This means that for as long as the equipment is utilized with high efficiency and few quality rates, then the OEE is high even if your equipment suffers from many leaks and half of its sensors, protective devices and gauges are not working. My message here is that both the primary and secondary functions of the equipment must be satisfied and working since there are cases where the failure of a secondary function may deem more serious and critical than the failure of the primary function just like in this case. A pump can provide a discharge rate based on its rated capacity, but the pipes are terribly corroded in such a way that leaks are present or the tank itself is leaking.

Case 2: Molding Machine with an OEE of 96%: Again, in the semiconductor industry that manufactures IC where I used to work many years ago, there is a Mold Station. Molding is the process of encapsulating the device in plastic material or molding compound that will be heated to cover the chip inside to protect it from moisture. The molding compound is heated until it melts and flows into the cavities, containing the lead frame for encapsulation. During my time in this semiconductor, we have lots of this equipment with different types. This molding equipment produced the highest OEE in the plant, 92%, 94%, some reaching up to 97%; since we were on the TPM Office, we were very happy with the results. One day, my boss told me to make a surprise audit on one of the molding equipment since he has some doubt on the integrity of the data because the OEE was always high, and so I went to their station to observe if it gave us the true data on OEE. Obviously, the equipment performed very well with no failures and breakdowns, and there was not even an operator in the equipment since it is running smoothly without any problems. I observed for an hour until I got bored and returned back to my office. The following day, I went again and observed for an hour. Still, the molding machine was running perfectly until I asked the maintenance to open the molding machine's lower compartment. Whoa, I was shocked by what I found out. The molding machine was flooded with oil. And there was an oil leak all over. They never addressed the leaks, but their OEE was great. This means that this machine with a high OEE produces output perfectly but fails in its secondary functions once again.

If you have a car, the primary function of a car is to take you from one place to another, but we do not just make our judgment on purchasing the car based on the primary function since we also need to consider its secondary function. This means that a car may be capable of traveling, but the car's headlight is busted, or the car which is supposed to have 4 doors have only 1 door, the car's wiper is not functioning, or the side mirrors are missing. These are called secondary functions. Like industry's equipment and machines, they also contain many secondary functions and what is important is that both primary and secondary functions must be working on the equipment.

[4]From the book of John Moubray RCMII, he quotes: There is often a tendency to focus too heavily on primary functions when assessing maintenance effectiveness. This is a mistake because trivial secondary functions often embody bigger threats to the organization if they fail rather than the primary functions of the equipment. As a result, every function must be considered when setting up the maintenance effectiveness measures and targets.

The OEE, as defined above, only relates to the primary function of any asset. This is misleading because, as in the case of the gasoline storage system, every asset, machine tool

---

[4] Moubrey, John, **Reliability-Centred Maintenance II**, Butterworth-Heinemann Page 302 - 304

included have many more functions than the primary function, and each of these functions has their own unique performance standards. Consequently, OEE is not a measure of the overall effectiveness at all but only a measure of the effectiveness with which the primary function of the asset is being fulfilled. From page 304, Reliability-Centered Maintenance by John Moubray. So think again, if you have equipment with a high OEE yet failed in many secondary functions, then you are in a dangerous situation as there might be implications or consequences and these failures are just waiting for the right time to happen.

Figure 3.14: Function Loss and Function Reduction Breakdown

3.9: Lessons on OEE

OEE is used to prioritize bottleneck equipment to improve productivity. Since OEE is focused on the equipment's primary function, which is the output, it is also important to check the secondary functions of the equipment if they are working or not. Although JIPM states that to pass the TPM requirements, an 85% OEE should be achieved. It does not mean that we have achieved success. Remember that OEE is composed of 3 components and that these 3 components must be satisfied. Remember that every piece of equipment consists of both primary and secondary functions that are likewise important and needs to be satisfied as well, and there are cases where the failure of secondary functions is as important as or even more important than the failure of the primary function itself. This means that equipment can have a high OEE even if it has a severe oil leak, or its protective devices are malfunction. Hence, both primary and secondary functions on the equipment must be satisfied in this case. Except for breakdown losses which will be handled by the Planned Maintenance, the best people to impact OEE will be the Focused Improvement pillar of TPM since they will be addressing the different losses on the equipment. When calculating a meaningful OEE, consider the following;

- Make a thorough classification of what constitutes a Planned and Unplanned Downtime. (Example, no raw material, inventory).
- Decide the formula to be used for availability and be consistent.
- Decide the ideal or rated UPH or cycle time to be used on each individual product especially if the machine is not dedicated
- Since OEE is focused on the primary function which is the output, it is also important to check the secondary functions of the equipment
- Except for the Breakdown Losses, the pillar that will impact OEE is not Autonomous or Planned Maintenance but it is the Focused Improvement.
- When calculating the OEE of a line or process, just focus on the bottleneck equipment first.

3.10: Total Equipment Effective Performance (TEEP)

Although we never used this indicator and never heard of this during my TPM days, TEEP is similar to OEE, but the main difference is that TEEP will be considering the Available Time to be at 365 days a year and not the industry Available Time, in which there are days in which the industry is not totally operating. The value of TEEP will always be lower than OEE which makes it more challenging. The formula for TEEP will be the same for OEE, however, we will be multiplying OEE with either the availability or utilization depending on which formula for availability you are using. Therefore, TEEP will be equal to:

- TEEP = OEE x Utilization

OEE will measure its productivity based on the company's calendar time. This means that if the plant is running at 335 days in one year, then this will be the available time. While TEEP will consider 365 days whether the plant is running or on a holiday. The only difference is that if our factory is located on Mars, then the available time will be 687 days since that will be the total number of days in a year on Mars. Although TEEP will provide an idea of how much a factory can actually produce in a 365 day, 24 hours period. Although some consultants may think of this as a cool measurement, this can only be true if your operators and maintenance are robots. If they are not and your operators are human beings, then you need also to consider the circadian pattern, even equipment needs to rest during their Preventive Maintenance schedules, what more with human beings.

Therefore, when your industry will be measuring these indices, start with your Human Resources and tell them to hire operators who are willing and able to work 365 days without any Vacation Leave or Sick Leave, and even if the operators agree on this, then we also have by-laws from the government depending on the country and company that one of the benefits of every employee is to be paid for their sick leave and vacation leave. Meaning, if your industry denies this, then your industry will have a problem with the government and labor laws.

3.11: Take Quiz on OEE

1. OEE will address the secondary functions of the equipment.
   a) True
   b) False

2. The primary measure of TPM is OEE or Overall Equipment Effectiveness.
   a) True
   b) False

3. According to the author of this book, once the OEE of the equipment achieved 85%, it is already considered to be world-class.
   a) True
   b) False

4. Besides the development of SMED or Single-Minute Exchange of Dies, Shigeo Shingo also developed what we called Mistake Proofing or Poka-Yoke.
   a) True
   b) False

5. Errors in automated handling equipment where workpiece flow stops operator resets and machine run is called Minor Stoppages
   a) True
   b) False

6 When the machine's availability is low, one possibility is that the machine is suffering from many minor stoppages.
   a) True
   b) False

7. OEE is originally designed for manufacturing plants or those that manufacture goods and products.
   a) True
   b) False

8. In the calculation on utilization, when the equipment is not loaded this will be considered as a Planned Downtime.
   a) True
   b) False

9. Set-up time is considered as a Planned Downtime or Non-Machine Related Downtime.
   a) True
   b) False

10. The best TPM Pillar that will impact a reduction in minor stoppages will be the Quality function of the organization.
    a) True
    b) False

11. A minor stoppage that has already consumed 15 minutes due to no available technician will automatically be converted to breakdown.
    a) True
    b) False

12. The losses incurred on Performance Efficiency will be minor stoppages and design speed loss.

a) True
b) False

13. The goal of improving design speed loss is to bridge the gap between the current operating speed and the design speed of the equipment.
    a) True
    b) False

14. Function Reduction Breakdown is when a specific function or part had failed in the equipment, but the equipment is still capable of running and providing its primary function.
    a) True
    b) False

15. The best people to measure and monitor utilization will be the maintenance function.
    a) True
    b) False

Note: Refer to Appendix B for the answers.

Chapter 4

Understanding MTBF, MTTF, Failure Rate, and Reliability

> *Many say that reliability cannot be improved by doing maintenance. My wisdom on this is that maintenance is two folds, which means that we may be talking about the tasks, or we may be talking about the people in our plant. If we consider maintenance as a task, then I agree that reliability cannot be improved but can only be sustained. But when we consider maintenance as a human being capable of modifying parts with inherent design weaknesses, then reliability can be improved.*

4.1: Clarity on When Do We Declare a Breakdown or Failure

Before we can discuss these maintenance measurements and Mean Time Indicators, we need to be specific and explicit on when do we declare a failure or a breakdown. As discussed in Chapter 1, **failure** means the inability of an asset, or equipment to perform its required function. The failure of a component can be viewed as terminating its life. **Breakdown** is a lost time due to equipment failure that may or may not cause downtime. This will be the case if the system or component has a redundancy. It also refers to machine-related downtime, unplanned downtime, tooling failure, unplanned replacement of spares, and components. Equipment failure refers to an event in which the equipment cannot accomplish its intended function and objective. It may also mean that the equipment stopped working, is not performing as desired, or is not meeting its target expectations. Breakdown and failure are almost similar, as failure will lead to a breakdown. A failure can occur, such as an increase in vibration, which is a potential failure and can eventually end up in a failed state, which is a functional failure. In this case, the equipment finally breakdown.

The word breakdown is a very tricky and complex word. The definition of breakdown and when to declare an equipment breakdown varies from one industry to another. Maintenance people must be precise, consistent, and unanimous on when will we declare a breakdown or not on our equipment and assets. Should maintenance and other organizational functions will not be consistent on having a clear and precise definition of breakdown, then our OEE calculation and other indices will either be wrong, compromised, or will not tell us the exact losses in our equipment, machines, and assets. The problem with many industries is that they make their own rules, standards, policies whatnot neglecting to understand that they compromise their calculation on OEE and other maintenance indices. An example of this confusion is a clear distinction between the term breakdown and minor stoppages. In many manufacturing industries, when a minor stoppage occurs, and there is no available technician for 10 minutes or more, they will automatically declare it as a breakdown which should not be the case. If a piece of equipment has 2 breakdowns and 10 minor stoppages in a week, then the denominator for MTBF which is the frequency of breakdowns now becomes 12 instead of 2, not to mention that we also add the total amount of downtime due to minor stoppages in the Unplanned or Machine Related Downtime for the denominator. In this case, we now have a lower value since we combined both the minor stoppages and breakdowns in the MTBF calculation.

In my 4th book on Cutting-Edge Maintenance Management Strategies, I explained in detail that the word failure is technically broad in its sense; therefore, before measuring the number of failures or breakdowns occurring in the equipment, maintenance needs to be precise, consistent, and crystal clear on what will eventually constitute a breakdown or not. What is important for maintenance and for all levels of the organization is to have a consistent definition of the word failure or breakdown. The question is, when do we precisely declare that a failure or breakdown has occurred in our equipment? This is very important because if this will not be done, then every single person may have their own definition of failures and breakdown. This will have a big difference in several KPI's such as MTBF. A safety officer can declare the equipment in a failed state if it is leaking oil excessively. His reason for declaring a failure is to avoid any unsafe conditions where a person can slip or lose his balance when he steps on the oil and slip. In this case, the equipment is still running and delivering the goods. For both production and maintenance, the leak may not be declared a breakdown immediately but not for the safety person. One of the sensors on the machine malfunctioned, and maintenance bypass the sensors for the equipment to run because they need to ship this lot today and the sensors are a non-stackable item in the storeroom. Do we declare a breakdown on the sensor? One of the pressure gauges was broken, in which the consequences of the failure can end up in an environmental consequence. Do we declare a breakdown on the broken pressure gauge? Try answering the following questions in Section 4.2 for a while and check out the answers in Appendix B of this book. The purpose of this quiz is for the reader to determine if this will be included in tracking the frequency of breakdowns or not. Another important reason why we need to be precise and consistent in our definition is that there will be indices that will be affected. For example, in calculating MTBF, the formula will be equal to the operating time divided by the number of failures; will the following be included in the number or frequency of failures? Another important thing to consider is that only function loss will be considered as a breakdown. All function reduction will be excluded as a breakdown. When tracking the different Mean Time Indicators such as MTBF, will you include the following in Section 4.2?

4.2: Take Quiz on Will You Consider This A Breakdown or Not?

To be clear on the instructions, the denominator on MTBF will be the number of breakdowns or failures, will you consider the following cases as a breakdown or not and included them in the denominator of MTBF?

1. A function loss breakdown where the machine totally stops. An example is that the gear fractured caused the machine to stop.
   a) Considered as a Breakdown
   b) Considered not a Breakdown

2. Function reduction breakdown where machine still runs. An example is that the emergency stop button of the equipment is no longer functioning but the machine is still capable of delivering the products needed.
   a) Considered as a Breakdown
   b) Considered not a Breakdown

3. A scheduled Preventive Maintenance was performed on the equipment which lasted for 48 hours of downtime, which is a Planned Downtime.
   a) Considered as a Breakdown
   b) Considered not a Breakdown

4. Failure of a pump with a standby unit that runs automatically, if we speak about the failure of the duty pump or component.
   a) Considered as a Breakdown
   b) Considered not a Breakdown

5. In item 4, if we speak about the whole system where the pump is just a part and the system is still operational.
   a) Considered as a Breakdown
   b) Considered not a Breakdown

6. An assist, error, or minor stoppage, which is already consuming 30 minutes because of no available maintenance technician to perform the correction.
   a) Considered as a Breakdown
   b) Considered not a Breakdown

7. Random breakdown of an electronic part that stopped the equipment.
   a) Considered as a Breakdown
   b) Considered not a Breakdown

8. Infant mortality failures or failures encountered right after a major overhaul or outage where the operator cannot operate the equipment after the PM crew endorsed the equipment back to operations.
   a) Considered as a Breakdown
   b) Considered not a Breakdown

9. Unscheduled repair, replacement, or overhauling of parts
   a) Considered as a Breakdown
   b) Considered not a Breakdown

10. A virus from equipment prevented the operator from starting the equipment, which took the IT and maintenance 24 hours to get rid of the problem.
    a) Considered as a Breakdown
    b) Considered not a Breakdown

11. Detection of a potential failure such as vibration on the machine detected through Vibration analysis.
    a) Considered as a Breakdown
    b) Considered not a Breakdown

12. In item 11, after detecting a potential failure due to excessive vibration, the maintenance performed corrective maintenance and replaced the part that is about to fail. The downtime was recorded to be 1 hour.
    a) Considered as a Breakdown
    b) Considered not a Breakdown

13. Software problems in the equipment in which the machine runs intermittently and then suddenly stopped and the operator reset the machine, and then it runs. After a few minutes, the same thing happens.
    a) Considered as a Breakdown
    b) Considered not a Breakdown

14. Hidden failures of protective devices such as sensors and alarms in which the equipment is still operating and delivering.
    a) Considered as a Breakdown
    b) Considered not a Breakdown

15. Failures of function reduction breakdown where the machine is still running and capable of providing the primary function of the equipment.
    a) Considered as a Breakdown
    b) Considered not a Breakdown

16. A compressor fails in such a way that it was not able to provide compressed air to 100 pneumatic equipment, if we speak about the compressor, then it is
    a) Considered as a Breakdown
    b) Considered not a Breakdown

17. In item 16, if we speak about the 100 pneumatic equipment that stopped since there was no supply of compressed air from the compressor
    a) Considered as a Breakdown
    b) Considered not a Breakdown

18. Waiting time due to no available parts and spares to repair the machine.
    a) Considered as a Breakdown
    b) Considered not a Breakdown

19. A failure in which a component or spare will be involved
    a) Considered as a Breakdown
    b) Considered not a Breakdown

20. The machine is stopped since it is producing defects and rejected products due to the poor quality of the raw materials recently delivered.
    a) Considered as a Breakdown
    b) Considered not a Breakdown

21. A blackout occurred which caused the maintenance in charge a delay of 20 minutes in switching the generator of the plant. If we speak about the production equipment that stopped.
a) Considered as a Breakdown
b) Considered not a Breakdown

22. A hacker hacked one of the robots in an automotive manufacturing operation causing it to malfunction and injure someone.
a) Considered as a Breakdown
b) Considered not a Breakdown

23. In one semiconductor industry, four out of twelve sensors in a Wirebond machine were bypassed by a technician. The machine is still operating and producing the products. Do we declare a breakdown on the bypassed sensors?
a) Considered as a Breakdown
b) Considered not a Breakdown

24. In a manufacturing plant, the operator shut down the equipment and took his lunch for around forty-five minutes.
a) Considered as a Breakdown
b) Considered not a Breakdown

25. Air leaks from the pipes leading to the compressor, causing pneumatic equipment to produce short stoppages frequently. If we speak about the pneumatic equipment, do we declare a breakdown or not.
a) Considered as a Breakdown
b) Considered not a Breakdown

26. A set-up and conversion were done on the equipment which took 45 minutes, however, the maintenance forgot to remove his Allen wrench after conversion. When the operator runs the equipment, the pick-and-place was damaged which took another 1 hour to replace. If we talk about the additional 1 hour to replace the pick and place, do we consider this a breakdown or not?
a) Considered as a Breakdown
b) Considered not a Breakdown

27. The cutter was replaced due to broken teeth since it was producing defective products. Do we consider this as a breakdown or not?
a) Considered as a Breakdown
b) Considered not a Breakdown

28. The force on the pick and place was not sufficient enough causing the product to slip and fall which caused an error on the equipment the operator picked the product and restart the machine and the machine runs again.
a) Considered as a Breakdown
b) Considered not a Breakdown

29. The P-F curve of the bearing was predicted by the Vibration Monitoring Analysis to be 45 days, however, there was a delay in the delivery which caused the equipment to fail while waiting for the delivery of the spare. The bearing arrived 24 hours after the bearing failure occurred. Although the failure was predicted, do we consider this a breakdown or not?
a) Considered as a Breakdown
b) Considered not a Breakdown

30. An equipment with a built-in Artificial Intelligence was no longer following the command of the operator. The maintenance decided to shut down the equipment. Do we consider this a breakdown or not?
a) Considered as a Breakdown
b) Considered not a Breakdown

31. If we replaced an item or spare such as an impeller during a Preventive Maintenance schedule since it has already eroded, technically it is already in a failed state since the flow rate has dropped by 10%. Do we consider this as a breakdown or not?
a) Considered as a Breakdown
b) Considered not a Breakdown

32. A machine is producing defects and was stopped by the operator and the cause of the defect is due to the raw materials. Do we consider this as a breakdown or not?
a) Considered as a Breakdown
b) Considered not a Breakdown

33. Machine A is causing excessive vibration, which caused Machine B (nearest to Machine A) to fail. If we speak about the failure of Machine B do we consider this as a breakdown or not?
a) Considered as a Breakdown
b) Considered not a Breakdown

34. 15 minutes is lost every time the operator starts his machine since the machine is in the process of warming up which is considered as a start-up failure. Do we consider this as a breakdown or not?
a) Considered as a Breakdown
b) Considered not a Breakdown

35. After conducting extensive Preventive Maintenance on the equipment, the operator is having a hard time running the equipment due to infant mortality failures. Do we include infant mortality failures as a breakdown?
a) Considered as a Breakdown
b) Considered not a Breakdown

Refer to Appendix B for the answers.

4.3: Confusion Between Breakdowns and Minor Stoppages

For Oil and Gas, Power Plants, Distribution Plants, and similar industries, this would not be an issue, since minor stoppages do not happen in these industries, but for manufacturing plants, it's a different story. During my TPM days in the semiconductor industry, I was assigned to handle the TPM Planned Maintenance pillar. We organized a group of people from every single department to be part of this company-wide initiative. We also included the facilities and utility section of the plant in our Planned Maintenance journey. Our initial requirements were to agree on what KPIs and maintenance indicators we should track initially. The number of breakdowns was one of them, so we started to measure the number of breakdowns on our first pilot equipment. It was a simple task to ask but indeed very confusing. There were eleven departments, to be exact, and each department was composed of two members. One was from the front of line, and the other represented the end of line. We had around 24 members during that time, and our meeting was every week. Since all of these departments produced IC packages except for the Facilities and Central Lead Finish or Plating section, it was very common to have similar equipment types in each station of a semiconductor plant. One of the stations in front line was the Wirebond Station. This station could have a minimum of around 100 to 200 wire-bond machines, and we had around 10 departments in the plant with a wire-bond station.   A wire-bond is a machine responsible for adhering or welding a thin wire to a bare semiconductor die-chip pad and the other end of the wire to a conductive pad on a substrate. The wire is usually made of 99% pure gold and is usually very thin, often 0.001 to 0.0013 of an inch. Anyway, so much for the introduction and back to the breakdown part of the story. All members of the Planned Maintenance were instructed to select an initial machine to undergo our Planned Maintenance activities, and one of the instructions was to provide the number of breakdowns on their pilot machine that would undergo Planned Maintenance for the past 3 months so we could have a sort of benchmarking data. When they submitted the data, I was shocked by the results. A couple of wire-bond from different stations recorded breakdowns of 3 to 8 times in a month, but what was shocking was that some stations recorded breakdowns of around 30 to 50 times a month. This was where I sensed that something was wrong with the data.   So what was wrong?   They included the short stoppages, assists, errors, and interruptions in the number of breakdowns.

The frequency of breakdowns and failures are easy to identify for oil and gas plants, mining, metal plants, power plant industry, and so on but relatively difficult in some cases for manufacturing plants. ***The words failure and stoppage are two different things.*** When a minor stoppage, error, or assist occurs on the equipment, nothing actually failed. The machine just stopped. The thing is if a breakdown occurs, the maintenance repairs the equipment. For minor stoppages, there is nothing to repair since nothing failed, the maintenance corrects the error and restarts the machine. These errors and stoppages occurred mostly for automated equipment or if your machine contains many electronic parts. For example, a pick and place lift the product and the product falls and an error occurred on the machine. In the majority of cases, the machine will just be reset and turned on once again. The problem in most cases is that due to the separation of activities between operator and maintenance, operators do not know what to do as they have not been taught by maintenance. The operator will wait for the maintenance to address this minor stoppage. This type of equipment loss must totally be separated from breakdowns especially when the industry is calculating MTBF as one of their indices and KPIs. If 1 breakdown and 10 assists occurred during the shift, then these errors or minor stoppages may be included in the frequency or number of breakdowns which is the denominator for MTBF which will provide an inaccurate and lower MTBF of the equipment. My advice for some manufacturing plants was that before measuring the number of breakdowns you have, maintenance people should agree on what would constitute and what would not constitute a breakdown. Therefore, for the succeeding weeks, we deliberated and discussed this matter thoroughly with the members. There were a lot of debates and arguments, just like passing a bill in congress and the senate, but finally, we resolved the matter and arrived at a consensus among the Planned Maintenance members regarding what to include and exclude as breakdowns, and here was what we agreed upon.

What We Included as Breakdown

- Breakdowns caused by poor set-up and conversion where the maintenance stops the machine replaced parts which are damaged as a result of poor set-up and conversion
- When the technician stops the machine due to breakdowns caused by defects and reworks such as worn-out punch, thereby producing product defects
- Function loss breakdown due to no availability of spare parts
- Unscheduled repair and overhauling of equipment
- Unscheduled replacement of parts
- Failure and breakage of tooling
- Run to Fail items that provide little or no consequences at all either a redundancy is in place
- All cases of Function-Loss Breakdown

What We Excluded as Breakdown

- Breakdowns caused by a material related problem
- Actual conversion and set-up or the process of changing from one product to another
- Machine stoppage caused by interruption assists, and minor stoppages resulting from jamming of products due to dirt and so on. (Japanese call this chokotei)
- Machine downtime caused by PM Schedule such as overhauling of parts
- Machine Downtime caused by scheduled replacement of parts as reflected from the tool algo schedule
- Repairs attributed by Predictive Maintenance findings
- Machine stoppage caused by quality audits
- Downtime caused by Facilities stoppages such as no air, no power, water, etc.
- PM Schedule, the shutdown of equipment, all planned downtime
- No inventory and all non-machine related downtime
- Intermittent stops on the equipment such as assists and errors
- Parts and items that have worn out and replaced during a Preventive Maintenance
- Failures of secondary functions such as protective devices, leaks, broken gauges, sensors where the machine is still running and providing the primary function
- All cases of Function-Reduction Breakdown

Although the readers may or may not agree on the lists we provide on what to exclude and include in the breakdown, what is important is that the list generated is a result of a consensus among the members of our Key Planned Maintenance team. I think the message I would like to emphasize here is simple. Each of us may have our own understanding and definition of failure or breakdown. A failure may have a different meaning from a safety point of view. Suppose the safety person finds something that can sense danger or accident, such as an oil leak in the equipment; in that case, he can declare the equipment to be in a failed condition even if the equipment is still capable of functioning. Maintenance can declare equipment to be failed if there is some form of functional loss failure in the equipment, while from an operation's point of view, the equipment has encountered a failure if it simply stops running, even if this is a minor stoppage. What is important is that before measuring breakdowns and failures, maintenance and other people from the organization must agree on what to include and exclude in tracking breakdowns. The key here is being consistent and having only one definition for all maintenance involved.

The frequency of breakdowns and failures should not only be the sole measurement or indicator that must be used in the maintenance department. In fact, in some cases having a low breakdown frequency is not actually an indication of having a good performance. The consequences of failure matter most. I have several experiences to prove this. Again, let me give you an example based on my actual experience. One department recorded a breakdown of at least 10 times a month. The other department records only one breakdown a month. When asked which department is performing better, you might be tempted to say that it was the department experiencing only one breakdown. *If you think you are right, this is where you are so wrong!* It is actually the department with ten breakdowns that are performing better. The breakdowns the department encounters have little effect on the consequences of failure, while the department with only one breakdown encountered was from the Facilities and utility section of the plant. A single breakdown was encountered in one of their sub-stations that paralyzed certain sections of the production line from operating. The effect of failure on the department with more breakdowns was only confined to that equipment alone, while the failure of one substation in a manufacturing plant can stop at least 500 or more equipment on the production floor. So do not be impressed with departments having a low breakdown rate unless you are convinced that the consequences of their breakdown are minimal. Bhopal India encountered one breakdown on one of their tanks in the form of a leak. It killed more than 16,000 people for that single failure alone.

Based on my experience, I believe that this is one big problem for industries, especially for manufacturing plants. One rule that manufacturing plants have is that if an assist, error, or minor stoppage is experienced on the equipment or machine and no maintenance is currently available at the moment, they will be inclined to include this as a breakdown and not as a minor stoppage. This means that if a minor stoppage is encountered on the equipment and it took maintenance more than 6 to 10 minutes or more to fix the problem, then this will be converted into a breakdown loss. Usually, it will just take seconds or a few minutes to correct a minor stoppage, but the downtime will include the time for the maintenance to respond. This means that if it took 1 minute to correct a minor stoppage and 30 minutes of waiting for the maintenance to respond, the total downtime will be 31 minutes which is caused by the minor stoppage. Breakdowns and Minor Stoppages are entirely different. Assuming that the equipment experienced 3 breakdowns and 20 minor stoppages last month, then the denominator for calculating MTBF will now be 23 instead of only 3 which will provide a lower value. If the industry insists on including these minor stoppages then just call this **MTBFAMS** which stands for Mean Time Between Failure and Minor Stoppages. What an ugly Acronym!

4.4: The Word Mean Explained

More than a dozen mean time indicators are used, and the list continues to grow depending on which is used by different industries for different purposes. Before we discuss the most common Mean Time Indicators, let us define the word Mean. Mean refers to the average. If I am teaching 15 people in the class and ask each of these people to weigh themselves on a weighting scale which I place outside the room, then perhaps after getting the results of each of these people, I will be having the following value.

Average Weight = (65 + 72 + 90 + 82 + 68 + 120 + 75 + 95 + 60 + 88 + 69 + 71 + 74 + 98 + 78) kg / 15
Average Weight = 80.33

After the last person completed weighing, I have a value of 80.33 kg. This means that this is the average weight of the 15 people in my class. It does not mean that each person weighs exactly 80.33, some will weigh less and some will weigh more since we are just speaking about the average. Therefore, if we have an MTBF value of 160 hours in a month, it does not mean that we will be experiencing a failure every 160 hours, since we are simply discussing the average time of failure. Assuming that we experienced 3 failures or breakdowns, then this can mean that 1 failure was experienced after 60 hours, another failure occurs after 80 hours and the last failure happened after 40 hours. Hence, it is wrong to assume that the exact time of failure will happen every 160 hours which is unlikely to happen since just like the weight of the 15 people, we are simply talking about the average or what we called the mean.

4.5: Mean Time between Failures (MTBF) Explained

By definition, MTBF is an average measure of reliability that the equipment will run without failing. The most common units used will be in hours. Its origin can be traced from the US Military Standards (MIL-STD-217), and since then, it has been widely used in other applications in various types of industries. The original Reliability Prediction Handbook was MIL-HDBK-217, the Military Handbook for the Reliability Prediction of Electronic Equipment which was published by the Department of Defense, based on work done by the Reliability Analysis Center. The MIL-HDBK-217 handbook contains failure rate models for various electronic systems and components such as integrated circuits, transistors, diodes, resistors, capacitors, relays, switches, connectors, and so on. These failure rate models are based on the field data that can be obtained for a wide variety of parts and systems. MTBF can also be defined as the average time between two successive failures, and it is a measure of the time the equipment is operating until the time it encounters a failure. By empirical testing or allowing a part to fail which is a form of destructive testing, the length of performance or the functional life of a population of items can be divided by the total number of failures to achieve the MTBF of a part or component. By the word mean, we speak about an average amount of operating time between the occurrences of breakdowns or failures that require repair.

MTBF will be easy to compute if the equipment is suffering only from breakdown losses, however, if a machine is experiencing other losses, these must be excluded in calculating the MTBF value of the equipment. Several equipment losses will also constitute a downtime such as minor stoppages, set-up, or conversion, defect losses, start-up losses, and so on, what is important in calculating the MTBF value is to consider only the downtime caused by breakdown only.

MTBF is also defined as the average time between two successive failures and therefore is a measure of the trouble-free time. MTBF is an average measure of reliability that a device will run without failing. It is a reliability engineering term that means the average amount of operating time between the occurrences of breakdowns that requires repair. It means the average amount of time between two successive failures and is based on historical data or estimated by the vendor, which is used as a benchmark for reliability. MTBF trend will be, the higher the value, the more reliable the machine or asset is performing. Depending on how you intend to measure MTBF since this can be measured on a system-based level, sub-system based, component-based, equipment-based, several groups of machines, sub-assembly based, or even to the spare part level itself. The formula for MTBF will be as follows:

$$\text{MTBF} = \frac{\text{Operating Time}}{\text{Number of Failures}}$$

Where: Loading Time = Available Time – Non-Machine Related Downtime
Operating Time = Loading Time – Machine Related Downtime

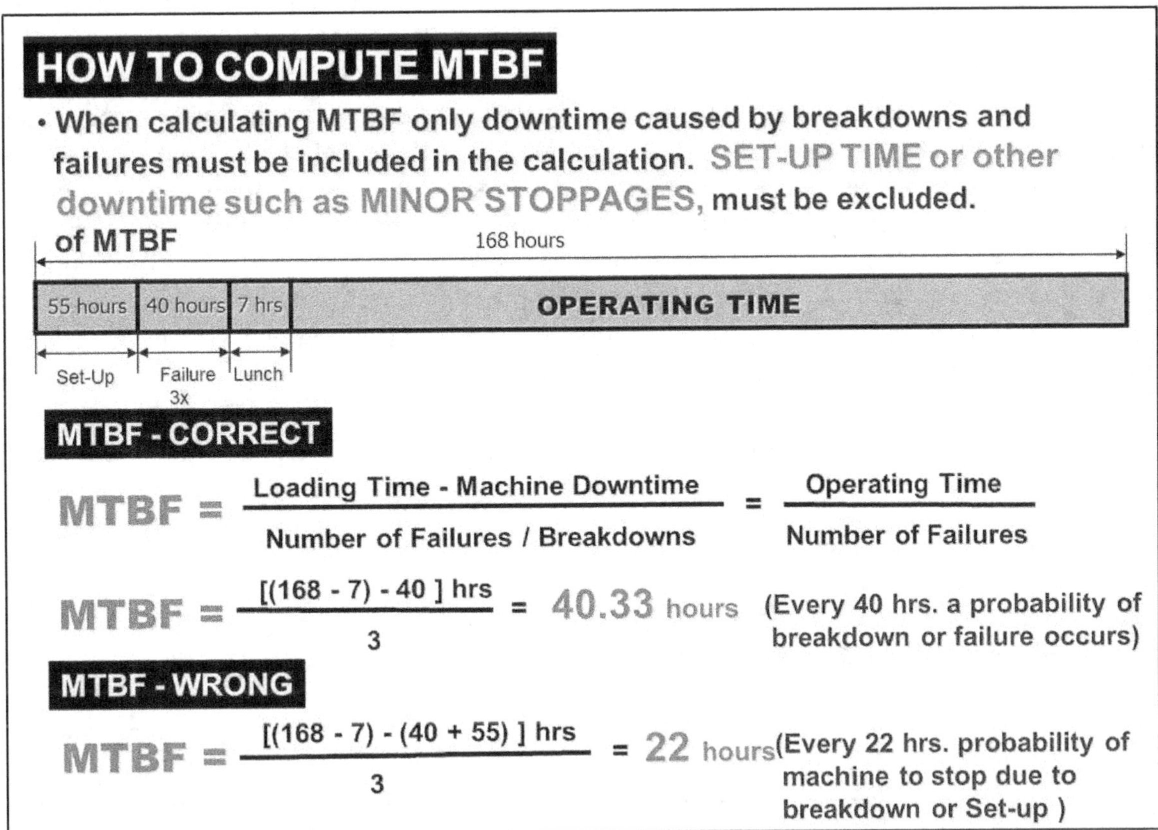

Figure 4.1: Only Downtime Due to Breakdown Should Be Included in MTBF

Since the denominator in the formula is the number of failures, the question is raised, what will be the MTBF in the case when there is no breakdown or failure? An MTBF of infinity will be obtained. This simply means that there is nothing wrong with the formula on MTBF. We can either prolong the duration of MTBF until a breakdown occurs or when there is no failure, we can assume a denominator of 1 to obtain a value. If you obtain an MTBF of 168 hours in one week, then that is the highest MTBF you can ever achieve on the equipment in a given week. The only downtime due to equipment breakdowns should be included in the MTBF calculation of the operating time. In manufacturing, breakdown and failures will not only be the cause of the downtime. Downtime can also be experienced during the changeover, minor stoppage, defects, and reworks. It must be made clear that downtime due to set-up and conversion, as well as short machine stoppages, assists, errors, defects, and other equipment losses must be excluded in the MTBF calculation. This must be made clear in an attempt to calculate a true and meaningful MTBF calculation. The acronym MTBF means that we are only interested in the average time the machine will stop due to an event of a failure or breakdown.

Another important point, I would like to consider is that when considering the denominator which is the frequency of failures, we are only interested in the failures or breakdowns that stopped the equipment from performing which is the function loss breakdown. There are 2 types of breakdowns which are the function-loss breakdown and the function reduction breakdown. Function Loss Breakdown or failure of the primary function is a situation in which the failure occurs, the equipment will totally stop and will definitely halt operations. RCM termed this as a failure of the primary function. Function Reduction Breakdown is also termed as a loss or failure of a secondary function. This is a situation where a failure occurs, but still, the equipment is running. This means that the equipment is still capable of operating and producing the needed output. Hence, when you are driving your car and your seat belt is not working, then you have a case of a function-reduction breakdown. If the police stop you since you are not wearing your seat belt and start writing a ticket and bribing you, then this is a case of function corruption. Anyway for industries, an example will be a machine that has an oil leak or sensors that is being bypassed. All equipment has both its primary and secondary functions that we must be aware of. Again for the record, let me reiterate that there are cases where the consequences of the failure of a secondary function can be far much worse than the failure of the primary function. This means that when calculating MTBF, we should only consider those Function Loss breakdowns and not the function reduction breakdown. The rule of thumb to consider is when the maintenance tasks come first before the failure, then we have no breakdown, however, when the failure happens first and we repair the machine, then there is an actual breakdown on the equipment.

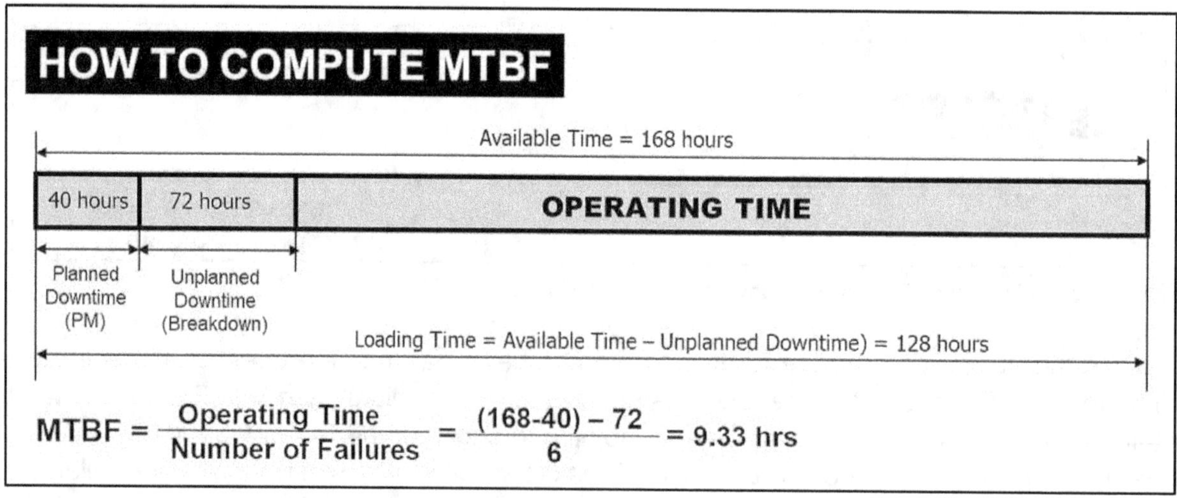

$$MTBF = \frac{Operating\ Time}{Number\ of\ Failures} = \frac{(168-40) - 72}{6} = 9.33\ hrs$$

Figure 4.2: How to Compute MTBF

Given:
• Available Time = 168 hours
• Machine Downtime due to Breakdown= 72 hours (Also called Unplanned Downtime)
• Non-Machine Downtime due to PM = 40 hours (Also called Planned Downtime)
• Loading Time = Available Time – Non-Machine Downtime = (168 – 40) hours = 128 hours
• Operating Time = Loading Time – Machine Downtime = (128 – 72) hours = 56 hours
• Frequency of Failures = 6 times

• MTBF = [(168 – 40) – 72)] hours / 6 hours
• MTBF = 9.33 hours

In figure 4.2, the MTBF is equal to 9.33 hours. So what do 9.33 hours actually mean? Most people are confused about interpreting this figure and will say that when it reaches 9.33 hours of operation, the equipment will fail at that precise period. Remember as we mentioned previously that the word "mean" refers to an average amount of time; therefore, this is interpreted as a probability or possibility of failure, but this is not the case in almost all instances. The failure can occur before or much longer than 9.33 hours. So the correct way of interpreting this figure is by stating that there is a probability of failure when it reached 9.33 hours of operation.

Reliability predictions are used to compute the predicted failure rate or the Mean Time Between Failures (MTBF) of its system. MTBF is usually being expressed in terms of hours. For example, if the system has an MTBF of 1000 hours, this means that, on average, the system can experience one failure in 1000 hours of operation. By using accepted standards for modeling failure rates of components, we can calculate our equipment's failure rate which is the reciprocal of MTBF. The goal is to make sure that the asset's MTBF is within its acceptable limits. If your prediction analysis shows low MTBFs, this means that you can expect your system to fail more often, and you may need to take steps to improve it. By making changes to your design, for example, maybe lowering temperatures or stress levels, the MTBF can improve and expect better performance. Only losses due to breakdown should be included in the MTBF calculation. Other equipment losses such as set-up and minor stoppages must be excluded in calculating MTBF. Breakdowns and failures not only on mechanical parts but on electronic, electrical, and electro-mechanical parts that caused the equipment to stop will be considered when calculating MTBF.

4.5.1: How is MTBF Calculated

MTBF can be calculated in many ways depending on how the industry wants to use this Mean Time Indicator. If you plan to use MTBF as one of your indices, we need to understand how your plant intends to measure MTBF. MTBF can be used to compute the following in figure 4.3.

In figure 4.4, when calculating MTBF for a group of machines, we can either get the Total MTBF, Average MTBF, or Percentage MTBF. For calculating MTBF for a given process or line, the equipment, or machine with the lowest MTBF will serve as the bottleneck equipment depending on the losses of the equipment. This means that improving the bottleneck equipment or machine means increasing its MTBF. We can measure MTBF as follows:

• MTBF by Critical Component to determine on an average when a critical component will fail

• MTBF by Sub-Assembly to determine which sub-assembly fails frequently on a machine

• MTBF by Machine to determine the MTBF of an individual machine and identify the part or spare that frequently fails. Anything that fails on the machine will be included.

• MTBF by Group of Machines: To determine the MTBF of several machines and identify both the highest and lowest MTBF value.

• MTBF by Process or Line to determine which process in the manufacturing fails frequently and to identify the bottleneck or constraint machine in a process.

• MTBF by Item or Spare to identify the part or item that frequently fails in the equipment. To determine the average time a particular item or spare will fail.

• MTBF of the Plant to determine the average failures for all equipment in the plant.

• MTBF by System or Sub-Systems to identify the system or subsystem with the lowest and highest MTBF value so that we can compare their performance.

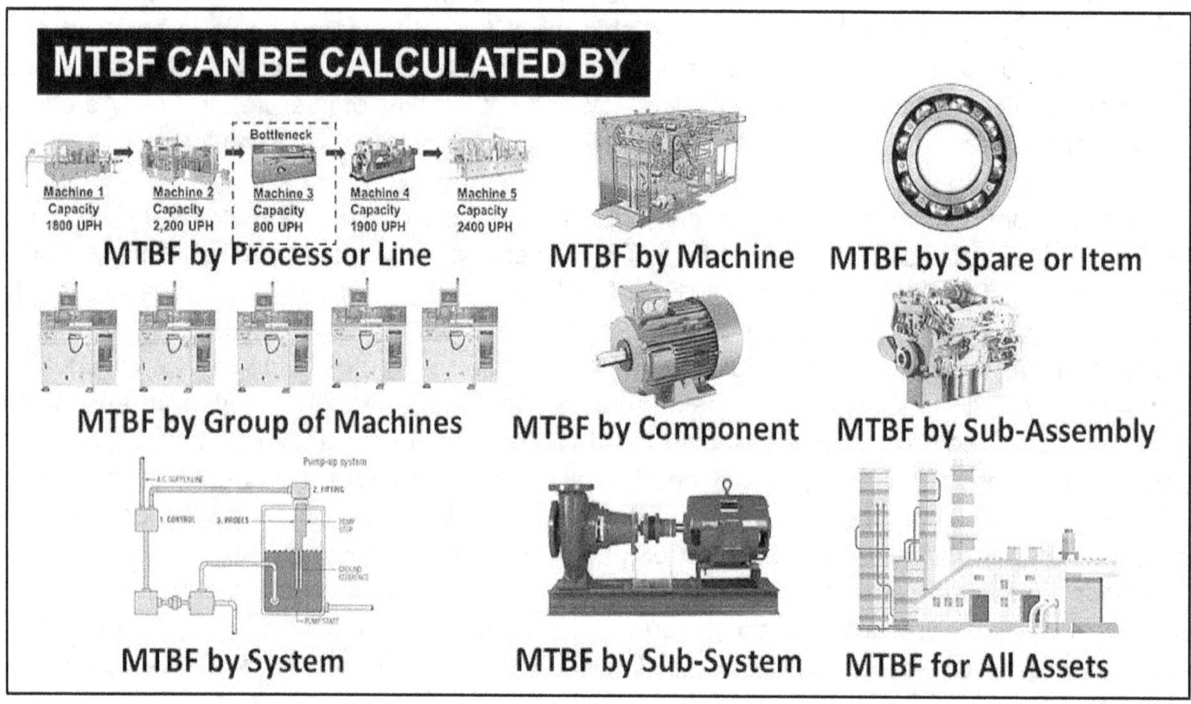

Figure 4.3: How MTBF can be Calculated

Again, what is important in calculating the formula will be to differentiate what will or will not be included as a breakdown. Accuracy and consistency are important in calculating these indices as people may have different interpretations on what to include or exclude in the MTBF calculation. Another important factor to consider is to separate the Planned and Unplanned downtime. In calculating MTBF only machine downtime caused by breakdowns and failure should be included. Function-Reduction breakdown or failure of the secondary function should not be included in the frequency of breakdowns. Performing Preventive Maintenance tasks of replacement of a part that has deteriorated will not be considered as a breakdown while performing a repair on an item or part that failed will be considered in this case as a breakdown on the equipment. There should be a clear boundary on this.

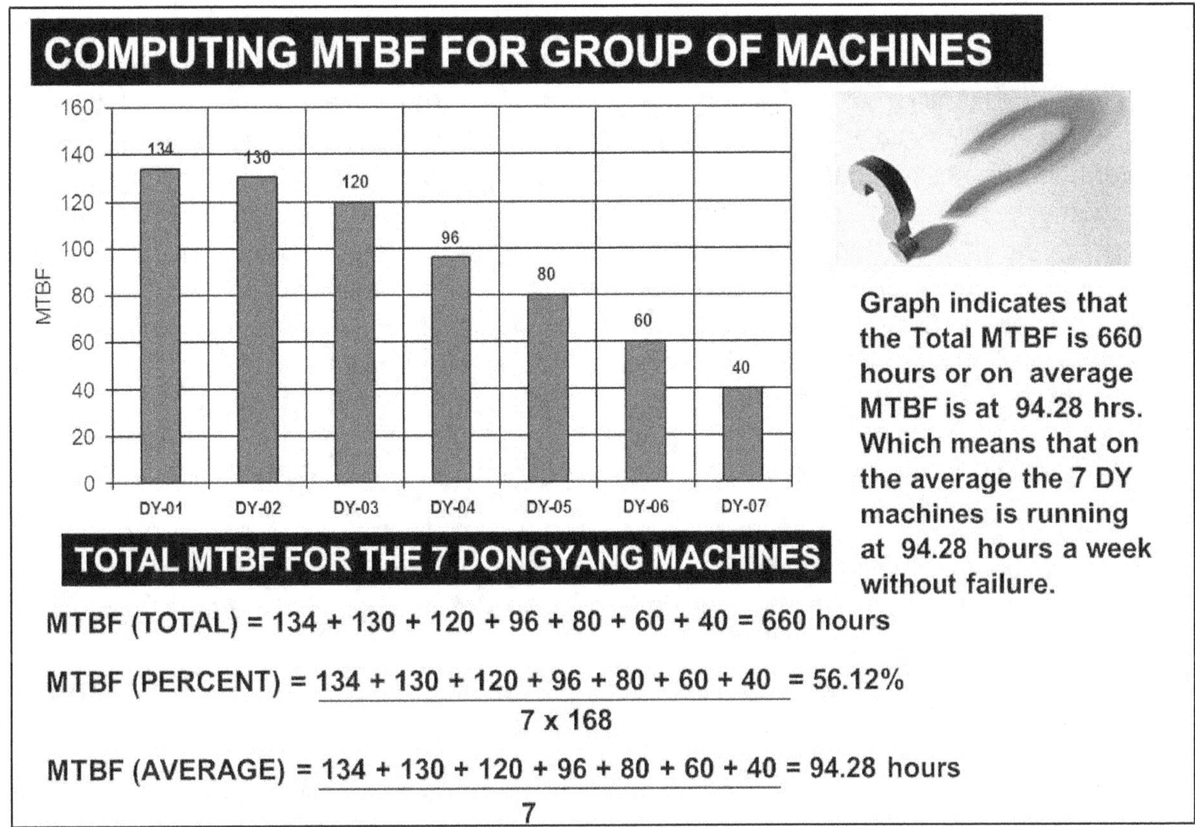

Figure 4.4: Calculating MTBF for Group of Machines

4.5.2: Where Can We Use MTBF

MTBF measurement has a lot of uses that can benefit maintenance in industries depending on how we want to measure it. Listed are the following where we can use MTBF.

Comparing Vendors: In many industries I know, Purchasing people often award the parts to a particular vendor or supplier with the lowest cost. MTBF can be used to compare two or more identical parts from different vendors. When we sub-contract some repairs of our equipment to some of our local vendors, one of the justifications in awarding the contract will be its MTBF. A part or spare with a higher MTBF may require a higher initial cost, but lower life cycle costs can be achieved in the long run if we consider the cost of doing repairs, costs of spare parts, overtime costs, overhead costs, replacement costs, and downtime costs, just to name a few. The part with the higher MTBF will definitely have a lower cost in the end. A part or component with a higher MTBF will yield a better return in cost savings in the future. Although some vendors would use the term MTTF or Mean Time To Fail.

Determining the Interval for Functionality Inspection: Functionality Inspection or Failure Finding Interval (also called Detective Maintenance) should only be done for hidden failures such as protective devices and redundant functions to avoid the chances of multiple failures. Multiple failures are stated as to where both the Protective Device and Protected Function are both in a failed state condition. Hidden failures will only become evident if the ones that they are protecting also fail in operations. Protective devices are placed on the equipment for a particular reason, and that is to protect something in the equipment. That something is called the protected function. An example of a protective device is a lighting arrester placed on a transformer. The function of the lighting arrester is to divert the current to the ground when the lighting will strike instead of the current going directly to the transformer causing damage. However, there are cases that when the lighting will strike and absorbed by the lighting arrester, it can damage the arrester permanently and lose its continuity to function. Therefore, maintenance tasks called functionality inspection or Failure Finding Tasks are used to check if these protective devices in our equipment are still functioning or not. The interval for determining Failure Finding Tasks or Functionality Inspection will depend on two factors, the desired availability and the average life or Mean Time Between Failure (MTBF). If the failure of a protective device is less critical then, we can aim for a lower availability but for failures of protective devices that are considered critical, then we should aim for much higher availability. Note that the higher the availability, the more frequent will be the Failure Finding Tasks or Functionality Inspection.

Availability for hidden function	99.99%	99.95%	99.90%	99.50%	99.00%	98.00%	95.00%
FFI Interval (% of MTBF)	0.02%	0.01%	0.20%	1.0%	2.0%	5.0%	10%

Figure 4.5: Table for Failure Finding Tasks Interval

Let us assume that an emergency stop button has an average life (MTBF) of 10 years. What will be the interval of performing a Failure Finding Tasks or Functionality Inspection?

If Availability = 95%, Failure Finding Interval (FFI) = 10%, MTBF = 10 years
- FFI = [(10 years) x (365 days / year) x (10)] / 100
- FFI = 365 days or Once a year it should be inspected

If Availability = 99.90%, Failure Finding Interval (FFI) = 0.20%, MTBF = 10 years
- FFI = [(10 years) x (365 days / year) x (0.20)] / 100
- FFI = 7.3 days or Once a week it should be inspected

If Availability = 99%, Failure Finding Interval (FFI) = 2%, MTBF = 10 years
- FFI = [(10 years) x (365 days / year) x (2)] / 100
- FFI = 73 days or can be rounded to every two months

Note that if we go for the lowest availability at 95%, the emergency stop needs to be checked once a year. This means that inspecting the emergency stop once a year will guarantee us 95% availability that it will be working with a 5% chance of failure. However, if we increase the availability to 99.9%, the chances of having the emergency stop available will be very high with 0.1% chances of failure. In this case, if we go for 99% availability, the FFI will be every 73 days with a 1% probability of failure. I think I will be happy with this. My recommendation is that if the protective device is critical, aim for higher availability. If the protective device is less critical, then we can aim for a much lower availability. Use this table to calculate the frequency of inspection for protective devices and redundant functions. Again, Failure Finding Tasks and this table will only be used for hidden failures.

MTBF / MTTR / BDO ANALYSIS CHART

STATION : Wirebond Station   Machines Included: WB 001, WB 002, WB 003, WB 004, WB 005, WB 006, WB 007, WB 008, WB 009

WW	1	2	3	4	TOTAL	WW	5	6	7	8	TOTAL	WW	9	10	11	12	13	TOTAL
DT (hrs)	12	15	10	16	53	DT (hrs)	20	13	5	3	41	DT (hrs)	18	15	14	13	6	66
BDO	4	5	3	5	17	BDO	6	4	2	1	13	BDO	6	4	4	2	1	17

(January) _(February)_ _(March)_

ITEM	MRO SPARE PART AFFECED	BDO TRACKING
1	Heater Block	IIIII - IIIII - II
2	X-Y Table	III
3	Clamp Left Low Side	IIIII - IIIII - IIIII- IIIII - II
4	Work Holder	IIIII-II
5	Clamp Righ Low Side	IIIII - IIIII - IIIII- IIIII - IIIII - IIIII
6	Spindle Motor	IIIII - IIIII - IIIII - IIIII - III
7	Servo Vale	IIIII - IIIII - IIIII - II
8	Cutter Blade	IIIII - IIIII - IIIII
9	Linear Bearing	IIIII - IIIII - IIIII - IIIII - IIIII - III
10	Window Clamp	IIIII - IIIII - IIIII - IIIII - II

ILLUSTRATE FLOW CHART PROCESS

Wirebond Machine

MAINTENANCE INDICES SUMMARY

	Jan	Feb	Mar	Apr	May	Jun
MTBF	36.41	48.54	45.53	48.92	58.73	40.68
BDO	17	13	17	13	11	19
MTTR	3.12	3.15	3.88	2.77	2.36	3.53

	Jul	Aug	Sept	Oct	Nov	Dec
MTBF	21.26	24.48	17.53	80.75	92.57	137.17
BDO	31	27	47	8	7	6
MTTR	0.42	0.41	0.34	3.25	3.43	2.83

$$MTBF = \frac{Loading\ Time - Machine\ DT}{BDO}$$

MTTR = Machine DT / BDO

LEGEND : DT (HRS) - Downtime in hrs   BDO - Breakdown Occurences

BDO	47	8	10	12	8	9	BDO	27	7	6	6	8	BDO	31	5	8	8	10	BDO
DT (hrs)	16	2	3	4	3	4	DT (hrs)	11	3	2	3	3	DT (hrs)	13	2	4	3	4	DT (hrs)
WW	TOTAL	39	38	37	36	35	WW	TOTAL	34	33	32	31	WW	TOTAL	30	29	28	27	WW

(September) _(August)_ _(July)_

Figure 4.6: MTBF / MTTR / BDO Sample Form for Spare Parts

Forecast Spares Needed for a Given Period: The use of MTBF and MTTF can be used as a baseline for understanding what parts seem to fail in our equipment. This data can provide both the storeroom, maintenance, and purchasing a clear idea, which parts seem to fail frequently in the equipment and what parts need to be stocked in the MRO Storeroom. By knowing what parts failed, we can subject them to the MRO Decision Diagram or algorithm which I have written in my book on MRO Spare parts to determine if they need to be stocked in the storeroom or not. The benefit of the MTBF analysis is that this is based on the actual usage consumption of maintenance compared to EOQ and min-max where the usage declared by the storeroom is actually based on the number of items withdrawn from the storeroom. I have developed a simple form in figure 4.6 on MTBF (Note that BDO means Breakdown Occurrence), where we can see clearly the spare parts that failed in the equipment. This sample form is the summation of failures for nine wire bonder machines. Depending on how large the industry is, most industries have several types of equipment. This form can be used for single or multiple machines or can be used per equipment type.

This form in figure 4.6 can be used per equipment type for a group of machines. If a plant has 15 types of equipment, then you will be needing 15 of these forms on the MTBF / MTTR /

BDO Analysis form. This single form is composed of nine wire-bonder machines, which are of the same equipment type. DT stands for downtime, and BDO stands for breakdown occurrences. In one year, we have 52 weeks; hence the form is equally divided into 52 weeks. Items 1 to 10 are the spare parts that eventually fail on the 9 wire-bond equipment. At the same time, both the DT and BDO are the summation of the total breakdown occurrences and downtime of the nine machines. In workweek 1 (WW1) in figure 4.6, we have a total downtime of 12 hours and a total number of breakdowns of 4 and so on. The MTBF, MTTR, and BDO will be calculated automatically on the lower right portion of the MTBF analysis form. You can use this form if CMMS or EAM does not exist currently in your plant. What is important is to have some actual figures for tracking failures and breakdowns. If a CMMS or EAM software is available, we can discuss with our IT or those responsible to have a second level on MTBF which corresponds to what particular part failed a well as its frequency, in which the MTBF can be calculated per equipment type and what are the failures experienced.

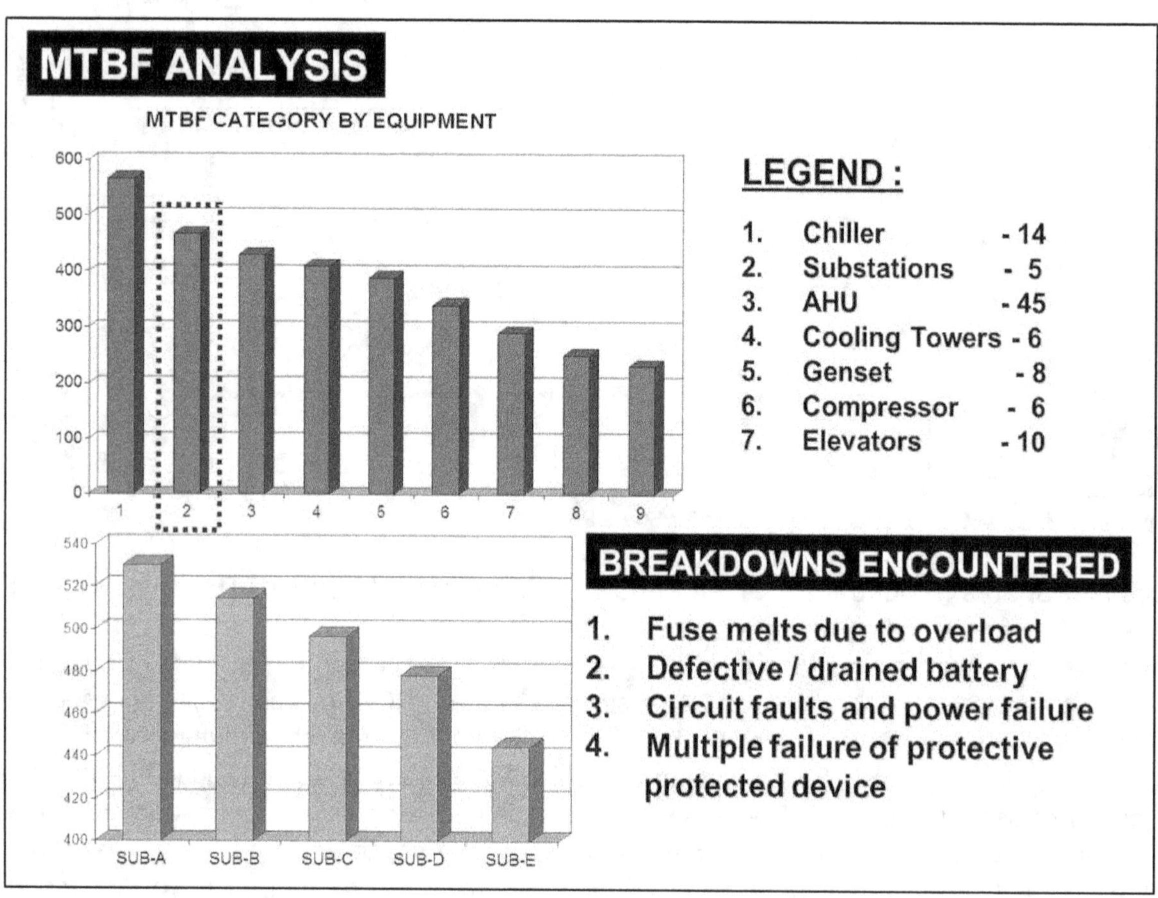

Figure 4.7: MTBF Analysis

For Conducting MTBF Analysis: MTBF is used to determine the main contributor as to why equipment keeps on failing. A group of maintenance people can conduct an MTBF study to determine what parts of their equipment keep on failing and to analyze the root cause of failure on that part so that they can modify or redesign the part with inherent design weakness. Just like in figure 4.7, when we click on item 2, it will show the MTBF of all substations in the plant, with substation E having the lowest MTBF value of around 445 hours. By moving to the next level, we have a list of failures experienced that resulted in a low MTBF on sub-station E. We can provide a Pareto analysis and prioritize 20% of failures, constituting around 80% of the problems. In figure 4.7, the breakdowns experienced by Substation E are determined, hence, corrective actions can be generated to increase the MTBF of Substation E.

4.5.3: Limitations of MTBF

Although there are many uses and benefits of using MTBF, it has also its limitation. MTBF should not be used as a baseline in determining the frequency of overhauls and replacement in the Preventive Maintenance schedule because we are merely talking about the average time of failure. A common misconception about MTBF is that it specifies the average time when half of the components and items will fail. Since this is just an average, it is not suitable for predicting or forecasting the time of failures for an individual part, component, or item due to its inherent variability concerning time. Collecting failure data to calculate MTBF in determining the maintenance interval for overhauls and replacement is wrong. It should not be done since MTBF is merely an average, and failures fall predominantly into three patterns, age-related, which make up to 20% to 30% of all failures, and the bigger portion will be that of both random and infant mortality failures, which constitute to around 70% to 80% of all failures. And for age-related failures, it is not MTBF but rather the remaining useful life that is significant when attempting to determine the best frequency of replacement or maintenance overhaul. The frequency of doing Preventive Maintenance replacements and overhauls should be based on the useful life and not on the average life of the part, component, or item.

In figure 4.8, let us assume that in a span of 9 years, 100 bearings failed. During the first year, 5 bearings failed, between the first and second year 15 bearings failed, between the second and third year 10 bearings failed, and so on. In this case, the bearings failed at different intervals which indicates that the failure is random. Hence, we cannot use MTBF to determine the frequency or interval of replacement since MTBF is just about the average. If we conduct and replace the bearing every 2 years, then our PM will be costly, while if we replace the bearing on the 7th or 8th year, then we can expect a lot of failures before the Preventive Maintenance is due. Determining the interval for replacement on Preventive Maintenance should be based on the useful life and not on the average life of the part. If a part can fail at different intervals, then we can use Predictive Maintenance in this case and base the replacement on the potential failure of the item.

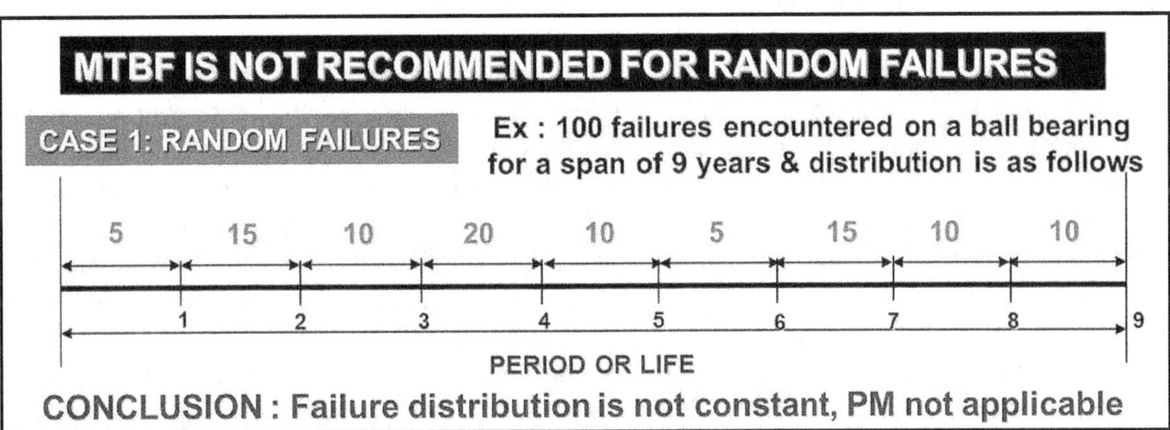

Figure 4.8: MTBF is Not Recommended for Random Failures

Vendors' MTBF may vary from how we used the part or item. Field conditions where the part is subjected for MTBF study vary enormously, from their surroundings and environment, corrosive atmospheres, ambient temperatures, relative humidity, moisture, air cleanliness, all of which will affect the MTBF of the same component depending upon the conditions in which the part is being subjected. Therefore, MTBF value may vary depending upon the conditions of the environment. Example: An electronic board will have a shorter lifespan if the environment where it is placed has high contents of sulfur and acids and will definitely last longer on a cleanroom facility such as a semiconductor plant.

MTBF may vary up to ten times, depending on the installation site, environment, and, most especially, how the machine is operated. Field conditions vary enormously, from the power supply voltage to corrosive atmospheres, ambient temperatures, relative humidity, moisture, air cleanliness, all of which will affect the MTBF of the same component depending upon the conditions in which the part is being operated. Hence, vendors' MTBF may not be the same if the operating conditions will be different. A bearing manufacturer may declare a useful life of 5 years on the bearing they are selling, however, our experience tells us that more than the majority of the bearings just lasted for 1 year, or even less. In this case, we cannot conclude that the vendor is lying about the bearing's useful life on how the vendor arrived with the figure. We cannot litigate the bearing vendor in court even if you got the finest lawyers in town, since the environment in which the bearing was subjected for the life data study and your environment may not actually be identical. If you are a coal power plant in which your location is beside the sea then that perhaps is one factor on why the life of the bearing is shorter compared to the dictated vendor's life, not to mention the load the bearing absorbs in your equipment and how the maintenance performs their lubrication, its frequency, and other factors.

4.6: Mean Time to Fail (MTTF)

Suppose you buy a 30 or 40-watt bulb and place it in your house. The bulb lasted for 3 years or 26,280 hours. What is the MTTF or Mean Time to fail for the bulb? If you answer 3 years, then you are correct. Let us say that this is the first time that you have replaced the bulb in your home, and the last time you replaced it was exactly three years ago. What is the MTBF or Mean Time between Failure of this bulb? Again, if you say 3 years, then you are correct. Very good! My next question, if both MTBF and MTTF are 3 years, is MTBF the same as MTTF? *If you answer yes, then this is where you got it all wrong!*

Technically speaking, MTBF should be used only on parts or components that can be repaired, while MTTF should be used for parts and components that are said to be non-repairable items. However, in most cases, MTBF is commonly used for both repairable and non-repairable items since this is the more familiar or popular term. When a bearing inside a motor fails, we replace the failed bearing; since the bearing is severely damaged, it cannot be repaired nor restored. In this case, we are speaking about the MTTF of the bearing. However, once the bearing has been replaced, the motor can be used again. In this case, we are speaking about the MTBF of the motor. Although both MTBF and MTTF trends would be the same where both mean time indicators, require, the higher the value, the better. MTTF refers to parts that cannot be repaired, while MTBF refers to assets and equipment that can be repaired.

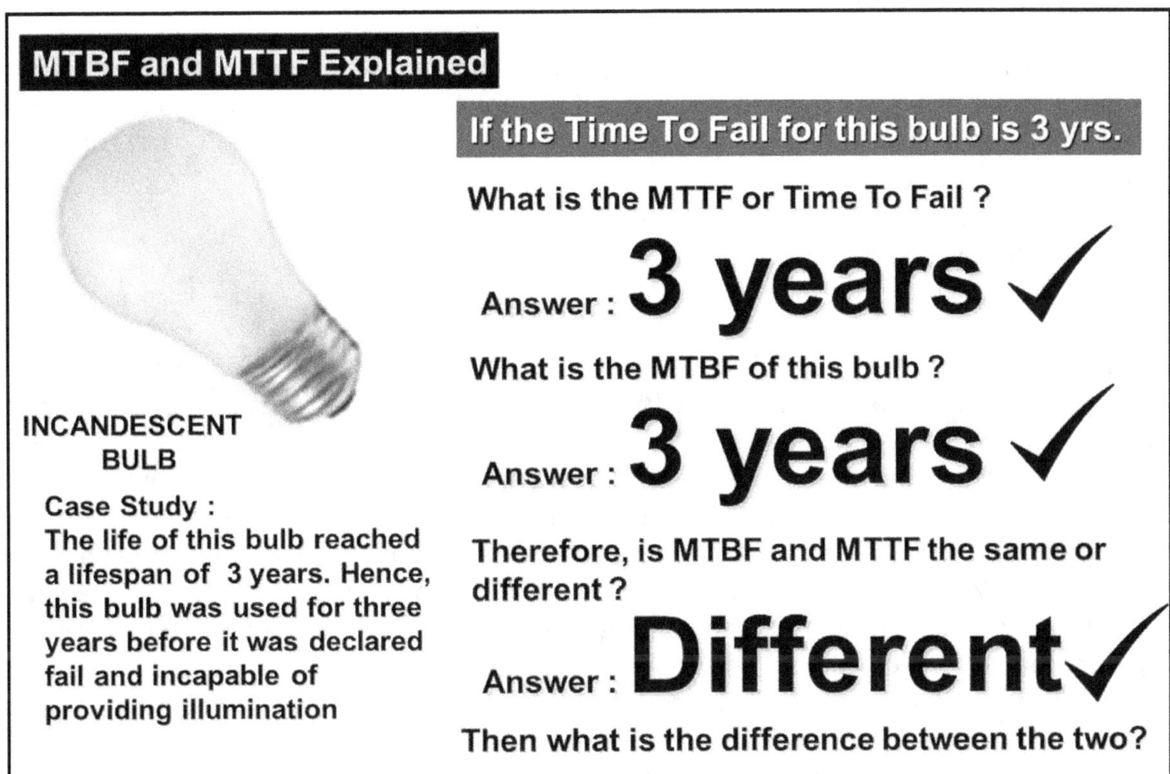

Figure 4.9: Is MTBF and MTTF the Same or Different

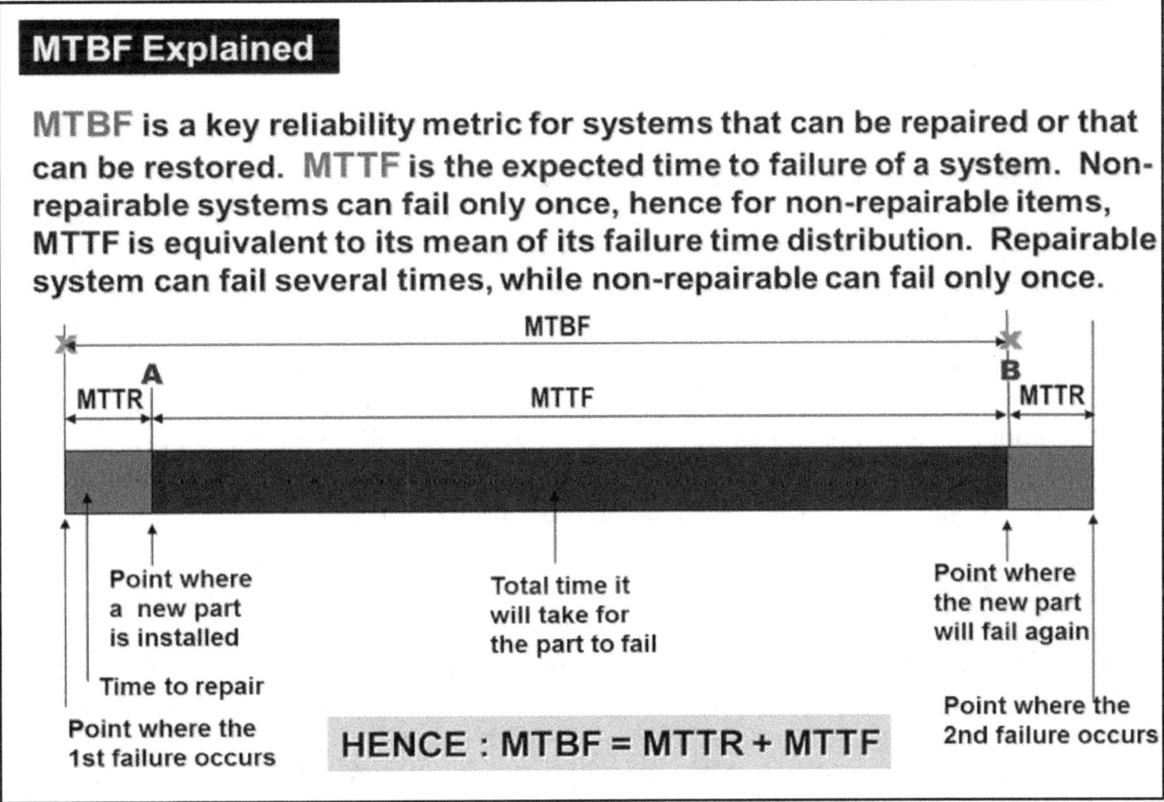

Figure 4.10: MTBF and MTTF Explained

This means that in the case of the bulb that lasted 3 years, the MTTF will be equal to exactly 3 years. MTBF will be equal to the MTTF plus the repair time which is the MTTR. Assuming you have a spare bulb and replaced the busted bulb which took you 5 minutes to get the spare bulb and replaced it, then the

• MTBF = MTTF (Mean Time to Fail) + MTTR (Mean Time to Repair)
• MTBF = 3 years + 5 minutes.
• MTBF = 26,280.08 hours

I have a Sony Vaio laptop which I purchased in 2009 and am still working up to this point of writing this book (well-maintained). In 2015, when I opened it, there was absolutely no power, and I got an upcoming training event since this laptop has been my source of bread and butter for living. I went to a Sony Service center and had a technician diagnose the laptop; after a few hours of waiting, he said that the motherboard had been damaged since 3 IC's failed. I told them if they could replace the IC that failed, and they responded negatively, but they said that I was lucky since they have a spare motherboard for this laptop if I wanted to. Since I needed my laptop so badly, I agreed with them and left my laptop. After a couple of days, they called me, and I went to their service center once again, and now my laptop was working. The laptop was repaired. Hence, if we speak about the laptop, then we speak about MTBF since it was repaired, but the motherboard will never be repaired for life. Therefore, if we speak about the motherboard, it will never be repaired, so it will be the end of life for the motherboard. This is where we used the term MTTF or Mean Time to Fail. In layman's terms, if you got sick and got the **Covid 19** virus and end up in a hospital, after some medication, you got well, then that's MTBF, meaning you were repaired. On the other hand, if you got worst and your heart stops pumping, then that is the end of your life, so that will be MTTF. For people, MTTF is also the same as Rest in Peace.

MTBF (Mean time between failures) for a repairable product is the average length of time between failures that requires repair. Therefore, for both MTTF and MTBF, the longer the period between failures, the more reliable the component or asset will be. MTTF (Mean time to fail) is the life test data collected, which can be used to calculate the average length of time between failure of a non-repairable component that is occurring, such as in the case of a light bulb. Therefore, when we speak about MTBF, we also consider the repair time since MTBF = MTTR + MTTF. The trend for both MTBF and MTTF would be the same, where both mean time indicators require, the higher the value, the better. Again, MTTF refers to parts, which cannot be repaired, while MTBF refers to systems or equipment that can be repaired. The reciprocal of MTBF and MTTF is the failure rate, which is the probability of failure.

4.7: Failure Rate

Failure Rate is the expected rate of occurrence of failure or the number of failures in a specified period. It is usually expressed by the Greek letter λ (lambda) and is often used in reliability engineering studies. The failure rate of a system or equipment usually varies concerning time. This means that as the equipment age, the failure rate increase as it nears the end of its life cycle period. The failure rate is typically expressed in failures per million or billion hours. It is simply the reciprocal of MTBF. It is also the probability of a failure in a stated unit of time. Like MTBF, the most common unit to express the failure rate is failures per hour. As the Failure Rate increases, the shorter the MTBF will be, and the greater the chances of failure to occur in the equipment. Although most will prefer to use MTBF instead of failure rate as the figures will be much easier to relate. This means that if we have an MTBF of 1 year, the probability of failure is at 8760 hours which is easier to interpret instead of 0.000114155 failures per hour, yet both of these denote the same meaning.

If a pump has a Failure Rate of 0.001 failures per hour, the probability of failure in a year can be calculated as follows:

Failure Rate = 0.001 / hour x 24 hours / day x 365 days / year
Failure Rate = 8.76 this will be the probability of failure in a given year

To determine the average repair time, we need to determine the time it takes for the maintenance to repair the equipment. If it takes an average of 2 hours to repair the pump and the failure occurs only once a year, then the MTTR or Mean Time to Repair will be:

MTTR = 2 / 1
MTTR = 2 hours

If the failed component was the impeller, and it is said to be non-repairable but replaceable, therefore the MTTF or Mean Time to Fail for the impeller will be:

MTTF = MTBF - MTTR = (1000 - 2) hours
MTTF = 998 hours

Failure Rate is the probability of a failure in a stated unit of time. It is the inverse or reciprocal of MTBF. For example, the MTBF of a roller bearing is estimated to reach 1000 running hours of operation. What is the failure rate of the bearing? What is the probability of failure after running for 800 hours? This means that as the Failure Rate increases, the shorter will be the MTBF and the greater the chance for a failure to occur.

FAILURE RATE	MTBF (hrs)	AVE. FAILURE FOR 800 hrs
0.0010	1000 hrs	0.8
0.0015	667 hrs	1.2
0.0020	500 hrs	1.6
0.0025	400 hrs	2.0
0.0030	333 hrs	2.4

Figure 4.11: Failure Rate and MTBF

If the MTBF of a Television set is at 200,000 hours. The Failure Rate which is the reciprocal of MTBF is given as 0.000005 failures per hour which means that after watching Netflix on TV for about 200,000 hours of television shows, it is expected that your TV set will encounter a failure besides your eyes getting so tired.

4.8: Reliability Explained

Reliability is the probability that the equipment will perform its specified function under specified operational and environmental conditions under a given and specified time. Reliability can be said as the probability that no failure will occur throughout a prescribed operating period. According to Bazovsky, the modern concept of reliability in popular language is the capability of equipment not to fail or break down in operations. When equipment works well and performs to do its job for which it was designed to perform, such equipment is said to be reliable. Reliability is the probability that an item will operate without failure throughout a specified interval. For example, if we schedule the next week's production, the equipment's reliability or probability will be our main concern. I have trouble accepting Bazovsky's definition of reliability, since how about if there are no failures but the speed of the machine was reduced, can we declare it to be reliable? Since reliability is a probability, we need to have some historical data on how the equipment performed, perhaps within the past couple of months, to better understand how the equipment will run in the future. This means that to know the future of the equipment, we need to know the past on how the equipment performed. One problem with Production Planning is that it assumes that the equipment will run without a single failure or downtime, which, in reality, is hard to achieve. Second, breakdowns or failures are not the only types of equipment losses that can be experienced on the equipment. What is important is determining the type of losses the equipment is suffering and assigning the correct indicator to measure its performance. My point of view, in this case, is that failures and breakdowns must not only be the indicator to determine if equipment or asset is reliable or not since failures and breakdown is just one part of the losses. This means that if a piece of equipment experienced no breakdown, the MTBF is high as well as its reliability, but if the equipment's speed is reduced by 15% would we declare it to be reliable? The problem with the definition of reliability is that it assumes that failures and breakdowns are the only losses on the equipment which is not actually the case.

If I explain reliability in the simplest way I can, assuming that the reader is in his youth or high-school days today, and you have your current girlfriend, but if we think about our life 20 years from now your current sweetheart may or may not be your future wife. What I am saying is that reliability is about your future wife because when we talk about reliability it is about the future. How will my equipment run for the next 10 hours, for the next 10 days, for the next couple of months, for the next six months, and so on? To know the future on how the equipment will run, we need to know the past on how the equipment performed.

In figure 4.12, let us say that we want to determine the reliability of this equipment for the next 10 days, and reliability is at 90%; this means that there is a probability that the equipment will run 90% of the time with a 10% chance of failure. This will answer how your equipment will function in a given set of times, but again this calculation is only a probability to give us an idea of how the equipment will run for the next 10 days. To know the future on how our equipment will operate, we need to have a background about the pasts and asks questions such as:

• What is the MTBF of this equipment for the past period?

- What is the Failure Rate of this equipment for the past period?
- How well did the equipment perform over the past couple of years?
- What failures were mostly encountered?
- Were we able to eliminate the failure?
- What are the probability of failure this time?
- Is the failure likely to occur in the equipment?
- Are we prepared for the failure this time?

Figure 4.12: Formula for Reliability

4.8.1: A Deeper Meaning of Reliability

Technically speaking, if we use Bazovsky's definition and formula of reliability, we can increase the reliability of our equipment and assets. This is a wrong concept since the inherent design reliability of any equipment is given at 100% on the assumption that the equipment is running without any single failure, breakdown, design speed loss, or other downtimes which in reality is not a possibility. As these failures and breakdown losses are reduced, the reliability percentage increases. If the reliability of the equipment was at 60% last month and this month it becomes 85%, it does not mean that reliability actually improved. The reality is we are not actually increasing the reliability in this case since it is already given at 100% at its design stage, and we are just sustaining it by avoiding unnecessary failures and losses in the equipment. Achieving 100% reliability in the real world is not possible since there are many losses to address. One of them is a Planned Shutdown. We cannot operate equipment without any form of scheduled Preventive Maintenance. When we perform Preventive Maintenance, we need to shut down the equipment.

I have heard others say that reliability cannot be improved by any means of doing maintenance. Well, my wisdom on this is that maintenance is two folds, which means that we may be talking about the maintenance tasks, or we may be talking about the maintenance people in our plant. If we consider maintenance as a task, then I agree that reliability cannot be improved but can only be sustained. But if we consider maintenance as a human being capable of modifying or redesigning parts with inherent design weaknesses, then reliability can definitely be improved.

I have a friend, his name is John, and our friendship goes a long way ever since we were in high school in Don Bosco Makati, in the Philippines. His father owns a shipping business and has several cargo ships. They were rich. One of those ships was called MV Sea Raider, which was my first taste of work experience way back in 1986. One day, John called me and asked for some favor if I can help him. I said ok and asked him what he wanted me to do. John said that he bought a tag boat and wanted to convert it into a fishing boat. He said that he salvaged the old anchor winch of his father's ship, which was the MV Sea Raider, where I used to work, which was already retired and scrapped. John's idea was to use the anchor winch and attach it to the net to catch some fish. Since this was where I used to work, it has some form of sentimental attachment with me. I accepted his offer to help him. We went to the shipyard, salvage the old anchor winch, and we started to open the gearbox of the old anchor winch. I started to measure the different gears inside the old winch. His tugboat crew installed the winch, placed a net, and tested it. It was a slow winch as it was originally designed for the anchor of the ship. The other crew in the fishing boat were laughing at us and one guy even said that even the most stupid fish on the sea will not be captured by the fishing net as he has plenty of time to escape.

The diameter of both the motor gear (D1), which is the driving gear and the driven gear (D2), will be modified. After studying and providing some calculations on the gears, I have provided five options regarding the size of the driving and driven gear, which were D1 and D2, respectively. We also calculated the time for the net to reach the ship when released up to one mile in length. The original design of the winch will take 5.34 hours to reach the ship when it was released one mile from the tug boat.

After one week, I submitted my calculation to my friend John. I provided him several options as to the size of the gear ratio as in figure 4.13. He told me that he would be going for the ratio of 1:25 or option 4 with an rpm of 71.70. Based on the calculation, if this gear ratio will be used, then the time for the fishing net with a distance of one mile to return to the ship will be reduced from 5.34 hours to 1.23 hours. Finally, he went to a machine shop and has the two gears, which are the driven and driving gear, fabricated to the exact diameter he wanted for gear D1 is the driving gear, and gear D2 for the driven gear. The machine shop fabricated the gear, and after a couple of weeks, it was done. They installed the revised gears on the winch and have it tested. During their trial run, John called me; I remember I was in Malaysia conducting some maintenance training, and he said that if I could come. I said I was out of the country. He said that the revised gears have been tested, and the crew and other shipping boats were amazed by the new speed of the anchor winch. He told me that they caught so many fish during their first run using the modified gears. The crew went home with their families that night, and I think they ate as many fish they can. I was happy to hear about that. I think in this case, the reliability of the anchor winch to deliver and catch fish was increased, and I rest my case on that. The winch was originally designed for the anchor of the ship and modified to catch fish, and it did catch some fish and I think that's all I have to say about that.

Option	D1 (mm)	D2 (mm)	Gear Ratio	Torque (foot - lbs)	RPM	Time hours (1 mile)
Original	35	190	1:5.43	3190.76	26.46	5.34 hours
Option 1	112.5	112.5	1:1	1760.50	89.36	0.98 hours
Option 2	75	150	1:2	3524.83	44.70	1.97 hours
Option 3	90	135	1:5	3644.95	59.57	1.47 hours
Option 4	100	125	1:1.25	2203.54	71.70	1.23 hours
Option 5	100	135	1:1.3	2292.45	68.73	1.27 hours

Figure 4.13: Summary of Options for Gear Modification for D1 and D2

Equipment reliability can only be increased by going beyond maintenance and modifying the capacity of the equipment. Once the capacity of the equipment is modified, the reliability increase, but again, the maximum value for reliability we can get is 100%, and we cannot go beyond this unless a second modification takes place. Running the equipment at 100% reliability during its entire lifespan is simply an impossible task as there are equipment losses and downtime, both planned and unplanned, that cannot be eliminated. If a spare part wears out and was replaced during a scheduled Preventive Maintenance, technically it failed but this failure will not be considered in the MTBF calculation since we have performed a task before the failure.

What if there is a schedule change, as the equipment needs to process a different product as in the case of several manufacturing plants. This could not be done in a split of a second, since there will be some changeover or conversion that will be done on the equipment that will include changing the parameters, changing some jigs and dies to produce the next product. This process, called set-up, changeover, or conversion, will be done when the equipment is stopped. The same goes true when we conduct some Preventive Maintenance on the machine. This means that the inherent design reliability cannot be 100% or even surpassed. We cannot reach a perfect output every time we operate on our equipment due to the losses, and not all losses can be eliminated. This means that if we want to zero out the downtime due to set up then we need to release some of our clients so the products being manufactured will be the same but this will definitely reduce the revenue of the industry.

Figure 4.14: Why Reliability is Difficult to Achieve in Industries

Here is the difficult part, reliability is not only limited to our physical equipment and assets. If we speak about asset management, it includes everything, including our company's name, logo, brand, and trademark. In this case, reliability is no longer controlled nor managed by maintenance. It now becomes a c-level responsibility, and the plant needs to remain in business. As said previously, reliability is the responsibility of everyone in the organization because reliability has something to do with everything we do. This will include having a reliable way for human resources on the process of hiring the right people, locating reliable suppliers by purchasing, having a reliable storeroom that can be dependable when maintenance needs parts, on having reliable operators in the plant that can operate, from the computer or laptop we use. I can go on and on with this one. Everything needs to be reliable, that is why reliability is not just a maintenance nor a reliability responsibility. Still, it is everybody's responsibility in the plant. The difficult part is that company has silos, and I have seen and worked in various industries in which their overall vision and mission comes in conflict with other functions of the organization which must not be the case.

The overall corporate success depends on the reliability of the company's products, processes, services, equipment, and assets. This will always be a top customer concern. A company's reputation is closely related to the reliability of its products and services. A reliable product may not dramatically satisfy a customer but an unreliable product will definitely affect a customer. Always remember that it will just take 1 person to let others know that your product is unreliable and this dissatisfaction will be spread to more people to let others know that they are not satisfied with your product or service. When a customer is dissatisfied with your performance then he can simply walk away and find somewhere else that can deliver. If we go to the dictionary, the word reliability means to be trusted or being trustworthy and dependable.

Maintenance is considered as the tasks which are the activities performed to preserve, conserve and sustain our equipment and assets. Safety, Quality, and Reliability is the effect of doing the correct maintenance tasks on our equipment and assets. This can only happen if these 3 groups do not work in isolation but in constant communication, cooperation, and collaboration. Safety, Quality, and Reliability must be given equal weight and importance. That is why among the 3 there should be no first as each of these priorities are connected and equally important. If these 3 can just work as one, they will be considered a powerful force to reckon with, and have a greater chance of dealing with industries problems.

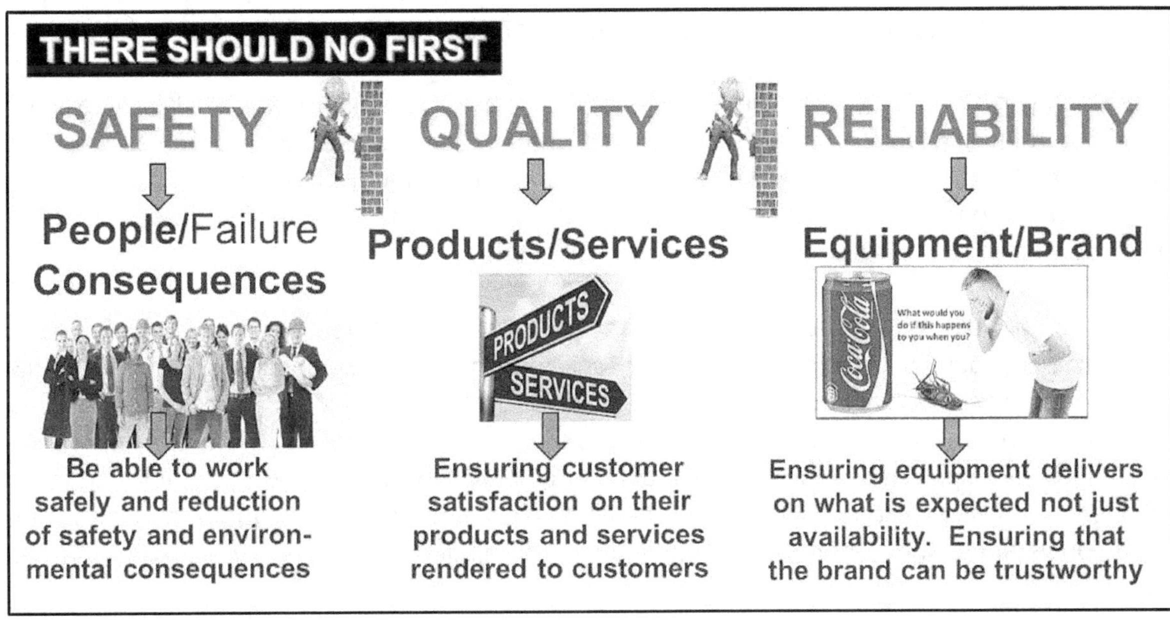

Figure 4.15: Why There Should be No First

Let me share with you the forward message of my dear friend, R. Keith Mobley on my 4th book Cutting – Edge Maintenance Management Strategies that truly enlightens me towards a deeper explanation about reliability.

Rolly is totally unique, and perhaps the most passionate person that I have met, especially when it concerns Reliability and Maintenance. We have been sharing philosophies for the past year, and while I sometimes have trouble following his logic, it is clear we share far more than we disagree on. I admire his passion, his energy, and his absolute commitment to helping others in their journey towards reliability. At times, our philosophical discussions had become heated. Still, in the end, we have always come to a consensus that both could accept. This book can be difficult to follow at times, but there is knowledge to be gained when you take the time to really read and understand what he is trying to help you understand.

As you read his book, remember that reliability is more than Mean-Time-Between-Failure. Reliability is universal, dimensionless, and immeasurable; it either is or is not. It, or the absence of it, surrounds us every minute of every day and is ingrained into every facet of our personal, family, social, and professional or business life. Reliability manifests itself as the consistency of purpose, dependability by always meeting commitments as well as the expectations of others; being steadfast and focused on goals, objectives and creating a positive future; and being unfailing in meeting responsibilities. At first glance, these appear to be human characteristics, and they are. Still, they also apply to physical assets as well as the business and work processes that all enterprises depend upon each day.

Those that continue to believe that reliability only pertains to physical assets and is measured by Mean-Time-Between-Failures (MTBF) cannot see that the reliability of the processes, procedures and practices that govern asset design, procurement, operation, and maintenance are the variable that determinate that result in the measurable MTBF, which is really the symptom of unreliability. Without reliable business, work processes, and employees who execute them, physical asset reliability is impossible. Instead, the volatility and instability of the unreliability of the infrastructure, not the assets, determine their reliability.

Understanding the true magnitude of reliability defines a reliability leader. It is also what differentiates reliability leaders from their peers. It enables them to lead their organizations into a brighter future, one that would not have happened without their leadership. Reliability is an essential, foundational part of a viable enterprise's life cycle asset management process. Still, it must be more than just reliable assets as defined by MTBF.

Capital or tangible assets are the heart of any enterprise that relies on its ability to produce or manufacture products or as in the case of facilities, to maintain conditions that occupants require. They are the engine that drives the revenue stream but can quickly become a demanding drain on the enterprise's cash flow and net operating profit when not effectively managed. Reliable processes, procedures, and practices are the framework, the infrastructure that binds the entire enterprise operations, including asset management together. It provides stability, consistency, and dependability that is crucial to all aspects of the organization. *From R. Keith Mobley*

Achieving a World Class Reliability in the industry is far more difficult than having World Class Maintenance since the challenge, in this case, is that reliability is everyone's responsibility in the plant and that the drivers should be the C-level people. This cannot be a sole maintenance function since the fact lies that there are a lot of factors to consider where maintenance has no direct control of the situation. But the C-level people can **<u>ONLY</u>** drive reliability if they truly know what it means. C-level and Decision-makers in the plant must take a fraction of their busy time to understand what reliability is and how it affects their business

4.9: An Introduction to Weibull Analysis

Waloddi Weibull was born on June 18, 1887, in Sweden and died on October 12, 1979, at Annecy, France. Among his lifetime works, he has written many books on strength of materials, fatigue, the rupture in solids, and the Weibull Distribution which he developed in 1937. The

Figure 4.16: Waloddi Weibull

Weibull distribution is about Life Data Analysis which is used to determine the probability and capability of parts, components, or systems to perform their desired functions for some time without failure. Weibull distribution is one of the most widely used distributions for failure data analysis also known as the Life Data Analysis of a component

The Weibull distribution is often used to describe the lifetimes of parts. These can be light bulbs, capacitors, disk drives, ball bearings, etc. When several parts are put on test, these parts will not fail at the same time. This tool provides a means to quantify the effect that various design options will have. Weibull distribution can be used in analysis to predict failure rates and to provide a description of the failure of parts and equipment.

Weibull Distribution Analysis Will Include

• Plotting the data and interpreting the plot
• Failure forecasting and prediction
• Analyzing failure mode and causes
• Evaluating corrective action plans
• Implementing Engineering change and modifications
• MRO Spare parts forecasting
• Calibration of complex design systems
• Maintenance planning and optimal replacement strategies

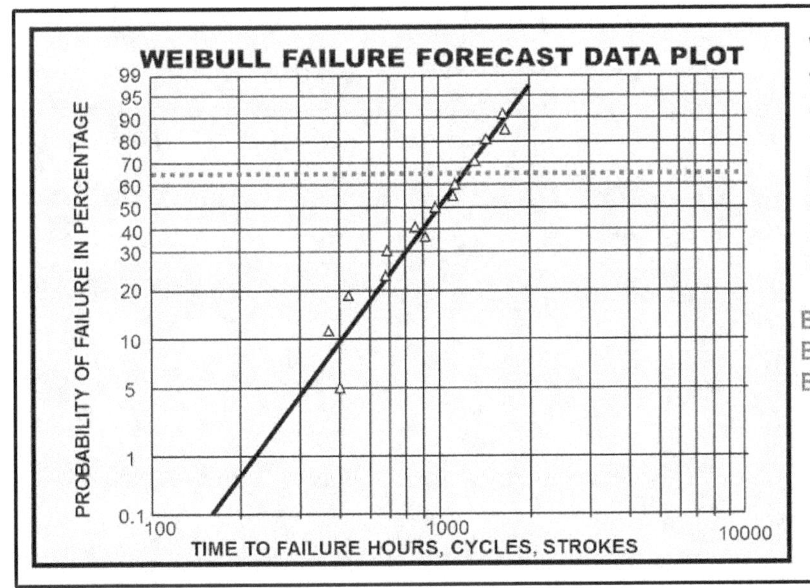

Figure 4.17: Weibull for Forecasting Failures

Life Data Analysis

This is the time, a given product, part, or component has operated successfully or the time it operated before it failed. It can be measured in hours, miles, cycles, strokes, minutes, or any means. One of the important aspects in the study of Life Data Analysis is to determine what constitutes a failure and to be clear when the product has actually have failed. In the study of Life Data Analysis and Failure Forecast, the output of the analysis will always be an estimate. Its purpose is to accurately estimate the time the said component will probably fail or rupture. Weibull Analysis studies the relationship between the life span of a component and its reliability by graphing life data for an individual failure mode on a Weibull Curve.   Weibull Analysis is most often used to describe the time to failure of parts such as light bulbs, ball bearings, capacitors, disk drives, printers, or even the lifespan of people.   Failures include cracks, fractures, fatigue, deformation, corrosion, infant mortality, or wear out. Another significance of the Weibull Analysis is that it provides a simple and useful graphical plot. The Horizontal Scale or X-Axis is the measurement of life or age in miles, hours, cycles, strokes, etc. while the vertical line is the cumulative percentage that will fail.

Weibull plot is useful for maintenance planning. If BETA is less than or equal to 1, then it tells us that schedule overhauls and inspection are not cost-effective since the failure is considered infant mortality. A Weibull curve indicates that a 1% probability of failure will occur after 160 cycles making the failure random and a 5% probability for failure to occur after 400 cycles if the horizontal axis is in cycles indicates that this is a case of age-related failure. The advantage of Weibull Analysis is that it provides moderately accurate failure analysis and failure forecast with extremely small data samples, making solutions possible at the earliest indication of a problem.   It is extremely useful in the study of mechanical, chemical, electrical, and electronics failures.   Perhaps the primary advantage of Weibull Distribution Analysis is to provide reasonably accurate analysis and failure forecast with relatively few samples which are critical when the failure has safety or extreme costs.

Chapter 5

MTTR, MTTS, and MTBA Explained

> *There is no question about having highly skilled people in our maintenance team. The goal of MTTR is to reduce the repair time, but the real goal of maintenance is to analyze failures by addressing the root cause of the problem. Remember, maintenance is not measured by how fast they repair but by how they were able to analyze the failure.*

5.1 Mean Time to Repair (MTTR) Explained

When a failure occurs, it is critical to restore the equipment as quick as possible. Typically, much of the repair time is spent on identifying what exactly failed. Most of the time, the traditional trend will be to repair the equipment and never get to the root cause of the problem. Although repair time should be performed at the shortest possible time, our goal will be to put back the equipment in its operating state of condition. For failures that repeatedly keep repeating themselves, the best strategy to adopt will be to address the real root cause of the problem and prevent it from recurring again on its own.

MTTR (Mean Time to Repair) is the average time required to repair an equipment. It is defined as the average time required in repairing the equipment divided by the total breakdown occurrence. Other terms used to designate MTTR are Mean Time to Restore or Mean Time to Recover. The trend for MTTR will be the lower, or the shorter the time to repair, the better. Improving MTTR means shortening the time required to repair the machine. This (Mean Time to Repair) is also the average time required to perform corrective maintenance or repair for all of the removable items in a product, component, or system. It analyzes how long repairs and maintenance tasks will take in the event of a system failure. MTTR may be defined as the time it will take to bring a failed system back to its available or operating status again. When the motherboard in our computer fails, and it takes you 3 hours to purchase a new motherboard and one hour for the technician to repair and put it back in operation, the total time to repair the computer is 4 hours, however; the MTTR for the motherboard that failed will be forever.

$$MTTR = \frac{REPAIR\ TIME}{FREQUENCY\ OF\ FAILURES}$$

Many assets, systems, and components can be repaired when it fails in operation. MTTR varies depending upon the difficulty of repairing a failure. It is also the measure of a system's ability to be restored to a specified condition when maintenance is performed by personnel having skill levels using the prescribed procedures and resources. What is important in calculating a true and meaningful MTTR is to consider only when the machine is down due to the event of a failure or breakdown and not including other machine downtimes such as set-up and conversion and most commonly assists and errors encountered in the equipment. A true and correct MTTR starts at the time of failure and continues until the system is back once again for operations, regardless of whether a system part or component will be available or not. MTTR is also difficult to estimate since one must consider a variety of repairs.

MTTR is also defined as the average time necessary to troubleshoot, remove, repair, and replace a failed system component. Activities during the repair are usually composed of;

• Fault Isolation: Time associated with those tasks required to isolate the fault to the item.
• Spares Acquisition: When a part or item is needed, the maintenance checks the system if the part or item is available in the MRO Storeroom.
• Disassembly: Time needed to replace the part or item that failed or items identified during the corrective maintenance process.
• Interchange: Time correlated with the removal and replacement of a faulty replaceable item or suspected faulty item.
• Reassembly: Time correlated with closing up the equipment after interchange is performed.
• Alignment: Time correlated with aligning the system or replaceable item after a fault has been corrected.
• Checkout: Time correlated with the verification that a fault has been corrected and the system is operational.

Other organizations may be tempted to use MTTR as one of their key points in their Performance Appraisal for maintenance. I would not recommend that. Why? Because it is difficult to compare the repair time between two people since repair varies depending on the magnitude of the work involved. Replacing a complete engine cannot be compared to someone who is just tightening a bolt. That would be quite unfair. I know of two technicians; one technician repairs the equipment so fast, but the problem is that the breakdown keeps on repeating after a few hours, but boy, he is fast. While the other technician takes his time to repair the equipment, he ensures that it will not fail again due to the same problem. So how can we compare the two in terms of MTTR? I do not even know why this Meantime indicator was invented in the first place. It contradicts what maintenance is all about. There is no question about having highly skilled people in our maintenance team. Technically speaking, the goal of MTTR is to reduce the repair time, but the real goal of maintenance will be to understand the cause of the problem and prevent its recurrence. The goal of maintenance is to preserve the equipment and not to become heroes and fast at repairing. As the Chinese once said, if our maintenance people become so good at repairing failures, then something is definitely wrong with our maintenance organization since they forgot to analyze the root cause of the problem, but if we expect a different result from the same things that we are doing, it's just not possible. The Chinese called this "Insanity."

MTTR and MTBF are limited to considering predictable failures of parts, systems, and components for operational-related causes. Equipment failures caused by war, vehicle collision, fires, terrorism, force majeure, Acts of God, and sabotage are generally ignored in the calculation of MTTR. MTTR can be used to track down the level of skills for maintenance and technicians in performing repairs so that we can monitor which people need improvement. Although we need maintenance that is both effective and efficient in performing repairs, again, let me reiterate that this is not the goal of the maintenance department. When failure keeps repeating itself, maintenance must slow down and take the time to analyze and address the root cause of the problem. If we really want to understand why the failure keeps on recurring, take things slowly, and start analyzing the failure. Figure 5.1 indicates the actual MTBF graph initiated by our Planned Maintenance team. As we achieve near to zero breakdowns on our 22 PM Pilot Machines, we also improve our MTBF or Mean to Between failures, hence, we have a good trend on our MTBF Indices. This was the actual data we submitted to JIPM in 2001. Just simply reminiscing the good old times in our Planned Maintenance journey.

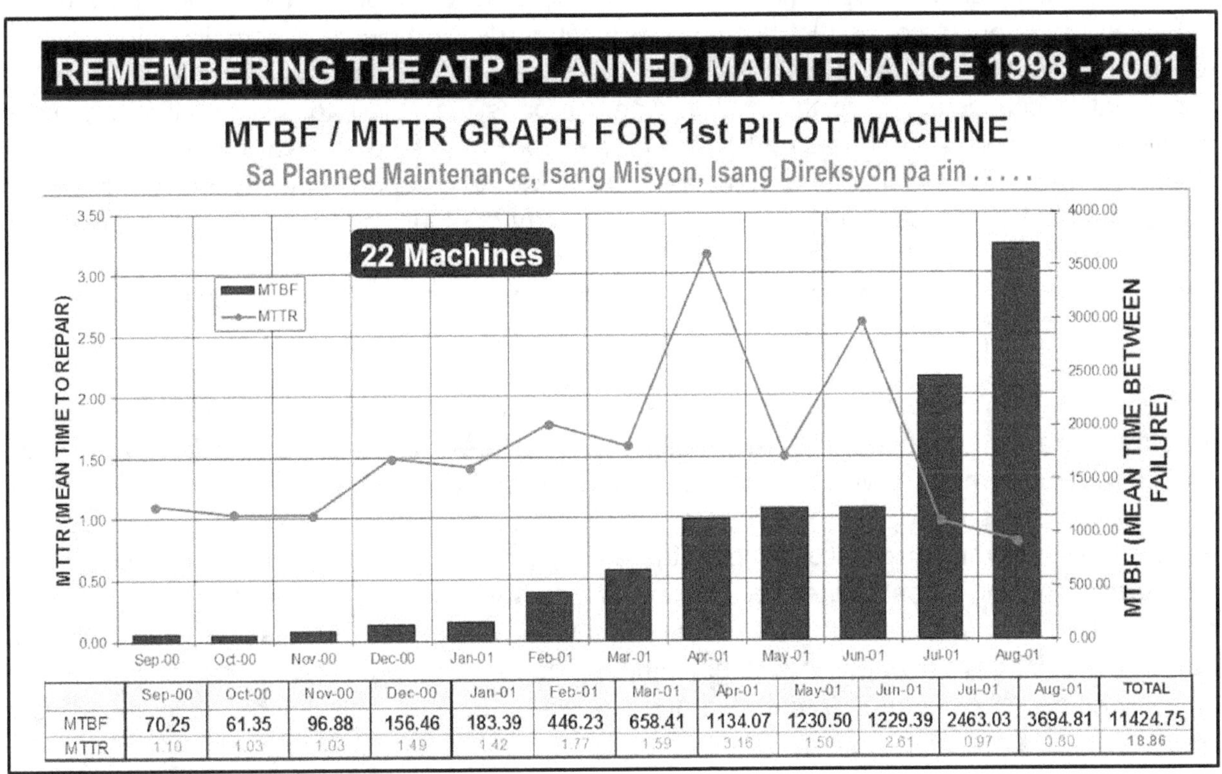

	Sep-00	Oct-00	Nov-00	Dec-00	Jan-01	Feb-01	Mar-01	Apr-01	May-01	Jun-01	Jul-01	Aug-01	TOTAL
MTBF	70.25	61.35	96.88	156.46	183.39	446.23	658.41	1134.07	1230.50	1229.39	2463.03	3694.81	11424.75
MTTR	1.10	1.03	1.03	1.49	1.42	1.77	1.59	3.16	1.50	2.61	0.97	0.80	18.86

Figure 5.1: Actual MTBF and MTTR Graph

5.1.1: Where MTTR Should Be Used

In most cases, when an experienced person retires or leaves the plant for good, his experience usually goes with him to the grave. What industries will do is to hire a new breed of people with little or no experience or those who just recently graduated from college. Once these people are hired they will find their way into the plant, commit errors on repairs by experimenting on what works and what do not. This is how they finally earned their experience. Once they reach the age of retirement, the same cycle happens and everything will just end up in a Merry-Go-Round, which sometimes makes me dizzy. Most industries have not captured nor documented these people's experiences of performing this type of repair for a particular breakdown. Others keep their repair skills a trade secret for hero stuff purposes. What was his first step, then what, after that, what's next, and so on? When these kinds of things are in place in the plant, then these can be used to teach other maintenance people who are new to the plant so that maintenance will be guided on how to repair this type of breakdown and not to experiment all the time. I believe that this is where the true value of MTTR should be reflected. Most of us in maintenance admit that we still cannot get rid of the hero stuff where no one on that plant can repair the failure and finally when the boss called us late at night in our home telling us to go back to the plant to fix something that nobody can fix because you alone with all your sweat and tears have experienced this in dealing with this magnitude of failure, then you get a hero's welcome. This is entirely the essence of a reactive environment.

For a start, if your plant encounters a major breakdown, call on the most experienced person to repair it, then try to video the whole process of how he started and ended the repair. Once the video is completed, set up a meeting with the person who performed the repair, showed him the video frame by frame, and let him discuss what he did on how he accomplished the whole repair process from the start to the end. Write down what he said step by step. Once the document is completed, have the person who performs the repair to review it for any more addition or deletion. Once the step-by-step repair procedure is completed, encode this in your CMMS (Computerized Maintenance Management Software). Suppose a similar failure happens in the future; in that case, every maintenance can print this out and have a basic guideline on repairing this type of failure if it occurs again. This way maintenance doesn't need to experiment on how to repair the failure. And the good thing is that we have captured the knowledge of this person.

Perhaps if MTTR is being monitored in the plant and in a certain group of 10 people, Charlie is said to have the lowest MTTR in performing repair and troubleshooting the hydraulic systems, while Joe has the highest MTTR in the same category. What we can do is we can define the proper procedures on repairs based on Charlie's practices, which can easily be applied and followed by other maintenance people, thereby avoiding trial and error during repair and troubleshooting, or we can either partner Joe and Charlie so that Charlie can teach Joe from time to time while Joe can assists Charlie in performing the repair. After some time, if we can see Joe repairing the hydraulics himself without Charlie, it simply means that the knowledge Charlie got had already been transferred and passed on to Joe. This is what we actually want on our maintenance people.

On the other side, if the failure keeps on recurring on our equipment, we need to have a balance on when do we continue repairing the failure and finally conducting a thorough root cause investigation. Although I do not recommend performing a Root Cause Failure Analysis for every single failure or breakdown that can happen in the equipment, since that will be a very tedious situation for maintenance to do, rather what I am saying is that for failures that keep on repeating themselves, it is best to slow down and finally address the root cause investigation to prevent the cause of the problem from recurring on its own. The saying goes that when our people become really good at fixing failures, then something is definitely wrong with our maintenance organization. Why? What is wrong? It seems that they cannot seem to let go of the problem. Try to think why they became efficient and good at repairing. It is because the failure keeps on repeating again, and the problem itself keeps on resurfacing. When failure keeps repeating itself, maintenance must take the time to slow down, analyze, and investigate the problem's root cause. If you really want to understand why the failure keeps recurring, take things slowly, and start analyzing them. I believe that to be fast, we need to take things slowly. Having a low MTTR due to years of experience is not always good because this means that the failure simply keeps repeating.

People become good at repairing because the failure keeps on coming back. Having a low MTTR for most maintenance people due to having some repair or troubleshooting guides and procedures based on the experiences of these old-timers and maintenance people from the University of Hard Knocks will be of great value. When these good people finally retire, they know that they have left a legacy in their plant. MTTR or Mean Time to Repair can never be a measure of Reliability; when people become too good at repairing, it only means one thing: the failure keeps on repeating. I would rather take my time and learn from the failure rather than repeatedly repair the same failure. Before fixing the failure, my recommendation is simple: have the restoration team take photographs of the affected part or component of the equipment and secure the part that failed. Place them on a plastic bag and any other foreign parts that they can locate near the part that failed and endorse them to the slow team, which is the team that will investigate and analyze the problem. This way, the evidence is finally preserved.

I remember a student of mine who attended a course on the root cause and told me that after this training, once he reports back to work, he will create a Repair Menu. I asked him what it's all about and he said that this time when operators call him to repair a failure, he will just hand the repair menu, just like in a restaurant when you want to order the food you would like to eat. Once, the repair menu has been handed to the operator, the operator can choose what repair he wanted, a fast, or a slow repair. When I asked what's the difference, he told me, well you see Rolly, if the operator wants a fast repair, then that is all we would do, but if the operator wants a slow repair, then we will also take the time to perform a failure analysis on the failed item so that the failure will not repeat itself, but expect the repair process to take much longer. My thoughts were, this guy actually makes some sense.

5.1.2: Application of Precision Maintenance on Repairs

Precision Maintenance involves performing maintenance work in a consistent, precise, and industry-accepted way. If properly implemented, this means that maintenance should yield the exact same results, no matter who is performing the tasks whether the person is the most or the least experienced of the craft. Precision Maintenance is much more than merely having procedures on PM. Precision Maintenance must be the correct culture of the organization. The good thing about doing Precision Maintenance is that it can even eliminate the chances of Infant Mortality Failures seen during the start-up of assets right after a Preventive Maintenance shutdown or Preventive Maintenance Scheduled Overhauls are performed. It involves performing maintenance work in an accurate and precise manner, which means that the maintenance should provide the exact same results no matter who is performing the work, whether the work is done by the most or the least experienced maintenance craft in the plant. Originally, Precision Maintenance is designed for Preventive Maintenance tasks, but the way I see it, this can also be applied to corrective maintenance, repairs, and other tasks.

When different maintenance people perform the same PM or corrective maintenance on the same equipment, variations occur. These variations cause problems since they do not meet the requirements for the process. Precision maintenance help to ensure that equipment will be maintained to the highest possible standard so that the variations that can cause defects and failures can be reduced, or eliminated enabling equipment and assets to run at their optimized and peak level of reliability. Precision maintenance rebuilds machines and equipment to the highest standards so that fewer problems and errors can occur during operation. It is a matter of ensuring the important things for equipment and machinery health are done correctly and accurately.

The key to Precision Maintenance lies in the knowledge and skill of the maintenance craftspeople. Precision Maintenance will take time to achieve, but the benefits that can reap off are worth implementing. Doing Precision Maintenance will not only reduce downtime but will reduce the cost of performing maintenance. Training maintenance to perform, the correct tasks on the equipment will yield outstanding benefits for the plant. Organizations that can build a culture of Precision Maintenance will reap the rewards of a better bottom line, a safer workplace, and more productivity throughout the facility.

Achieving Precision Maintenance implementation, success, and sustainability requires a full commitment and advocacy from C-Level Management, senior leadership, as well as their supervisors and management roles. There should be a clear focus on setting written expectations for maintenance, engineers, planners, MRO Stores, and all contributing roles, including operations in achieving Precision Maintenance and reducing human errors, especially during Preventive Maintenance.

5.1.3: Survey on What is the Total Repair Time

Typically, when a machine fails, the operator will find someone who can repair it. Once the maintenance arrives at the failed equipment, they will diagnose the fault. If a part or spare is affected, he will leave the equipment temporarily to check the system and acquire the part in their stockroom. Then the actual repair will take place. After the part had been replaced, the maintenance will revalidate and make some test runs, then finally endorse the equipment back to the operator to continue production. When the maintenance or technician arrived at the failed equipment, usually, this is what will take place.

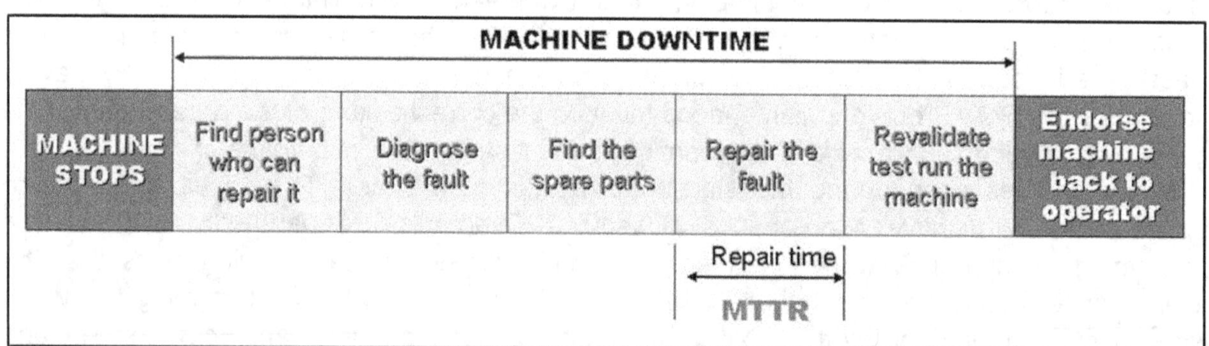

Figure 5.2: What is the Mean Time to Repair?

126

- Operations to prepare a Work Order and give it to Maintenance
- Maintenance will try to diagnose the equipment if a spare part is required
- If a spare is required, the maintenance will determine the codification or part number
- Check the system inventory of the part in the storeroom
- Travel time going to the storeroom
- Time for the maintenance to requisition the part in the storeroom
- Response time of storeroom to acquire the part requested
- Travel time going back to operations
- Time to repair the equipment begins

Question: If the part is unavailable in the stockroom and it took the Purchasing people one week to acquire the part where the machine was not used and idle for that period, and once the part arrived, it took maintenance just one hour to repair and replace the part; in this case, what is the total repair time? Is it 1 week and 1 hour or only one hour? I have made a mini-survey to a small list, and as expected, their response differs from one industry to another. Here are some of the answers they provided and how they responded. The question, in this case, what will be the total repair time? Will it be 1 hour or 1 week plus 1 hour?

a) 1 week + 1 hour (Total repair time is equal to the total downtime)
b) Only 1 hour (Total repair time is equal to the actual time the machine is repaired regardless of whether the part is available or not)

Response 1: It is on a case-to-case basis. When equipment fails, downtime for us is when the machine does not give a good product or there is no output. When we install an alternative part versus OEM, it is a risk we have to decide, equating the cost of additional repair to the profit we can make to produce a good product. *From Ryan Ben R. Sabilala, Philippine Overseas Working in the Middle East*

Response 2: Answer (B) only 1 hour. For record purposes, we usually put the actual time to repair, which is 1 hour. The waiting time for spares to arrive can be devoted to other activities. If materials and spares are not available, this will automatically be considered pending work orders and will be active upon arrival of the required materials. Hope this would help. *From Rusty Peralta working in a Power Plant, Philippines*

Response 3: Good afternoon Rolly, we would consider the following conditions. If the machine was restored to the following conditions where the speed before breakdown is the same and the quality before breakdown is the same, we consider only 1 hour. If we could not operate the equipment, then repair time will be equal to the total downtime. I hope this will help. *From Albert Floresca, working in an Automotive Plant, Philippines*

Response 4: Greetings! My answer to your question is a letter (B), only 1 hour. The reason for that is the maintenance will not consider the downtime due to the procurement of the parts considering that procurement is not under the maintenance function. We only consider the time upon the arrival of the said parts. *From, Kenn R. Rio, Head, Maintenance Group, Philippines*

Response 5: Rolly, good morning. My opinion on your question is the following, the repair time is only 1 hour. Since the rest of the machine downtime is just waiting time for the part to arrive. However, the machine downtime is 1 week + 1 hour. Thank you for considering me as one of your respondents. *From Fortunato G. Dequit, working in a power plant in the Philippines*

Response 6: Hi Rolly, my answer to your question is a letter (B). The total repair time is only one hour, which is equivalent to the actual repair time executed. *From Joemar Apolinario, working as a semi-conductor in the Philippines*

Response 7: Rolly, my answer is (A). Once you receive the permit to work, that is the time that the clock of downtime started. Once you receive the PTW (Permit to Work), they cannot use the equipment that has already been isolated. *From Joseleo Vicaldo, working in a power plant in, the Philippines*

Response 8: Hi Rolly, one-hour downtime plus one-week waiting time are two different scenarios. When we are talking about maintenance, particularly skills, the computation should refer to the letter B. The 1-week waiting time will become an assignable cause when trending the downtime. On the other hand, when we talk about OEE, I believe (A) is more appropriate. *From Anthony Bathan working in the semiconductor industry, Philippines*

Response 9: Rolly, I would choose (A). For me, repair time means exactly what it is because I consider the equipment under repair starting when it broke down until it is restored, regardless of delays in shipment, parts, etc. *From Mac Chan Lee, working in a power plant in, the Philippines*

Response 10: Hi Rolly, Good day; we measure repair time as the actual time in doing the repair. If the reason is the non-availability of parts, then the machine done is for different reasons. So the answer is (B). *From Charles D. Mendoza working in a semiconductor plant, Philippines*

Response 11: Repair time is 1 hour while downtime will be 1 week plus 1 hour. It's important to take both metrics into consideration to get a complete picture of how things can be improved. *From Sai Sharma*

Response 12: For me, it would be letter A. If the software gets the total downtime then the time to repair will be 1 hour plus 1 week and the MTTR of that month, for example, will be high. The time that you spent searching spare parts, tools, and so on is part of the time to repair too. *From Andre Eduardo Silva*

Response 13: Roly in principle it's letter A. The simple response is 1 hour repair time and 1 week and 1-hour downtime however there should be some consideration involved in the time spent purchasing the part as this is also a cost to the business. *From James Thomas*

Response 14: My answer is b. Because it is the only amount of time spent by the maintenance when repairing the equipment. Assuming that if it is a "stock item", maintenance should not shoulder the in-effectiveness of the purchasing group (in my plant, the purchasing is separate from the maintenance). But if the maintenance failed to forecast their parts or failed to establish their safety stock then my answer is letter A. The 1 week is part of the penalty of the maintenance by failing to do so. *From Wilfredo Bolatin, Reliability and Equipment Maintenance*

Response 15: Letter A, this is because when we are asking for the total repair time, we will not only consider the time which the technician or contractor will repair the machine itself. The consideration of how many days will be the downtime after the equipment failed is necessary to include. Even if the parts order is not part of our Operation or Facility Department, or it will fall from the Procurement side, still Operations or Facilities need to have a contingency plan that if ever our equipment will encounter problem or breakdown, what will be the necessary item needed to consider for it to lessen the repair time. That is why from the time the equipment encountered breakdown and purchasing of parts until the repair has been finished, for me that will be the calculation of the total downtime and repair time. *From Armie Magadan, Facilities*

Response 16: Hi Rolly, my answer on this is 1 hour because we should only be accounting for the true activities done in repairing the equipment. These repair activities will tell how easy or difficult the problem is and will tell how skilled or not our craftsmen are. This repair time will also align with the actual time spent during the activity especially if the nature of the problem can be restored logically in a short period. The unavailability of the spare parts which entails one week to procure is not directly caused by the problem of the equipment and should not be considered in computing MTTR. This will just create a misconception when it comes to correlating the MTTR to the true condition of the equipment as far as the problem is concerned. Furthermore, companies have inventory management policies to lower their inventory cost; and some spare parts do not have actual inventory stocks and on Order-Upon-Request (OUR) scheme especially for those slow-moving and high-cost parts. Hence, delays for these OUR parts are inevitable considering procurement lead-time. On the other hand, it is also best to highlight the downtime caused by the failure of not having the needed spare parts available on time. It can be included in the total repair time especially that equipment should be running as much as it should be. Having this long downtime will affect the productivity of the operations, which is worth highlighting. However, we need to ensure that data will help us and not confuse us. We need to have a concrete delineation of all downtimes that can be computed separately. After all, we can have as many KPIs as we can so that everything can be captured and will harmoniously help us to better manage our operations. Hope this will help and thanks for including me as one of your respondents. *From Miguel Gelig, Jr., Reliability and Equipment Maintenance Superintendent, SAP-PM Local Process, Dole Philippines International, Polomolok, South Cotabato, Philippines.*

Here are the responses of consultants whom I personally asked for their opinion that have participated in this survey question, and their choice is almost unanimous except for number 20.

Response 17: Rolly, answer (A), Repair time = Active repair time plus preparation time. The reason the machine is purchased is to produce and not sit idle. An airplane only makes money when it is in the air. A machine can only be in one of two states, working or not working. If we don't need it to produce in the second case, it is still available to work to treat it as work. All the rest of the time, it cannot work, so it is down for maintenance, so no $$$. An efficient maintenance system will minimize the preparation time and scheduled work, so we minimize the loss of $$$. Finally, maintenance is a Business Process, not a Department. *From Vee Narayan, Lead Author, 100 Years of Maintenance: Practical Lessons from Three Lifetimes, Industrial Press, U.K.*

Response 18: Hello Rolly, my answer is (A), a failure event must be seen regarding the total loss value and costs incurred to the business. Total repair time is a measure that means exactly what it says: The time from the second the machine stopped being available for production to the time the machine came back up to its required production rate. That includes commissioning time and hand-back time, removing tags, and completing permits. What is most important to the business is the full impact of a machine outage on production uptime. From the business, the repair took 1 week and 1 hour. That is the correct business measure to truly understand the impact of the business's outage. *From Mike Sondalini, http://www.lifetime-reliability.com, Australia*

Response 19: Rolly, my selection is (A) if the maintenance function objective is equipment ready to run, and the maintenance function has authority over resources required, diagnostics, preventive, spare parts, etc. If the maintenance function is only repaired failure, then b could be my answer. I like the maintenance goal defined as "equipment ready to run." However, the resource consumption required needs to be logical to the cost of downtime. There is no need to own an expensive spare part and pay interest $ when the part can be obtained in a short relative lost production time $. So, the maintenance function needs to be closely aligned with the value-added activity of production. *From Greg Peitz, Affiliate, Failsafe Network, USA*

Response 20: Interesting question Rolly, I am not an expert on maintenance metrics, but I always believe that you measure what you want to hold to a certain performance standard or to improve, and the measures need to reflect the actual problem. In this case, the downtime needs to be recorded, but what bucket do you put it in. Most places have reduced spare inventory, but if it causes a week delay, it must be visible that spares or inventory was the issue, not maintenance working on the repair. So my answer is total repair time is 1 hour, total downtime is 1 week + 1 hour, and the cause is inventory. *From Virginia Edley of SBK Consulting, LLC, USA*

Response 21: Rolly, anyway, I really don't have anything to add to either Greg or Virginia's response. I do like Virginia's inference that we measure what we're interested in improving. So, suppose we're interested in improving total availability. In that case, I'd opt for A. And as Greg said, if I'm only interested in maintenance response time, then I'd opt for b. Most people and organizations might be a bit too myopic and don't measure from a broad enough perspective. So, from the two options, I personally would opt for "A." Thanks for asking. *From C. Robert (Bob) Nelms, President, Failsafe Network*

My Response: I am quite surprised by the responses since there are people that actually worked in the same industry, yet they have different opinions and points of view. When a part is unavailable when it is needed, a stock-out occurs, and both downtime and repair time is prolonged. If a part is unavailable in the storeroom where it took purchasing one week for the part to arrive in the plant and it took one hour for the maintenance to repair the equipment, the total repair time will be equivalent to the time that the equipment was placed back in operations. This means that the total repair time is not one hour but 1 week + 1 hour. In my opinion, the total repair time will be 1 week and 1 hour, which includes the time the spare had to be on hand up to the time the equipment had actually been repaired. Hence, I choose (A) because spare parts management is one of the responsibilities of the maintenance function. If the part is not available when needed, then maintenance should be held accountable since we are the ones who control our spares. When the equipment is idle because a part or spare is not available, then the industry is not making any profit. My feeling on this is that those who voted for B (one hour), what I believe, is that this will be the case if maintenance people do not manage the storeroom. The consideration of the repair time should include the time the part or spare is acquired. This simple survey question is that if one industry opts to choose (a) or (b), MTTR computation will vary greatly. Considering the total repair time is only 1 hour, then we can have the following computation.

MTTR = Total Repair Time / Breakdown Occurrences.

If we select (B) as our answer as the total repair time, then the actual repair time in one week is just 1 hour. In one week, the MTTR value will be 1/1 or 1 hour, while the rest of the week, it is idle and not working since the spare is unavailable. The MTTR, in this case, is one hour in a week, but the rest of the days, the machine sits idle. Although the MTTR value is giving us a good figure since repair time is just 1 hour. In this case, we are just fooling ourselves with the numbers since the rest of the time, the machine is not functioning and is not making any revenue. On the other hand, if we consider both the waiting time for the spare, which is 1 week and 1 hour, to be the total repair time, then the MTTR value will be 168 hours in one week, giving us more truthful and realistic value. A true and correct MTTR starts at the time of failure and continues until the item is operational once again, regardless of whether a system part or component will be available or not.

5.1.4: Why MTTR and RCFA are the Opposite of Both Worlds?

Root Cause Failure Analysis simply tries to understand why something went wrong so people in industries can finally learn from the things that go wrong. It will identify the cause or origin of the problem based on the evidence unfolded by separating the facts from the other probable causes. It provides a methodology for investigating, categorizing, and eliminating the cause of safety, quality, reliability, and equipment-related problems and consequences. Every system, spare, or component failure happens for a specific reason. There is always a specific succession of events that will eventually lead to failure. RCFA will follow the cause-and-effect path from the final failure back to its origin. Therefore, Root Cause Analysis is a tool to better explain what happened, determine how the failure happened, and better understand why it happens.

Fast Team (MTTR) taking a ride in the Fast Train   Slow Team (RFCA) taking a ride in the Slow Train

When a failure occurs there are actually two teams working on this failure. First, we have the fast team. They are the restoration team and they will be responsible in bringing the equipment up and running in the shortest possible time. On the other hand we have the slow team. These people will be responsible for analyzing the problem and they will be taking a ride on the slow train.

Figure 5.3: The Fast and Slow Team

I have several friends who used to work for industries, and one of them is Charlie. If I recall him right, he had been working in the plant for a good number of years, 25 years to be exact, and is still working on that same old plant up to this point of writing. There is a lot of rotating equipment in their plant, and Charlie is one of the maintenance guys that maintain them. He used to tell me that when he was new at that plant, it took him several hours to fix the equipment experimenting on what to do because out there, you are usually on your own. When a bearing fails, it usually took him a couple of hours to replace the failed bearing. 25 years had passed, and today when the same bearing fails, it will just take him around 20 to 30 minutes at the most to replace it. Charlie's repair time improved. My question is, is that a good thing or a bad thing? The answer here is both good and bad. If we speak about MTTR or the Mean Time to Repair, then this is good. Charlie's MTTR improved a lot throughout the years he had worked in the plant by gaining experience. If we speak about the trend of MTTR, the lower, the better, then this is a good thing. On the other end, if we asked about Root Cause Failure Analysis, Charlie simply cannot find the cause as to why the bearing failed, since the failure just keeps on repeating itself because he never bothers to identify the root cause of the problem, and I believe that this is not a good thing. Why did Charlie become fast at fixing bearings? The answer is quite obvious because failure happens all the time, and that's about it.

Although we still have to consider the economic point of view in this situation, meaning if the consequences of failure will be limited to the cost of failure alone, or if the motor that failed has a standby motor or redundancy in place, making the failure less critical then MTTR can either be prolonged to allow us the time to gather evidence on the part or component that failed. However, suppose this is a stand-alone motor where the failure can affect some critical parts or the entire operations of the plant, then the repair time should be in the minimum and shortest possible time. This is where performing root cause will have its stall, or hang in the balance.

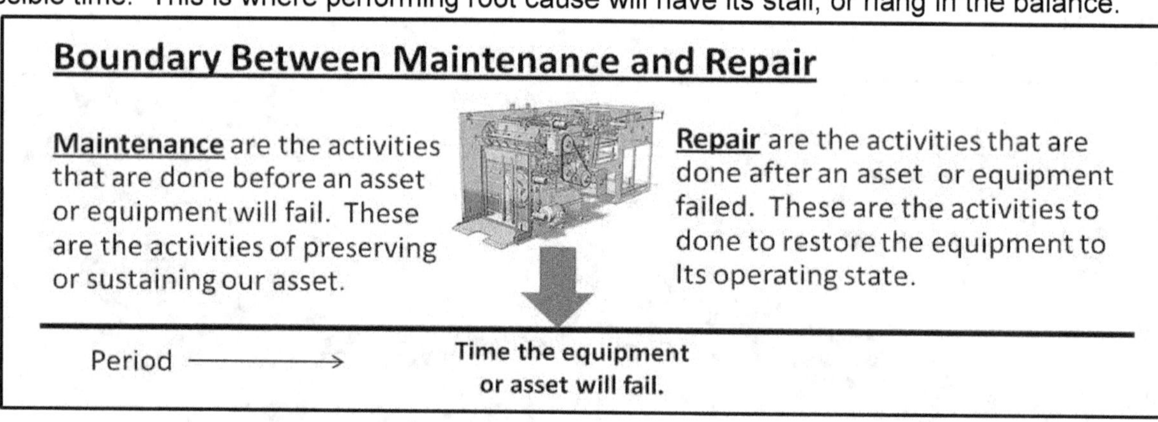

Figure 5.4: Why Maintenance and Repair are the Opposite of Both Worlds

When the equipment fails, there are actually two teams working on the failure. First, we have a fast team. The fast team will always be present in the plant in the shortest possible time. They will arrive at the scene of the crime in a matter of seconds. They will be the restoration team, and they will be responsible for bringing the equipment up and running in the shortest possible time. They need to do things fast, and they know that they are under pressure to repair, even if they did not cause the equipment to fail since most of the time, operations people would be watching them. These people know that the clock is ticking, and the longer they repair, they know that the equipment is not producing any revenue. The fast team will be taking a ride on a fast train. The restoration team knows that time is always of the essence.

On the other end, we have the slow team. These people will be responsible for analyzing the root cause of the problem, and they will be taking a ride on the slow train. They will take things rather on a slow basis by picking up debris, taking pictures initially on the failed equipment, spending time interviewing operators and people nearest to the asset during the time of the failure. However, when these people arrived at the failure scene, the fast team had already restored the equipment. In this case, the question is how on the planet can you perform a Root Cause Failure Analysis when the equipment is already restored? You just simply can't because all the shreds of evidence and traces of failure had been washed out in the first place. One of the most important things that I've learned in performing a thorough Root Cause Failure Analysis investigation is to freeze out the evidence, but in this case, the evidence had completely vanished into thin air. And most of the time, their managers complain that their maintenance people lack root cause analysis skills. Come on, give me a break.

Suppose we perform a Root Cause Failure Analysis investigation on the equipment or asset that failed, MTTR will be prolonged since we need to pick up whatever physical evidence can be found on the equipment before restoring the asset. I don't think operations people will like that very much. I know one Key Top Management person from a manufacturing plant who almost lost his job because one of their biggest customers almost pulled all their work in their plant when they cannot explain the cause of their equipment failure. The corrective action simply indicates that they repair the failure, but when the customer asked what caused the part to fail, they have no answer, and their customer was very unhappy and pissed about it.

Maintenance and repair are the opposite of both worlds. While others may think that maintenance and repair are the same. They are not, and we need to create this mentality with our people. Whether we agree or disagree, maintenance will involve cost, but looking at the brighter side, maintenance can be a profit center if we truly understand how to use them correctly. The question is do we learn from the failures? Industries will never learn from their problem by just repairing the equipment as it will just cause a recurrence of the failure. In my book on Investigating Root Cause Failure Analysis, I wrote that when a failure or breakdown is experienced on the equipment, the first thing to ask does this failure warrant a Root Cause Failure Analysis Investigation, if we answer no, then we proceed and repair the equipment, but if we unanimously agree and say yes, then the first thing we need to do is to preserve the physical evidence, since conducting a root cause investigation is 100% dependent on the evidence unfolded. Without the presence of evidence, we can just speculate, guess, or base our conclusion on experience which may or may not be the actual root cause of the problem.

5.2: Mean Time to Set-up

In manufacturing industries, there are pieces of equipment that are not dedicated and are used to process different products at different intervals. Although the same equipment will be used, dies, jigs and fixtures will be changed on this equipment since it will process a different product on the same equipment. Other industries called this change-over, conversion, or set-up time. It is also the time spent replacing these dies and jigs until the equipment is ready to start manufacturing the next batch of products. Usually, the equipment will not be running during the time of conversion. As explained in Chapter 2 of this book, set-up, conversion sometimes termed as change-over, is the time required to remove dies, jigs for one product, clean-up, prepare dies and jigs for the next product, re-assemble the equipment, adjust the equipment, perform trial runs and make further adjustments until the product of acceptable quality is finally obtained on the equipment. It is important to have a fixed period in getting the MTTS or Mean Time to Set up the machine to establish a clear trend of improvement in the set-up time.

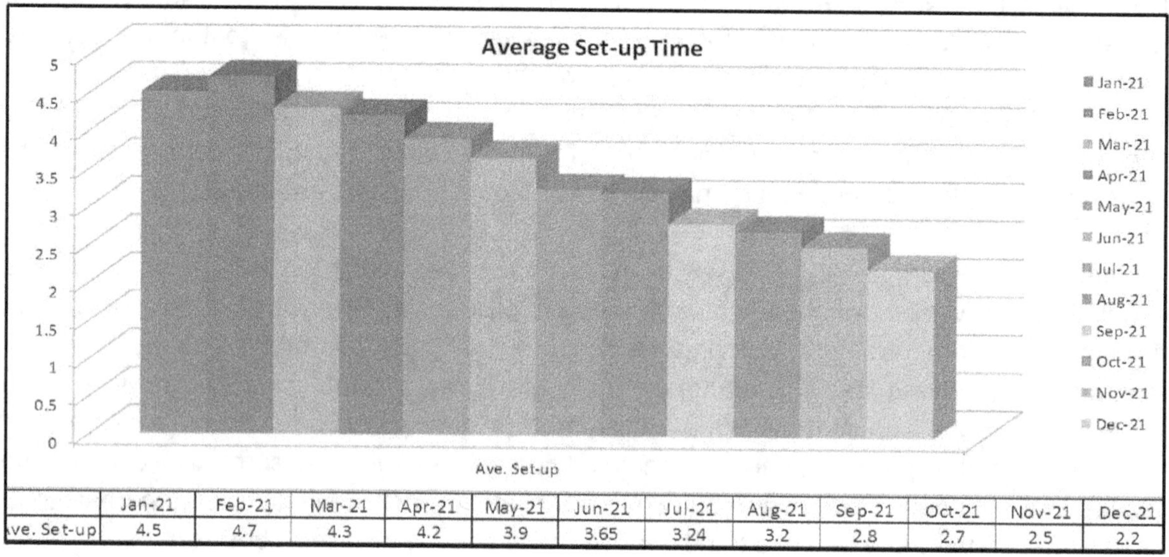

	Jan-21	Feb-21	Mar-21	Apr-21	May-21	Jun-21	Jul-21	Aug-21	Sep-21	Oct-21	Nov-21	Dec-21
Ave. Set-up	4.5	4.7	4.3	4.2	3.9	3.65	3.24	3.2	2.8	2.7	2.5	2.2

Figure 5.5: Monitoring Set-up Time in hours

$$\text{MTTS} = \frac{\text{SUM OF SET UP TIME}}{\text{FREQUENCY OF SET-UP}}$$

The trend we want for MTTS will be the lower, the better. If MTTS is calculated monthly, then it should be consistently reported on a monthly trend. It is recommended that to reduce the conversion or set-up time, what is needed is to perform an actual video covering how the set-up is usually performed and having a team study on what points of the set-up time can be reduced together with analyzing where does maintenance spend most of his time during the conversion. Another tip on shortening the set-up or conversion time is to distinguish between external and internal set-up time, which are the activities that can be performed while the last lot is being processed from activities that can be performed when the equipment will be finally stopped and converted. An example of this is when the machine is not yet converted, the maintenance or people in charge of conversion can get the tools needed, dies and jigs and place them near the equipment instead of waiting for the equipment to totally stop for the conversion. If 20 minutes will be spent on acquiring the needed items then we have reduced 20 minutes in the setup time. Application of SMED (Single Minute Exchange of Dies) and techniques developed by Shigeo Shingo will help reduce equipment set-up time. Set-up and conversion usually happen in manufacturing industries. The challenge in set-up time is to reduce the time to convert one product to another to 10 minutes or below. This is according to Shigeo Shingo's book on Single Minute Exchange of Dies (SMED). Remember that the longer the set-up time performed on the equipment, the fewer products we produced which means lesser revenue and profit for the industry.

5.3: Mean Time between Assists (MTBA) Explained

Assists, errors, or TPM called these minor stoppages are any unplanned interruption or variance from equipment specification that requires human intervention on the equipment. Japanese called this chokotei. If the time fixes an assist goes more than 6 minutes or more, many manufacturing industries I know will consider this a breakdown, which I totally disagree with. Causes of prolonged assists may include waiting time for the maintenance to arrive at the equipment, or it taking too much time to correct the assists. My take on this is that an assist is always an assist and is different from a breakdown. When a part breaks down caused by the assists, this is the time to consider it a breakdown.

Figure 5.6: Actual Meco Plating Machine

Meantime Between Assists (MTBA) is the average time the equipment performs its intended function between assists. It is also the product or the operating time divided by the number of assists; hence MTBA is the average time between assist on any stoppage of a machine caused by any unwanted operation. It is important to note that only downtime caused by errors and minor stoppages should be included in the machine or unplanned downtime.

For industries implementing TPM or Total Productive Maintenance, they are more familiar with the term minor stoppages or chokotei. A minor stoppage is an equipment stoppage due to an error in automatic handling, processing, or assembly of parts and workpieces. It sometimes occurs due to quality-related abnormalities. These are errors in automated processes where the workpiece flow stops, the operator resets, and the machine runs again. This usually happens on fully automated equipment with a lot of electronic parts.

In most manufacturing plants that encounter this type of problem on their equipment, it is quite tedious for the operator to record every single assist and error that can occur on a particular day since errors or minor stoppages occurred more frequently than breakdowns and is easy to correct. With the absence of software and system, it is unlikely that the operator can capture every assist that can occur on a particular piece of equipment; in this case, this is where an MTBA Snapshot can be taken down on that particular equipment. An MTBA Snapshot is an observation to get data and result at any given time, duration and quantity.

Procedure for Preparing MTBA Snapshots
• Step 1: Prepare MTBA Snapshot Form
• Step 2: Select equipment that has a high frequency of assists and errors

- Step 3: Sectionalize the equipment per station or sub-assembly and record all possible assists that can occur on each sub-assembly or station. Provide a code for each assist or error.
- Step 4: Perform an MTBA Snapshot (minimum of 2 hours) and write the duration, frequency, and type of assists that occur on the machine. The longer the time for taking an MTBA Snapshot the better.
- Step 5: Generate corrective measures and perform modifications
- Step 6: Horizontal Replication of the MTBA improvement to same equipment similar problems

MTBA SNAPSHOT FORM												
DETAILS OF ASSISTS	Frequency (Time in Seconds)										TOTAL ASSISTS	TOTAL TIME
	1	2	3	4	5	6	7	8	9	10		
Misloading on load	11	34	55	31	23	48	18	50	11	15	16	429
	33	58	35	17	20	15						
Mispick	17	21									2	38
Magazine stops	34										1	34
Strip jamming	60	19									2	79
Elevator jam	52	113	49								3	214
Magazine stop	51										1	51
Total Assists and Time in Seconds											25	845

Figure 5.7: Actual MTBA Analysis Snapshot Conducted on Meco Plating Machine

PROCESS	CURRENT UNFORSEEN ISSUES	CORRECTIVE ACTIONS	MP DESIGN NO.
LOADING	- Frequent turn-over of strips during change of magazine that cause mispick - Frequent alarm of mispick due to undefined setting of vacuum pressure - Early replacement of gripper plate holder for rotating cylinder	- Modify y-aligner arm by extending its arm to 3mm to cater upcoming strips. - Convert analog pressure to digital pressure switch - Change material from aluminum to ss plate	- CLF-00SP-MP-006 - CLF-00SP-MP-007 - CLF-00SP-MP-005
DEFLASH to DRYING	- Early deterioration of conveyor clips that cause strip jamming to processed cells	- Modify flywheel to non opening type that prolongs the pressure switch clip life span to reduce jamming	- CLF-00SP-MP-008
UNLOADING	- Frequent alarm and mispick due to undefined setting of vacuum pressure - Frequent turn over of strips during off loading	- Convert analog pressure switch to digital pressure switch. - Modify y-aligner arm by extending its arm to 3mm to cater upcoming strips.	- CLF-00SP-MP-007 - CLF-00SP-MP-005

Figure 5.8: Modifications Done on Meco 2 Machine to Increase MTBA

Below is an actual MTBA Snapshot Case Study we performed in the past

- Area: Central Lead Finish Station
- Machine: MECO Plating Machine)
- Machine type: EDF + EPL 2400S automatic load and unload strip to strip deflash and plating machine to process 2400 strips per hour.
- Package or lead count: ABC / 52 lead Integrated Circuit
- Area: Central Lead Finish (Plating)
- Machine Number: Meco 2
- Performed by CLF Planned Maintenance Team

- Date Performed: January 15, 1998
- Time Started: 10:00 am
- Time Ended: 12:00 pm
- Total Observation Time: 2 hours or 7200 seconds

Case 1: Computing for the MTBA

Given from the MTBA Snapshot from Figure 5.8
- Time Start (TS) = 10:00 am
- Time Finish (TF) = 12:00 pm
- Total Assists (TA) = 25 times
- Total Downtime (TDT) = 845 seconds

PHASE 0	PHASE 1		PHASE 2	
START	BEFORE	AFTER	BEFORE	AFTER
January 1998	February 1998	April 1998	May 1998	February 1999
4.24 min.	16.61 min.	22.54 min.	45.84 min.	115.74 min.

Figure 5.9: Before and After Data After Conducting MTBA Analysis

CASE STUDY 1 : MTBA SNAPSHOT REPLICATION

MECO	PHASE 1		PHASE 2	
	BEFORE	AFTER	BEFORE	AFTER
	FEB - 98	APRIL - 98	MAY - 98	FEB - 99
MECO 2	16.61 min	22.54 min	45.84 min	115.74 min
MECO 1	11.86 min	67.4 min	118.9 min	117.04 min
MECO 3	15.68 min	51.86 min	51.86 min	87.33 min
MECO 4	10.07 min	39.11 min	48.84 min	105.68 min

MTBA Before and After Data derived from CLF Planned Maintenance Activities 98-99

Following were actual data taken from CLF Plating Meco Machine as part of their improvement activities during their TPM's Planned Maintenance activities, as based from previous records, all MECO and TECHNIQUE machines passed Phase 2 Certification

Figure 5.10: MTBA Replicated to Other Meco Machines

Total Productive Time (TPT)
- TPT = Loading Time - Machine Downtime
- TPT = [(2 hours x 60 minutes / hour x 60 seconds / minute) - 845 seconds]
- TPT = (7200 – 845) seconds
- TPT = 6,355 seconds

Calculating for the MTBA and MTTA

- MTBA = Productive Time / Total Assists
- MTBA = 6,355 seconds / 25
- MTBA = 254.2 seconds
- MTBA = 4.24 minutes
- MTTA = Total Assists Time / Total Assists
- MTTA = 845 seconds / 25 = 33.8 seconds

This means that MECO 2 machine has a probability of experiencing an assist occurring every 4.24 minutes and the average time to correct the error MTTA (Mean Time to Assists) is 33.8 seconds. The following figure indicates the modifications done on the CLF Planned Maintenance team's plating machine which increased the MTBA of the MECO equipment. The following were the actual data taken from CLF Plating Meco Machine as part of their Phase 2 activities during their TPM Planned Maintenance activities. There were 4 Meco machines and since the other 3 Meco machines experienced many minor stoppages as well, the modifications were replicated to the other 3 machines as reflected in figure 5.10. Based on actual records, all MECO machines passed the Planned Maintenance Phase 2 audit and certification. The group was able to improve the MTBA of their equipment from 4.24 minutes to 115.74 minutes after conducting several modifications on the equipment that affected these errors. Before and after data derived from CLF Planned Maintenance activities 1998-1999.

Case 2: Computing for the MTBA

Total Productive Time (TPT)

- TPT = Loading Time - Machine Downtime
- Total Assists (TA) = 36 times
- Total Assists Downtime (TDT) = 2099 seconds
- TPT = [(2 hours x 60 minutes / hour x 60 seconds / minute) - 2099 seconds]
- TPT = (7200 – 2099) seconds
- TPT = 5101 seconds

Details of Assists	1	2	3	4	5	6	7	8	9	10	Assists	Time
1. Buffer Finger	33	58	35	17	20	15					6	178
2. Elevator Error for Loading	58	45	56								3	159
3. Magazine Need to Continue	33	15	80								3	128
4. Elevator Error Unloading	92	73	49	80							4	294
5. Buffer Error	40	48	49	50							4	187
6. Non Movement Rotating Motor	84	126									2	210
7. Mispick	50	55	45	60	87						5	297
8. Product Jamming	80	92	120	59	90	65					6	506
9. Product Fell	45	36	59								3	140
											36	2099

MTBA SNAPSHOT FORM

Figure 5.11: MTBA Snapshot Form

Calculating for the Mean Time between Assist (MTBA)

• MTBA = Productive Time / Total Assists
• MTBA = 5,101 seconds / 36
• MTBA = 141.69 seconds
• MTBA = 2.36 minutes

Calculating for the Mean Time to Assist (MTTA)

• MTTA = Total Assists Time / Total Assists
• MTTA = 2099 seconds / 36
• MTTA = 58.31 seconds

Just like MTTR for repairing failures and breakdowns, MTTA will be the term used to correct minor stoppages and errors. Remember that in MTTA, there is nothing to repair, in the majority of cases, we correct the error by simply restarting the machine. The goal just like MTBF is to increase the time for the error or minor stoppages to occur on the equipment. The longer the error occurs means that the machine is suffering less from these losses.

Key Points in Conducting MTBA Snapshot

• Make a distinction between what constitutes an assist and a breakdown.
• Make a record or list of each type of assist that is possible to occur on the equipment itself and provide a code for each error experienced.
• Minimum MTBA Snapshot time required should be a minimum of 2 hours, however, the longer the time the better. The person conducting the MTBA Snapshot should also be responsible for correcting the errors or stoppages that occurred on the machine.
• Other machine downtimes such as breakdown and conversion should be excluded from the Total Productive Time or Operating Time.
• Once the initial MTBA had been determined, recommend for improvement to improve the MTBA of the equipment. Once the MTBA increased, recognize the team for their effort.

5.4: Mean Units Before Assists (MUBA) Explained

Another way of tracking minor stoppages, or errors is through Mean Units Before Assists (MUBA). This indicates how many units were produced before an assist or minor stoppage is encountered on the equipment. There are both advantages and disadvantages of using this measurement which we developed during my time in the semiconductor industry.

Unlike MTBA which is only a snapshot where the minimum snapshot time required will be 2 hours, MUBA will be a continuous process for as long as the equipment is operated which will provide a more accurate figure on the frequency of assist which is measured in units. While MTBA will be based on time, MUBA will be based on the number of units produced before a minor stoppage occurs. The formula for MUBA will be as follows. The unit will be the number of units produced.

$$\text{MUBA} = \frac{\text{Number of Units Produced}}{\text{Total Assists}}$$

Code	Minor Stoppages Errors	Code	Minor Stoppages Errors
PPM	Pick and Place Malfunction	CE	Conveyor Error
MP	Mispick	PF	Product Falling
ML	Misloading	SE	Software Errors
MNC	Magazine Need to Continue	SM	Sensor Malfunction
PJ	Product Jamming	CS	Clogged Sensor due to Dirt
EL	Elevator Jamming	CC	Clogged Chute
BE	Buffer Error	ICA	Insufficient Compressed Air

Figure 5.12: Sample Codes for Minor Stoppages

MAINTENANCE INDICATORS AND MEASUREMENTS

No	Measurement Indicator	Code	Formula	Trend
1	Machine Downtime	MDT	n.a.	The lower, the better
2	Repair and Maintenance Costs	RNM	n.a.	The lower, the better
3	Breakdown Occurrences	BDO	n.a.	The lower, the better
4	Mean Time Between Failure	MTBF	Operating Time / Breakdown Occurrence	The higher, the better
5	Mean Time to Repair	MTTR	Machine Downtime / Breakdown Occurrence	The lower, the better
6	Overall Equipment Effectiveness (%)	OEE	Availability x Performance Rate x Quality Rate	The higher, the better
7	Availability (%)	AVAIL	(Avail Time – All DT) / Available time	The higher, the better
8	Utilization (%)	UTIL	(Load Time – Machine DT) / Loading time	The higher, the better
9	Performance Efficiency Rate (%)	PER	(Ideal Cycle Time x Output) / Operating time	The higher, the better
10	Mean Time Between Assists	MTBA	Operating Time / Frequency of Assists	The higher, the better
11	Mean Time to Fail	MTTF	MTBF - MTTR	The higher, the better
12	Mean Units Before Assist	MUBA	Total Units Produced / Frequency of Assists	The higher the better
13	The ratio of Maintenance Costs to Operating Costs (%)	MC: OC	Total Maintenance Cost / Total Operations Costs	The lower, the better
14	PM Achievement Rate (%)	PMAR	PM Tasks Completed / PM Tasks Planned	The higher, the better

Figure 5.13: Maintenance Indices and Trend

The disadvantage of MUBA is the difficulty in measuring this indicator if this will be done manually by operators since they will be spending a lot of time logging the products produced before an assist happens. Remember that the frequency of assist will occur more frequently than the frequency of breakdowns. However, if the equipment is fully automated, or a PLC is directly hooked up to the CMMS system, what is important is to provide a code for every single assist that can happen on the equipment as in figure 5.13. Every time an assist will be encountered, the operator just needs to enter the code of the assist or minor stoppage and it will directly be linked to the system. Both MTBA and MUBA can be linked to the system. It is important not to mix the codes for minor stoppages and breakdowns as this will provide inaccurate data and variations for MUBA, MTBA, and MTBF respectively.

Chapter 6

Preventive Maintenance Indicators

> ***Measuring the Meaningful Measures of Performance is important for any industry as downtime translates to loss productivity, loss revenue and loss of customer's confidence. Downtime in one business segment can easily have a direct impact on your customer's business. KPIs and measurements will allow us to reflect and improve the way we do things in our plant.***

6.1: Preventive Maintenance Explained

Before we discuss the different KPIs and indices for Preventive Maintenance, let us explain in detail what Preventive Maintenance is all about. Preventive Maintenance is a set of activities or maintenance tasks that is performed on the equipment and assets on a scheduled basis. The main goal of performing tasks on a scheduled basis is to extend the equipment's life and to assure its capacity in support of the industry's goals and targets. In Preventive Maintenance, the basic law to consider is that the cost of performing PM must always have to be lower than the cost of failure it is, meant to prevent. This means that if the maintenance planner estimates that the Bill of Materials (BOM) for conducting a quarterly scheduled Preventive Maintenance is $ 50,000.00 and the cost of failure it is meant to prevent is $ 15,000.00, then performing Preventive Maintenance is not feasible in this sense. It should be the other way around.

According to the definition of William Worsham, Preventive Maintenance is planned maintenance of the plant that is designed to improve equipment life and avoid any unplanned activity. PM includes routine cleaning, lubrication, adjusting, minor component replacement, and overhauls, to extend the life of equipment and facilities. Its purpose is to minimize breakdowns and excessive depreciation. In its simplest form, Preventive Maintenance can be compared to the service schedule of a car. The amount of Preventive Maintenance needed at a facility varies greatly. It can range from walk-through inspection of facilities and noting equipment deficiencies for later correction up to shutting down the equipment after a certain number of hours or after a certain number of units had been produced. The primary goal of PM is to anticipate the failure of equipment before it actually occurs. It is designed to preserve and enhance equipment reliability by replacing worn components before they actually fail in operation. Almost all industries have their regular and routine Preventive Maintenance that is being done in a timely and scheduled fashion. The interval for conducting industries' PM activities varies from one industry to another. While most industries will base their scheduled activities on calendar time, others would be using running hours, the number of volumes produced, cycle time, and there is even one industry I know which is a sugar milling industry in which they perform their scheduled Preventive Maintenance when all the sugar cane have already been harvested by the farmers.

Figure 6.1: Ideal and Complete Preventive Maintenance Activities

Developing a Preventive Maintenance strategy in any organization should begin with having an inventory of all equipment and assets in the plant. Once the inventory has been completed, the equipment is ranked accordingly based on its criticality. We can categorize all Rank A to be the worse, Rank B is in between the worse and good and Rank C will be considered as good machines. The planner in charge can use this machine ranking to prioritize which equipment and assets will be on the top list for Preventive Maintenance. The planner will write the detailed tasks, estimate the labor, bill of materials needed, spares, consumables, and the interval of performing the tasks. To make the PM more effective, we can coordinate initially with the Predictive Maintenance people to conduct a reading before initiating the actual Preventive Maintenance activities and we can call this a Pre-PM. The CBM or PdM team can meet with the PM and planners to discuss their findings on what parts, modules, systems to focus on. After the meeting, the PM tasks will be reviewed for any addition or deletion of tasks. The maintenance responsible for executing the tasks will be withdrawing in the storeroom the needed parts, consumables, and tools needed to execute their PM. Once the needed items had been withdrawn, then the PM tasks will be executed. Once the execution of the Preventive Maintenance tasks had been completed, they can call the CBM or PdM team once again to conduct a Post PM to verify if the readings improved. This Pre and Post PM is also important to minimize the chances of Infant Mortality failures. Once these PM tasks are completed, then the PM crew will close the report and a feedback session with the planner will be done to improve their next PM activities. The planners make the necessary adjustments based on the feedback and the next machine will then be scheduled for the next PM and the cycle continues.

6.2: Why We Need to Measure our Efforts on PM?

Almost all industries have their Preventive Maintenance activities, but many are not satisfied with the outcome. But regardless of how the industry performs its PM, it is important to measure the effectiveness of our activities on Preventive Maintenance. We simply just cannot perform PM for the sake of just doing it. This means that if our efforts on Preventive Maintenance are both effective and efficient, then we can experience fewer failures and breakdowns on our equipment and assets. Remember that when we execute a Preventive Maintenance activity, we are using resources and the industry's money. Therefore to gauge whether our efforts on PM are effective or not, we need to have some form of measurement. These measurements and indicators will also tell us if we are performing the correct tasks at the correct interval.

COMMON PM TASKS LISTS INCLUDES	
TASKS	**EXAMPLE**
1) Inspection	- Looking for leaks in a hydraulic system
2) Cleaning	- Removing dirt/debris on the machine
3) Tightening	- Tightening of bolts w/ correct torque
4) Take readings	- Recording reading on gauges
5) Adjustment	- Adjust tension on drive belt
6) Lubrication	- Check level & health of oil
7) Parts Replacement	- Replace worn out parts
8) Overhaul	- Scheduled overhaul of pump
9) Functionality Check	- Check sensors if working
10) Predictive Maintenance	- Routine thermal scanning of motors

Figure 6.2: Common Tasks Performed on Preventive Maintenance

Performing Preventive Maintenance includes different activities at different intervals. The PM tasks lists are the heart of the PM Program. The PM lists indicate what the maintenance needs to do, and how to execute it. In its highest form, the PM tasks list represents the accumulated knowledge of the maintenance workforce or those who will execute these tasks on what to do to avoid a failure on the asset. These PM activities and tasks have two major objectives, first, to extend the life of the asset, and second is to detect when the failure on the asset had begun its descent into a potential failure (not yet in a failed state) which is the start of wear-out mode. It is also the assumption of the PM design that when a problem is detected during inspection, the maintenance will respond with a corrective action to address the problem

These PM activities and tasks are assembled into lists and sorted by frequency of execution. They are directed on how the asset will fail. The rule is that these tasks should address the failure modes of the asset especially those with high consequences. But despite our efforts, there will still be failures and breakdowns even with the best PM structure. Our goal is to reduce, control and prolong the duration of the failure and convert the breakdowns that are left into a learning experience to improve the delivery of the maintenance function.

Preventive Maintenance is composed of different activities from mere inspection to overhauling the entire equipment. It also includes the replacement of parts, cleaning, and lubrication. For scheduled replacement and overhauls, this PM activity should only be done if the part or item has a wear-out pattern and the schedule of performing these tasks must be based upon the useful life and not on the average life of the part or item. This means that there should be a wear-out mode. The important thing is that the interval should not be too soon or too late.

In most industries, replacements and overhauls are based on mere guesses, but in most cases, they guess it wrong all along. PM uses the concept of JIC or Just in Case. This means that replacements and overhauls are done because our thinking is that it might fail again sometime if we do not do the tasks now. The problem here is that if a part is still working and it was replaced, then we are throwing something that is still useful. It means that we are throwing money away from our company, which is one of the reasons why doing Preventive Maintenance is costly. This is because the majority of industries are highly too conservative most of the time, in which industries try to select intervals that are way too short or doing maintenance that is intrusive or not needed. This means that we might overhaul equipment after one year in which the correct overhauling interval should be done after 5 years. These activities only cause our PM efforts to be more costly than ever.

Determining the correct frequency or interval of replacements and overhauls for Preventive Maintenance activities should be based upon the useful life and not on the average life of the part, which inhibits an age-related pattern. Parts that do not have a useful life should not be replaced on a time-based dominated frequency because their pattern is different. Only parts that will age because of stress-induced upon the part should be the subject for parts for PM replacement and overhaul. This simply means that the stress finally exceeds the strength of the material of the part. Aging is the process where certain parts of the equipment deteriorate because of stress-induced upon a part over a given period. Any part that conforms to this process will be the subject of Preventive Maintenance intervention.

In the absence of data and historical records, using our experience to guess the task interval is the only option. According to Anthony Smith, one proven method can be adapted to determine the correct task interval for overhauling and replacement, which is called the age exploration method. When we first try to overhaul or inspect the condition of the equipment on parts where aging and wear-out are thought to be possible and reveal that no wear-out signs exist, we can automatically increase the interval by 10% during the next schedule. Repeat this process until we can finally see the normal wear-out pattern on the part being inspected. If the equipment is due to be overhauled after one year and reveals that upon thorough inspection, there are still no traces of wearing or aging on suspected parts, then we can automatically increase the interval by 10% during the next overhauling schedule. This means the next PM overhaul will happen on the 13th month. If there are still no signs of wearing out, we increase it again by 10% until a wear out is evident on the parts.

6.3: <u>Machine Downtime Due to Breakdown and Frequency</u>

As discussed in Chapter 3, downtime has many causes and can be considered as Planned (Non-Machine-Related) and Unplanned Downtime (Machine Related downtime). Tracking this downtime should only be for a downtime caused by breakdowns and failures that will warrant a repair. This will be mostly expressed in hours as the best unit to use. Our goal definitely is to reduce both the number of breakdowns and downtime caused by these breakdowns on our equipment. It is also likewise important that all maintenance, technicians involved should have a unanimous and clear definition of what to include and not to include as breakdown. Only function-loss breakdowns will be considered in tracking breakdowns. This means that breakdowns that caused the machine to stop since something failed will be considered. Exclude function-reduction breakdowns in the tally which are breakdowns and failures that can occur where the equipment is still capable of providing the primary function. For example, if the emergency stop is in a failed state yet the machine is still capable of delivering the products, this would not be listed in the tally for breakdown. Figure 6.2 was our actual results on the number of breakdowns and downtime as we journey the Planned Maintenance teams which piloted 22 Rank A machines composing of different types across divisions and successfully reduced the breakdown recurrence through the 4 Phases of Planned Maintenance.

Our previous Planned Maintenance Committee is composed of 22 members from different departments in the plant for front and end of line stations. Facilities were also included since their equipment and assets are more critical. If one of the compressors failed, then expect several pneumatic equipment to stop in operations. These people were responsible for the implementation of Planned Maintenance activities in their respective areas. The Planned Maintenance team covers the following divisions from Plant 1 (P1) and Plant 2 (P2) which is composed of more than 10,000 employees. The organization and structure are mainly composed of membership from both front of line and end of line for each division including the facilities department. Most of the equipment we have were fully automated and operated 24 hours a day for 7 days a week. Our objective in implementing Planned Maintenance on a plant-wide scale is to dramatically reduce breakdown and improve the reliability of our equipment and assets through improvements and modifications. After our success in our First Pilot Machines, the PM teams formed the fan-out teams which were then composed of 172 machines from operations and 100 pieces of equipment from facilities totaling 272 machines. We then horizontally replicated the activities of the pilot team's Phase 1 restoration activities. Figures 6.2 and 6.3 were our actual collective efforts in reducing the breakdown rate of our equipment. Although the reader might think that this is an easy task which is not since before we can ever start the process, what needs to be improved is the mindset of our maintenance people.

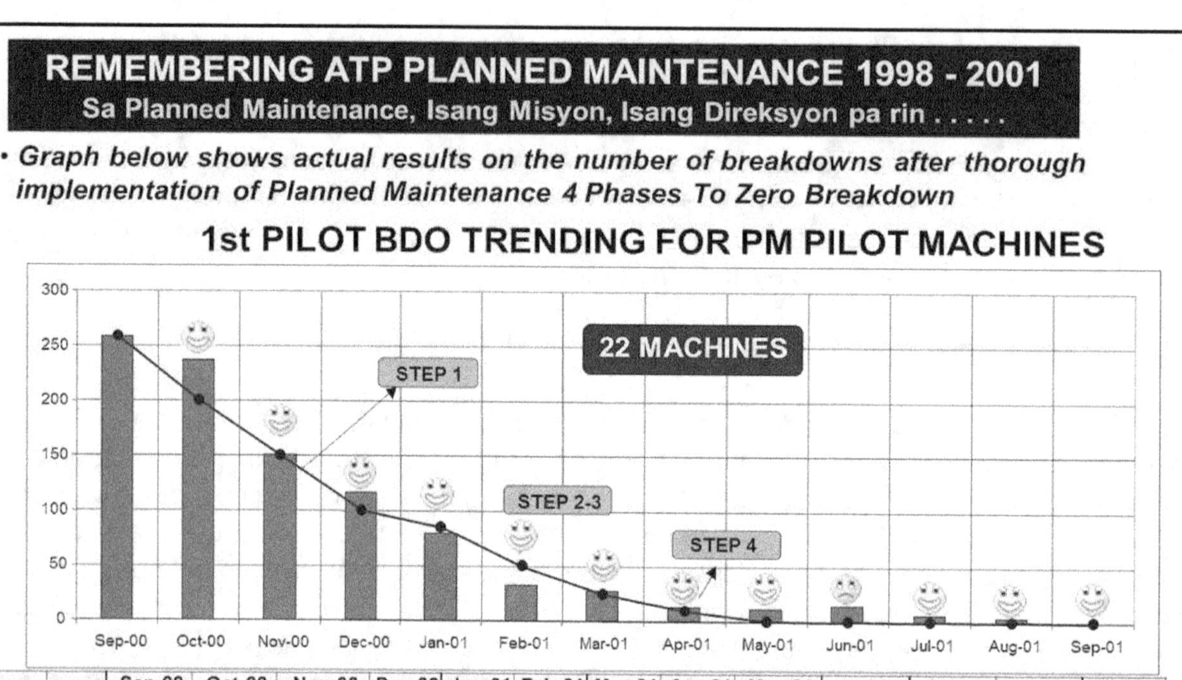

BDO		Sep-00	Oct-00	Nov-00	Dec-00	Jan-01	Feb-01	Mar-01	Apr-01	May-01	Jun-01	Jul-01	Aug-01	Sep-01
	PLAN	259	200	150	100	85	50	25	10	0	0	0	0	0
	ACTUAL	259	237	151	117	80	33	28	13	12	15	6	4	0

Figure 6.3: Actual Planned Maintenance BDO Tracking for Pilot

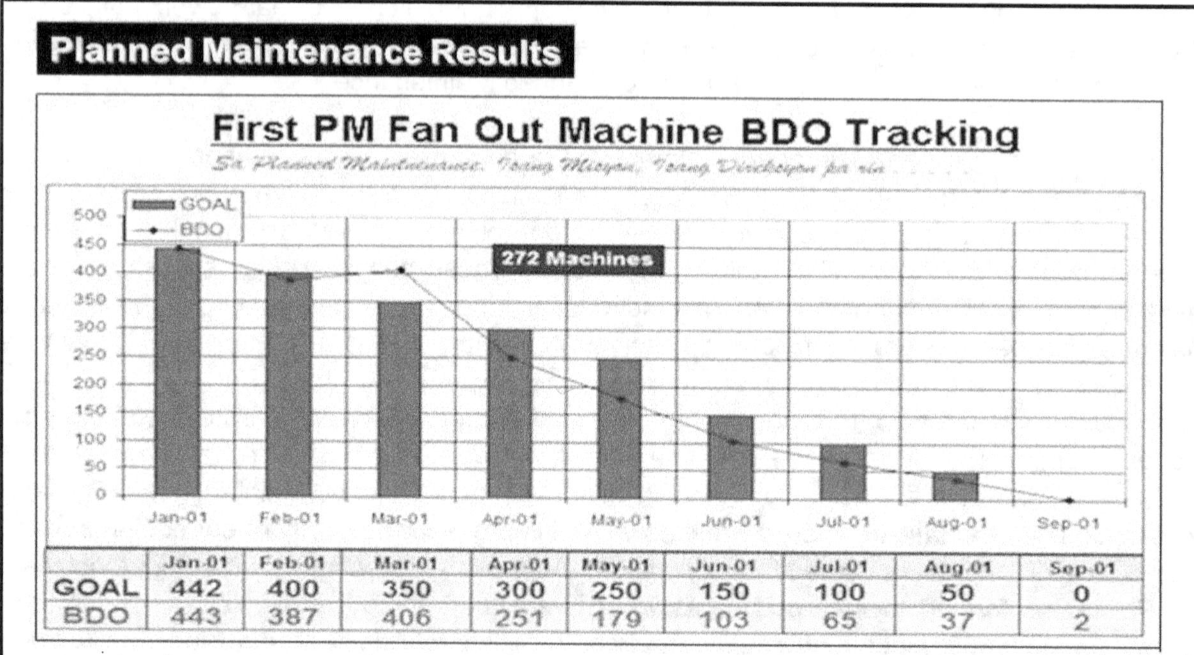

	Jan-01	Feb-01	Mar-01	Apr-01	May-01	Jun-01	Jul-01	Aug-01	Sep-01
GOAL	442	400	350	300	250	150	100	50	0
BDO	443	387	406	251	179	103	65	37	2

Next, the PM teams formed the fan-out team which then composed of 172 machines from operations and 100 from facilities totaling 272 machines then horizontally replicated the activities of the pilot team's Step's 1 -3 and checked other recurrence of breakdown

Figure 6.4: Actual Planned Maintenance BDO Tracking for Fan-Out

6.4: Preventive Maintenance Compliance

The amount of Preventive Maintenance needed at a facility varies greatly. It can range from walk-through around inspection of facilities and noting equipment deficiencies for later correction up to shutting down the equipment after a certain number of hours or after a certain number of units had been produced. This measurement is perhaps the most abused of all KPI's for maintenance. In an industry, a planner will provide the lists of maintenance tasks to be done on the equipment which will include the Bill of Materials, duration, manpower needed, MRO Spares for replacement, and in other instances will also include work done by third-party contractors. PM Compliance is a maintenance indicator that will measure how many PM tasks were actually completed and closed divided by the total number of tasks to be accomplished on or before the due date. The equipment will then be scheduled and another group of people will execute the Preventive Maintenance tasks. Hence, if we have 25 activities to execute in your Preventive Maintenance lists to be performed, let us say on a semi-annual schedule on a particular piece of equipment, and all 25 activities had been completed for a given period stated, before the due date, then we can say that the PM Compliance is at 100%, which is perfect. Measuring PM Compliance or having a very high percentage of being compliant on every activity on Preventive Maintenance is useless if the equipment still encounters many infant mortality and random failures in which maintenance performs emergency work right after a Preventive Maintenance had been initiated on the equipment. If we are doing Preventive Maintenance activities, yet the equipment is always failing, then something is wrong with how we execute our Preventive Maintenance tasks. Or perhaps, we need to overhaul not the equipment, but the maintenance tasks that we perform on Preventive Maintenance. All activities on Preventive Maintenance should and must address a particular failure mode. The formula for PM Compliance is as follows:

$$\textbf{PM Compliance } = \frac{\textbf{Number of PM Tasks Completed}}{\textbf{Total Number of PM Tasks Listed}} \textbf{ x100\%}$$

Performing Preventive Maintenance for the mere sake of complying will not improve the performance of the equipment. All PM activities must prevent a failure from occurring. If there are activities on the Preventive Maintenance lists that do not address any particular failure mode, then delete them in the PM lists since it is just a complete waste of time, money, and resources to execute such an activity. Remember the golden law on doing Preventive Maintenance. The law states that the cost of doing Preventive Maintenance must always have to be lower than the costs of failure it Is meant to prevent. If the Bill of Materials on performing this annual Preventive Maintenance is, around 100,000 US dollars and the cost of failure it is meant to prevent, is 10,000 US dollars, then it is a good idea to disregard the PM. Having a high percentage of our PM Compliance is next to useless if the cost of doing PM and replacing the equipment increases. If you have 25 activities to be done on the equipment and 15 of these activities mean replacing parts even if they are still working, then it serves no value whatsoever. PM compliance will be 100%, yet the cost of doing Preventive Maintenance almost tripled since we replaced the parts that are still functioning and we have not maximized the lifespan of these parts. PM Compliance can only be effective if we also measure and compare it with the emergency and repair hours on the equipment. If a PM activity does not need to be done, it is best to remove them from the list; otherwise, it might induce infant mortality failures into stable systems in the equipment.

Survey On Top Problems on PM (as of November 22, 2021)

1) Add on PM Checklists Syndrome - where PM checklists and activities seems to grow	377		6th
2) Infant Mortality Failures - where problems arises after a PM replacement and overhaul	405		4th
3) Replacement of good parts to conform with PM specs and procedures	308		7th
4) The Case of Random Failures - where random failures are included in the PM checklists	301		8th
5) Ageing workforce - nearing retirement	165		10th
6) Lack of training on the maintenance function	614		1st
7) Still Reactive and lot of corrective maintenance even with a sound PM Program	385		5th
8) Frequent reorganization in the plant - where new boss makes a new system	193		9th
9) Lack or poor documentation in PM	456		3rd
10) PM is waived - Operations wont give equipment for PM to cope with production	607		2nd

Figure 6.5: Survey on Top 10 Problems on Preventive Maintenance Revisited 2021

TASKS	INTERVAL	RESPONSIBLE	ACTIVITIES	FAILURE MODE
1	SHIFTLY	OPERATOR	Cleaning	Failure Mode A
2	MONTHLY	PM GROUP	Replace part B	Failure Mode A
3	MONTHLY	CBM GROUP	Overhauling	Failure Mode A
4	SHIFTLY	OPERATOR	Inspection of Belt	Failure Mode B
5	WEEKLY	MAINTENANCE	Routine Inspection	Failure Mode B
~~6~~	~~6 MONTHS~~	~~CONTRACTOR~~	~~Major Overhauling~~	
7	DAILY	MAINTENANCE	Inspection of gauge	Failure Mode D
8	DAILY	MAINTENANCE	Inspection of oil	Failure Mode E
9	WEEKLY	ELECTRICIAN	Routine Inspection	Failure Mode F
10	SHIFTLY	OPERATOR	Inspection of gauge	Failure Mode G
~~11~~	~~MONTHLY~~	~~PM GROUP~~	~~Replace part A~~	
			ADD	Failure Mode H

Figure 6.6: PM Tasks Should Address a Failure Mode

Another problem I see on these indices especially for manufacturing industries is that Preventive Maintenance is waived by Operations. Based on a survey I started in May 2009 until my last update last November 2020, this is the second biggest problem on maintenance in which lack of training comes first as in figure 6.5. When the equipment was subjected to maintenance intervention or scheduled Preventive Maintenance, operations or management frequently deferred or waived the equipment because the output is always the top priority and supersedes everything. When the demand for the product is high, or there were months when production is at its peak. At this stage, maintenance will find it very difficult or impossible to get the equipment for a scheduled Preventive Maintenance activity. As a result, more unexpected failures occur, and the feud between operations and maintenance gets worst and is pretty much alive. Maintenance ends up always on the losing end because the operation's decision prevails. Operations will give the equipment to maintenance only when the demand is low.

<u>6.5: Preventive Maintenance Effectiveness</u>

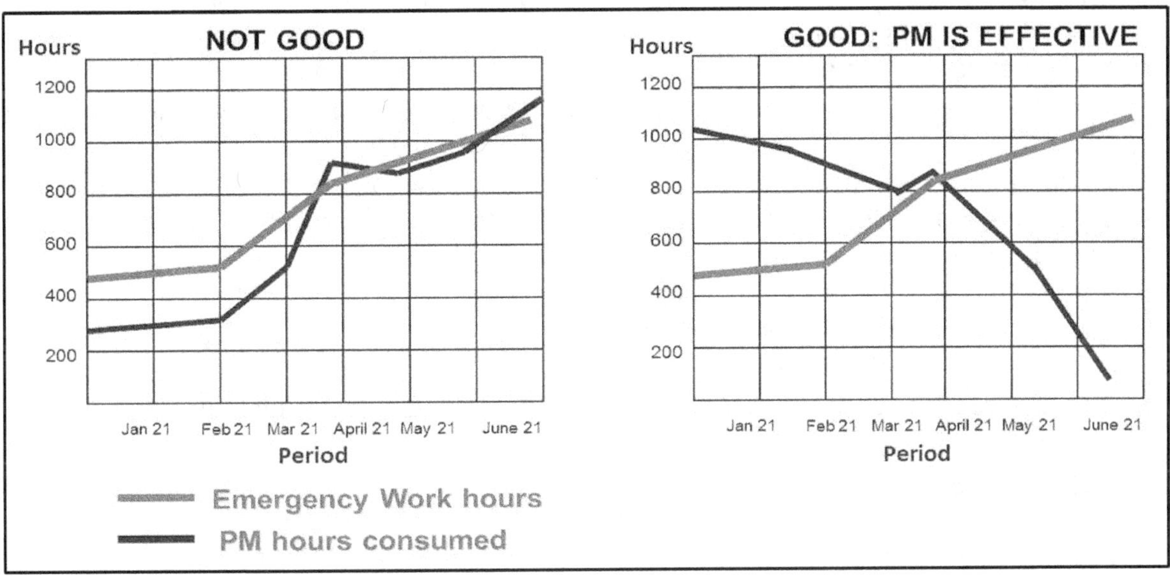

Figure 6.7: PM Effectiveness

As discussed in Section 6.2, PM Compliance is the most used and abused metric. Measuring PM Compliance or having a very high percentage of being compliant on every activity on PM is useless if your equipment encounters a lot of infant mortality failures and emergency work right after a PM initiative. However, a better approach is to we can compare these indices with the number of hours spent on emergency work. We can use these indices to determine if the Preventive Maintenance tasks we perform are effective or not. This means that if the PM Compliance is high, yet the number of breakdowns is also high, then something is wrong with how we perform Preventive Maintenance in the equipment since it should be the opposite. An effective PM strategy must reduce the amount of Reactive Maintenance in the plant by a margin of 20% or more. This means that if more time is spent on doing corrective and reactive maintenance compared to the Planned Activities, then we need to evaluate the maintenance tasks in our PM lists and how it is being performed. Therefore, to measure PM Effectiveness is to plot the number of hours spend on doing PM and the number of failures or breakdowns experienced after performing PM:

• Number of PM hours
• Number of hours of Emergency Work

If the Preventive Maintenance tasks were performed last January 15 to 18, 2022, the emergency worked hours should be measured starting from January 19 to February 18, 2022, and so on. Performing Preventive Maintenance for the mere sake of complying will not improve the performance of the equipment. All PM activities must prevent a failure mode from occurring, if not then delete those PM activities in your lists. The important thing to consider is that all PM tasks should address a particular failure mode.

6.6: Preventive Maintenance versus Breakdown Maintenance Ratio

Another way of measuring PM effectiveness is to compare the number of Planned Work Orders versus Emergency Work Orders but I would prefer to measure this in hours instead of the frequency. Although this will be the same for PM Effectiveness, the main difference is that this is based on the number of Planned Work Orders or Jobs generated compared to the number of breakdown or emergency repairs measured in frequency and not in hours. The word Planned in this case will include all activities on maintenance including Predictive Maintenance.

The only problem I see in this PM indicator is that it is not measured in time but on its frequency. This means that if I have 10 PM Work Orders that have been fully completed, this might look good on record. However, if we convert this into a time in which the 10 PM Work Orders took 10 hours to complete, and 1 emergency breakdown which took 48 hours to complete perhaps due to no available spare parts in the storeroom, then this is not actually a good performance from the PM Group. It would be better to use PM Effectiveness in hours to gauge the effectiveness of all the activities performed on Preventive Maintenance.

6.7: Maintenance Backlog

Maintenance backlog is a time indicator that consists of delays in performing the scheduled maintenance works, or those activities which have not yet been completed. This consists of pending scheduled planned activities allotted on the maintenance. The maintenance backlog to be executed may include emergency work, Preventive, Corrective Maintenance, Predictive Maintenance, or improvement activities. To measure this index is to consider all the tasks and activities included in the maintenance planning. The maintenance backlog will be equal to the total sum of man-hours for the planned, pending, and completed work orders, divided by the total available man-hours. Maintenance backlog may be expressed by the number of hours.

Although if we adopt this definition of maintenance backlog as those pending or delayed work where the due date had already been exceeded, then we might be missing something. There are various reasons for the delay in executing the work request such as;

• Delayed in the delivery of MRO Spare Parts
• Delayed in the delivery of consumables
• Operations waiving the Preventive Maintenance activities
• Maintenance performing other tasks

• Maintenance on leave
• Not enough manpower workforce on maintenance

One of the things that come to mind regarding the maintenance backlog is the maintenance tasks really need to be done, even if priority 1 is indicated in the work order, the question is, how do we define the priority in maintenance? The second question is, how about the interval? Are we sure enough to replace this part since the planner indicates to replace the part when it is still functioning? If we don't comply, then we have a backlog. My recommendation is before measuring this index, maintenance must have a clear definition of what to consider and what not to consider as a backlog and not just base everything on the due date of the work order but rather on the criticality of the tasks. This means that if there are two pending work orders on different equipment where a failure on equipment 1 will totally halt the whole operations, while for equipment 2 which has a back-up, then the priority should be given to equipment 1. Prioritization of executing the maintenance tasks should be based on the criticality of the equipment and the consequences of the failure if the tasks would have been neglected instead of basing the backlog on the due date of the tasks.

The criticality and consequences of the failure vary from one industry to another. A failure from a power plant may lead to a city blackout, where your plant will be on the nightly television news, and your company's lawyers are involved giving a lame excuse regarding the cause of the blackout. On the other hand, a failure from one carton box factory may mean a delay in the processing in which your boss eventually is asking you for the root cause of the failure all the time. Here are some cases to indicate if the failure will be critical:

• If the equipment is in an unmanned location
• If the failure can result in environmental consequences
• If the failure can result in safety consequences or can kill or injure someone
• If the failure can shut down the whole operations temporarily
• If the failure can permanently shut down the whole operations (loss of job)
• If the failure can result in industrial accidents
• If the failure can breach any known environmental regulations and laws
• If there is an absolute certainty that the failure cannot be detected
• If there are no redundancy or standby for the equipment
• If the cost of the failure will be extremely high and can result in a penalty
• If the company lawyers and insurance people will be involved in the failure
• The worst case will be if the media and those crooked old politicians will be involved

We can also base our judgment on the criticality table derived from FMEA in figure 6.8. If any of the activities specified in the tasks will lead to the above situations, then we can define a Priority 1 basis and without a doubt must be included in the maintenance backlog if the tasks had not been done on the due date provided. On the other end, for lesser critical failures, we can reschedule the date to a more convenient time and remove them from the backlog. Every failure has a specific set of consequences associated with it, and if the consequences are very serious, all efforts should be done to prevent the failure from occurring. Proactive maintenance has much more to do about avoiding or reducing the consequences of failure than it has to do with preventing or eliminating the failure themselves. Therefore, a proactive task is only worth doing if it deals successfully with the failure's consequences, which it is meant to prevent.

The maintenance backlog provides a list of maintenance tasks that must be accomplished over a given period. Although a small amount of maintenance backlog is generally acceptable. These maintenance backlogs must likewise be reviewed and prioritized if they really need to be performed or not. If a certain maintenance backlog is irrelevant, then it should be deleted from the lists. Prioritizing these backlogs and analyzing the causes for the delays will definitely improve the future scheduled maintenance plans of the plant. People involved in generating the activities such as planners must continuously be in constant communication with those who will execute the task itself. The idea is not to eliminate the backlog but to manage it effectively. The formula for Maintenance backlog is as follows:

Effects on	Criteria – Severity of Effects	Ranking
Hazardous without warning	Very high severity ranking – Affects operator, plant, or maintenance personnel, safety and/or affects non-compliance with government regulations	10
Hazardous with warning	High severity ranking – Affects operator, plant, or maintenance personnel, safety and/or affects non-compliance with government regulations	9
Very High Downtime or Defective Parts	Downtime of more than 8 hours or defective parts loss more than 4 hours of production	8
High Downtime or Defective Parts	Downtime of 4 to 7 hours or defective parts loss of 2 to 4 hours of production	7
Moderate Downtime or Defective Parts	Downtime of 1 to 3 hours or defective parts loss of 1 to 2 hours of production	6
Low Downtime or Defective Parts	Downtime of 30 minutes to 1 hour or defective parts loss of up to 1 hour of production	5
Very Low Downtime No Defective Parts	Downtime up to 30 minutes – no defective parts	4
Minor Effect	Process parameter variability exceeds Upper/Lower Control limits. Adjustment or other process controls need to be taken – no defective parts	3
Very Minor Effect	Process parameter variability between upper/lover control limits. Adjustments or other process controls needs to be taken	2
No Effect	Process parameter variability within Upper/Lower Control limits, adjustment or other process controls not needed or can be taken between shifts or at normal maintenance – no defective parts	1

Figure 6.8: FMEA Table on Criticality/Severity

$$\text{Backlog} = \frac{\text{Sum of All Planned, Pending, and Completed tasks}}{\text{Total Available Man-hours}}$$

Generally, these maintenance backlogs will always be present since performing maintenance is a never-ending process needed to sustain the equipment. They can also be tricky as any maintenance tasks are written by the planner with a due date that will always result in a backlog. Maintenance people must review all their tasks performed on their equipment and assets. First on the list will be the frequency of conducting the tasks, are we doing these tasks too much or not. For example, in building A, the window-type aircon needs to be cleaned monthly as indicated in the plan, however, if the room temperature doesn't fluctuate, then we can ask the planner to make this activity every quarter or even semi-annually.

6.8: Maintenance Costs

Although there is no distinct and universal standard on what maintenance cost includes as it varies from one industry to another. Maintenance cost is a universal and common measurement and indicator for all types of industries, unlike other KPIs.

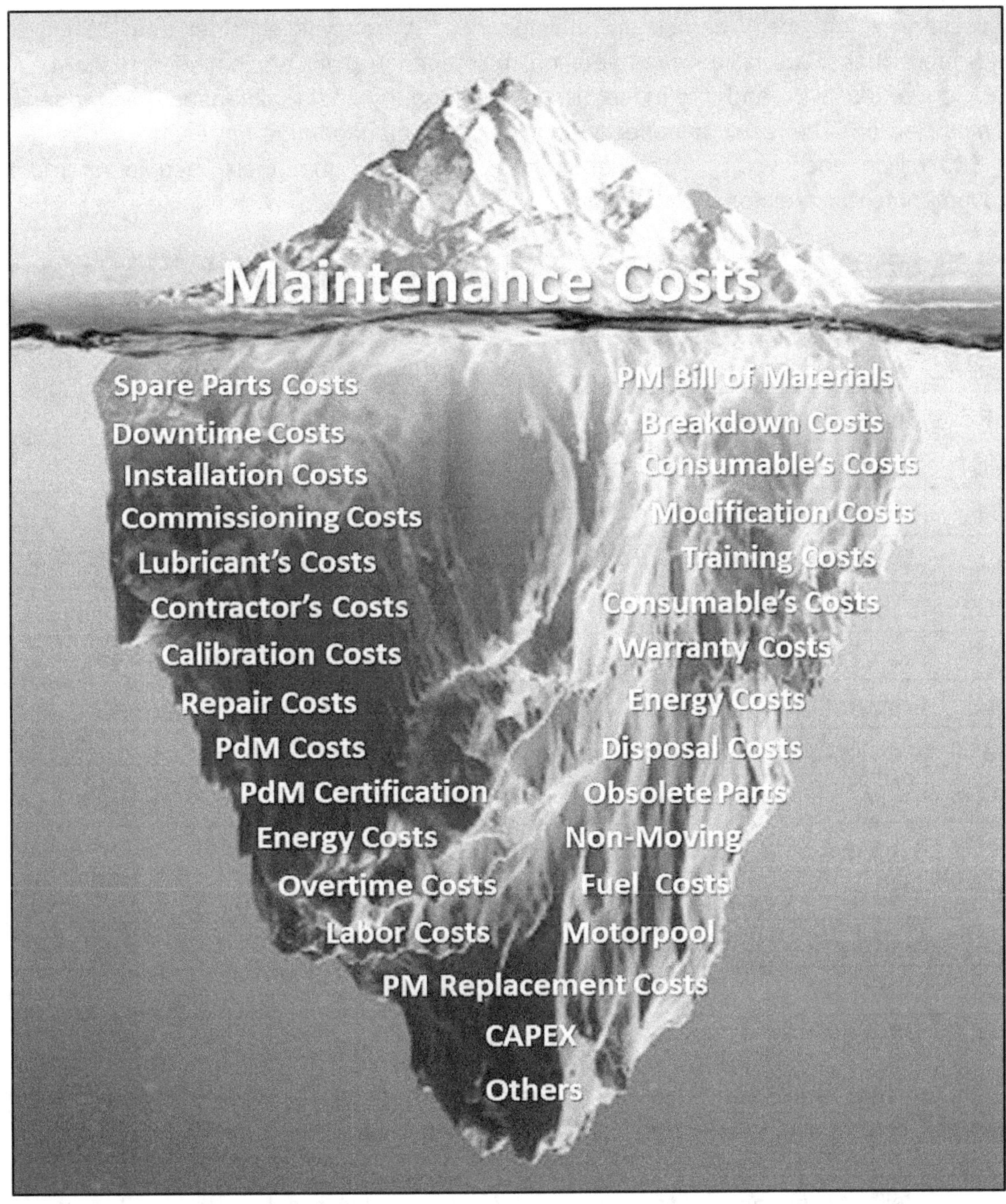

Figure 6.9: What Maintenance Costs Can Include

By definition, maintenance costs are any expenses incurred on the equipment, machines, or asset by the maintenance function to sustain and preserve it. It can be said as a major part of the total operating costs for all manufacturing and non-manufacturing industries. Depending on the type of industry, the Total Maintenance Costs can represent around 3 to 50% or sometimes more of the costs of products and services produced.

[5]According to my dear friend R. Keith Mobley in his book An Introduction to Predictive Maintenance, he quotes that recent surveys of maintenance management effectiveness indicate that one-third or 33 cents out of every dollar of all maintenance costs is wasted as a result of unnecessary or improperly carried out maintenance. When you consider that US industries spend, more than $200 billion each year on maintenance plant equipment and facilities, the impact on productivity and profit that is represented by the maintenance and operations becomes clearer. The result of ineffective maintenance management represents a loss of more than $60 billion each year. This statement truly signifies that there is a lot of room for improvement for the maintenance function.

Overall Maintenance Costs

Code	Costs Incurred	Code	Costs Incurred
SPC	Spare Parts Costs	PM BOM	PM Bill of Materials
MDT	Machine Downtime Costs	PM REP	PM Replacement Costs
IC	Installation Costs	MRO CC	Consumable Costs
CC	Commissioning Costs	MOD	Modification Costs
LUB	Lubricant Costs	TNE	Training and Education
CC	Contractor's Costs	WW	Warranty Costs
RC	Repair Costs	DIS	Decommissioning Costs
PdM	PdM Instrument Costs	OBS	MRO Obsolete Parts Costs
PdMC	PdM Certification Costs	NMP	MRO Non-Moving Parts Costs
EC	Energy Costs	FC	Fuel Costs
OT	Overtime Costs	MOT	Motorpool Costs
CAL	Calibration Costs	CAP	Capex
LAB	Labor Costs	OTH	Others

Figure 6.10: Coding Maintenance Costs

Although what to include in the lists of maintenance costs may vary from one industry to another, for other industries, each function or department have their own individual cost center,

---

[5] Mobley, R. Keith, *An Introduction to Predictive Maintenance*, Butterworth-Heineman Elsevier, 2002, Page1

especially for large organizations. Figure 6.9 represents what can be included on what is allocated on the maintenance costs. For example, I have clients in which the training costs for maintenance are under the budget of the training department or human resources,, while for other industries, the training budget is allocated per department. Hence, if a plant has several business units, then I need to prepare a Sales Invoice for each of these business units.

To visualize a clear representation of what the Overall Maintenance Costs include, it is important to provide a code for each of these costs individually such as in figure 6.10. Categorizing every cost incurred on maintenance can provide us a clear understanding of these maintenance costs individually and target our efforts to reduce what seems to consume where most of the money is spent on maintenance. For example, if the top 3 cost on maintenance includes lubrication, repair or corrective maintenance and non-moving parts, then we can have our people train on lubrication, MRO Spare parts, and deploy strategies based on the learnings from the training such as establishing an Oil Contamination Control Awareness and application of High-Efficiency Filtration System.

$$\text{Maintenance Unit cost} = \frac{\text{Total Maintenance Costs}}{\text{Number of Products Produced.}} \times 100\%$$

Ratio of Direct Maintenance Costs To Total Value-Added Costs

Sector	Percentage
Mining	**20 – 50 %**
Primary Metal	**15 – 25 %**
Manufacturing	**5 – 15 %**
Processing	**3 - 15 %**
Fabrication & Assembly	**3 - 5 %**

From John Campbell "Uptime" page 2

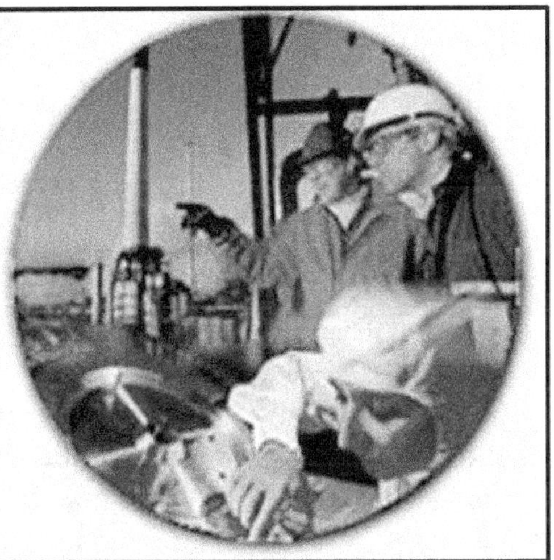

Figure 6.11: Ratio of Maintenance Costs to Total Value Added Costs

I have worked in a mining industry and this industry is very heavy on lubricants. Most of their failures and breakdowns are in a way attributed to lubrication. This means that assuming a gold mining industry produces 1 bar of gold at 20 kg in a day, which is equivalent to 705 ounces, and if the rate of gold is \$ 775 per ounce, the cost of a 20 kg gold will be at \$ 546,375.00. This means that the total cost of maintenance to produce this 20 kg of gold would be roughly around \$ 109, 275.00 to \$ 273,187.50, just for this single 20 kg of gold bar, which represents around 20 to 50 % for mining industries.

For manufacturing and industries that produce products, we can also compare the Total Maintenance Costs against the cost of products, which can range from around 3 to 50% of the total cost of maintenance. [6]Here is a table in figure 6.11, from the book of John Dixon Campbell on Uptime.

6.9: Should We Measure Wrench Time on Maintenance?

According to studies, most industries will have a wrench time of around 30 to 35%, yet this can be increased from 50 to 55%. Wrench time started its roots during 1910 when the IE (Industrial Engineer) conducted a time and motion study for the daily production operators in a plant. Although this practice is still being carried out today as industries hire IE (Industrial Engineers)to conduct this study. A smart operator upon seeing a person with a stopwatch approaching them will opt to go to a slow-motion mode because whatever the time the IE recorded for operators in producing a product will now become the standard and management are keen to question operators on why the target has not been hit.

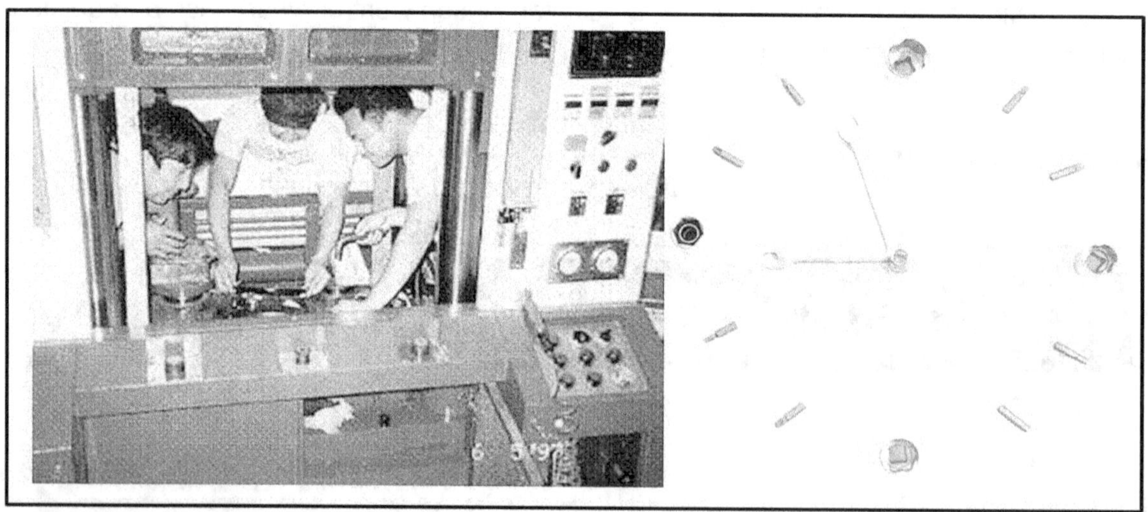

Figure 6.12: Wrench Time

Similarly, wrench time is an indicator that is used to measure how much time the maintenance people spend on doing actual maintenance work on the equipment. Others refer to this as the tool time or when maintenance is actually holding a tool which of course is not only the wrench. Although according to its definition, the wrench time will not include the travel time, work order assessment, tools organization, encoding report on the CMMS, locating spares in the system, attending meetings, task communication, working on an improvement, nor taking a break time.

In one of my training on World Class Maintenance – The 12 Disciplines, a couple of guys working in an automotive plant approached me during a coffee break and asked me if what they are doing on maintenance is right or wrong. I asked them precisely what they are doing so I can provide them my thoughts. One guy said that at the start of their shift, they need to get a form and write down all the activities they did on an hourly basis and at the end of the shift, they need to submit the form to their supervisors. I told them that supposed that I am your supervisor and

---

[6] Campbell, John Dixon, **Uptime, Strategies for Excellence in Maintenance Management,** Productivity Press, Page2

at the end of the shift, you wrote a 10 pages report stating what you have repaired every single hour, another person submitted a blank report with only his name on it with just a 1-hour activity on performing greasing on 20 rotating equipment. My question is, who performed maintenance more efficiently? Does it mean that if you repair an average of 5 machines every hour making your report very long indicates that your wrench time is better than the person who just performs a 1-hour greasing during his shift?

My thoughts may differ from other consultants but when we look up the word maintain in the dictionary, it means to sustain or to preserve something. Therefore, the term maintenance are the activities that are done before an asset or equipment will fail. These are the activities of preserving or sustaining our assets. Repair are the activities that are done after an asset or equipment failed. These are the activities to be done to restore the equipment back to Its operating state. Maintenance and repair are the opposite of both worlds. While others think that maintenance and repair are the same. They are not, since there is a boundary and a thin line that separates them, and we need to create this mentality with our people. Maintenance can be a profit center if we truly understand how to use them correctly. Just like MTTR, the question is, if maintenance becomes efficient in repairing failures, it means that we have not learned from the failure itself. Industries will never learn from their problem by just repairing the equipment as it will just cause a recurrence of the failure.

The problem with wrench time is that it will measure both Reactive and the Planned Work together. Does it mean that if what we did for the week is to repair failures all the time, our wrench time is high? The answer of course is yes, this means that if there are a lot of repairs done, the wrench time is high, which also contributes to a higher downtime and maintenance cost.

Another point I would discuss in wrench time is that doing improvements, performing root cause will not be included in the wrench time. The job of maintenance is not only to preserve but also to improve the reliability of their equipment and assets. Although other experts say that maintenance can only preserve and sustain their equipment since no amount of maintenance can really improve the reliability of the equipment. My take on this is that maintenance can be considered as 2 folds, we have the task and we have the people. If we consider the tasks, then I agree that maintenance tasks cannot improve reliability, they can only sustain it. But if we talk about the human side which is the maintenance people themselves, they can definitely improve the reliability of their assets by understanding the inherent design weaknesses of their equipment and adopting continuous improvement strategies and modifications. Although some industries would want to measure these indices, caution must be taken in measuring them, and I would strongly recommend including them for Planned Maintenance Work.

6.10: Percentage of Maintenance Cost to RAV

RAV or Replacement Asset Value is also called the Estimated Asset Value (EAV). It is the cost of maintaining the asset which is measured against the value of the asset. It can be said that as the percentage of the cost to replace the asset. As we continue to operate our equipment and assets, the value of our equipment depreciates. The main cause of depreciation is that our assets deteriorate over time. There will always be the subject of wear and tear for mechanical parts. Deterioration is given and will happen in our equipment and assets, the best that maintenance can do is to prolong its process, however, when the time comes, then the lifespan of the equipment had been reached, and the equipment will be decommissioned which finally end its lifespan. With this, the cost value of our equipment will be reduced based on the amount of time we use on our equipment and assets. The book value of equipment or asset refers to the book (from the Accounting of Finance Department) and is reflected on the industry's financial statement. This will be equal to the Total Cost of the Asset minus its liabilities. On the other hand, the Market Value will be the value or cost of the asset according to the stock market. This will be the cost of the asset when sold to the market.

If we speak about the Percentage of Maintenance cost to RAV which is expressed as a percentage, the lower its value the more effective we are in maintaining and preserving the asset. This can be used by maintenance managers and decision-makers to decide whether to still continue operating the equipment, modify it, or just retire the equipment for good and purchase a piece of new equipment. This is like owning a car for 5 years. This will be one of the deciding factors on whether we still continue using the car or just sell it at the current Market Value.

Imagine you and your wife (assuming you have just recently married and living independently) planning to purchase a car. Hence, you went to a car dealer and plan to purchase a brand new Toyota Vios 1.3 at PHP 681,000.00 or $ 13,266.00. You also heard from a colleague of yours that her friend is selling her Toyota Vios rush as she will be immigrating to another country. You and your wife went to see the lady and the car as well. The cost of the car was PHP 395,000.00. You learned that the woman just bought the car exactly 1 year ago and checked the speedometer which is at 12,050 kilometers. You interviewed the owner and asked about the expenses she incurred throughout and the owner said that she changed the oil, filter, car wash, and the consumable which is the fuel. There was no scratch on the car and it still looked brand new. You also checked the engine and seems everything is all good. So the question, in this case, would you purchase the brand new car or go on to purchase the 1-year-old car from the lady driver?

- Brand New Car   = PHP 681,000.00
- Lady Selling Car Rush   = PHP 385,000.00

- Difference   = PHP 286,000.00

- Lady Selling Rush Car Expenses
- Change Oil (just once)   = PHP   3,000.00
- Car Wash (Monthly)   = PHP   2,400.00
- Fuel (Consumable)   = PHP 11,685.00

- Maintenance Cost   = PHP 17,085.00

For industries, we need to define clearly what will be included in the Overall or Total Maintenance Cost since this can vary from one industry to another. Second, we need to be consistent in only including the cost that is spent on maintaining the asset during its period. This indicator, Percentage of Maintenance Cost to Replacement Asset Value can provide us information and decision on whether we need to maintain or replace the asset with a brand new one. The lower the value of this percentage, the better. This means that if your assets Percentage of Maintenance Cost to RAV is at 20%, then your maintenance is too expensive. Mostly the target value will be at 3%, but lowering this value means that we have lowered the cost of maintaining the asset and it would be more economical to keep the asset operating instead of selling it at a depreciated value and purchasing a new asset or equipment. If we compute for the Percentage of Maintenance Cost to RAV for the second car that we decide to purchase it will be equal to:

• Percentage Maintenance Cost to RAV (%) = (Annual Maintenance Cost ($) x 100) / RAV ($)
• Percentage Maintenance Cost to RAV (%) = (PHP 17,085.00 x 100) / PHP 385.000.00
• Percentage Maintenance Cost to RAV (%) = 4.43%

The Percentage of Maintenance Cost to RAV can also provide us information on which assets we need to replace, modify, and those that we need to maintain. If these indices will be included by the maintenance function, this should be discussed with the IT to integrate this if the plant currently has an EAM or CMMS software as this will be very tedious to do individually for all equipment and assets in the plant. Although we can have a decision on whether to continue operating or sell the equipment whose percentage will be 20% and beyond, what is important is we also need to consider the non-tangible or qualitative factors such as the following:

• Will the new equipment be easier to operate?
• Will there be less movement on the operator?
• Will the new equipment be more ergonomically convenient for operators?
• Will the training for the new equipment be easier than the old one?

Case Study: An OEM is proposing to management to purchase new equipment. The cost of the new equipment is $ 50,000.00, and the maintenance cost is projected at $ 8,000.00 annually. The cost of the current equipment is $ 65,000.00 and the average maintenance cost is $ 12,000.00. The Replacement Asset Value today of this machine is at $ 35,000.00. A TPM Planned Maintenance Team had concluded its prototype for evaluation on the current equipment and is proposing a modification cost of $ 5,000.00 which can increase the productivity by 30% and is projected to reduce their maintenance cost by 60%. It is forecasted that the demand in the next 3 months would be increased by 20%. Would it be feasible to buy the new equipment as recommended by the vendor or proceed with the modification by the Planned Maintenance?

Given:
• Cost of New Machine = $ 50,000.00
• Projected Maintenance Cost Annually for New Machine = 12,000.00
• Cost of Old Machine = $ 65,000.00
• Projected Maintenance Cost Annually for Old Machine = 12,000.00
• Projected Maintenance Cost Annually for Old Machine with Modification = $ 12,000 - (12,000.00 x 0.6)
• Projected Maintenance Cost Annually for Old Machine with Modification = $ 4,800.00
• RAV for the Old Machine = $ 35,000.00
• Modification Cost = $ 5,000.00

For the New Machine without Modification
- Percentage Maintenance Cost to RAV for New Machine = (Annual Maintenance Cost ($) x 100) / RAV
- Percentage Maintenance Cost to RAV for New Machine = $ (12,000.00 x 100) / $ 50.000.00
- Percentage Maintenance Cost to RAV for New Machine = 24 %

For the Old Machine without Modification
- Percentage Maintenance Cost to RAV for Old Machine = (Annual Maintenance Cost ($) x 100) / RAV ($)
- Percentage Maintenance Cost to RAV for Old Machine = $ (12,000.00 x 100) / $ 35.000.00
- Percentage Maintenance Cost to RAV for Old Machine = 34 %

For the Old Machine with Modification
- Total Maintenance Cost = [12,000 – (12,000 x 0.60)] = 4,800.00
- Percentage Maintenance Cost to RAV (w/ Modification) = (Annual Maintenance Cost ($) x 100) / RAV
- Percentage Maintenance Cost to RAV for Old Machine = $ (4,800.00 x 100) / $ (35.000.00 – 5000)
- Percentage Maintenance Cost to RAV for Old Machine = 16 %

The conclusion, in this case, is to proceed with the Planned Maintenance Modification as this indicates the lowest value regarding the Percentage of Maintenance to RAV and disregard the proposal of the vendor to purchase the new equipment.

6.11: KPIs for Predictive Maintenance

One of the differences between Preventive and Predictive Maintenance is the scheduled time on performing the tasks. Preventive Maintenance will base the tasks on a prescribed schedule, while Predictive Maintenance tasks are done once the Potential Failure is nearing its Functional Failure. In this case, corrective maintenance will be performed before it reached functional the failure. This means that if there are 10 potential failures detected and 10 corrective maintenance that had been initiated, then the Predictive Maintenance Percentage Rate is at 100%. This can be measured weekly or monthly.

$$\bullet \ \text{Predictive Maintenance Percentage Rate} = \frac{\text{Number of Potential Failures Detected}}{\text{Number of Corrective Maintenance}} \times 100\%$$

Although both cost savings and cost reduction will reduce the company's cost, there is a slight difference between the two. Cost avoidance focuses on taking actions that avoid incurring costs. For industries, these are measures to reduce their expenses. Any actions to avoid an increase in expenses are considered cost avoidance. Cost savings is the reduction from last year's spending on the same item. They will appear on the company's budget and in financial statements as a decrease in spending. This will be the distinction between cost savings and cost avoidance. An example is when C-Level people traveling internationally will no longer be traveling in business class but in economy class. If the business class ticket is at $ 1,000.00 and economy class is $ 500.00. Subtracting this, we have a cost savings of $ 500.00, every time a c-level executive travels. Or even more, if business meetings will be prohibited due to the new Omicron Covid Virus, and all international meetings will be held via Zoom or Ms Teams, then we have a 100% savings of 1000 USD depending on the cost of the plane tickets. Although there is a yearly maintenance cost, it may or may not be allocated precisely, hence, if this is the case, then we shall call this Cost Avoidance. Figure 6.17 is an actual Cost Avoidance Report our Predictive Maintenance generated as part of our TPM Planned Maintenance.

STATION	SMT1	ANNUAL TOTAL COST AVOIDANCE THROUGH THE USE OF THERMOGRAPHY								
DATE	EQPT ID	REPAIR DURATION		PRODUCTION DOWNTIME				Parts Cost in USD	Downtime Cost in USD	Cost Avoidance in USD
		MAN-HRS	$/MAN-HR	UPH	UTL	ASP	DURATION			
04/07/00	SOPM008	16 / 2	1.20	23,256	70%	0.0012	48 / 2	105.00 / 2.00	1,061.88 / 43.47	$1,018.41
04/07/00	SOPM002	16 / 2	1.20	12,912	70%	0.0026	48 / 2	105.00 / 2.00	1,252.19 / 51.40	$1,200.79
02/10/00	SOPM003	16 / 2	1.20	12,920	70%	0.0021	48 / 2	105.00 / 2.00	1,035.84 / 42.38	$993.45
07/08/00	SOPM007	16 / 2	1.20	12,920	70%	0.0021	48 / 2	105.00 / 2.00	1,035.84 / 42.38	$993.45
03/06/00	SOTE005	16 / 2	1.20	23,400	70%	0.0017	48 / 2	105.00 / 2.00	1,460.81 / 60.09	$1,400.72
03/13/00	SOTE007	16 / 2	1.20	23,400	70%	0.0017	48 / 2	105.00 / 2.00	1,460.81 / 60.09	$1,400.72
07/06/00	SOPM005	16 / 2	1.20	12,920	70%	0.0021	48 / 2	105.00 / 2.00	1,035.84 / 42.38	$993.45
05/14/00	SOPM010	16 / 2	1.20	32,559	70%	0.0012	48 / 2	105.00 / 2.00	1,436.98 / 59.10	$1,377.88
05/12/00	SOTE002	16 / 2	1.20	23,400	70%	0.0017	48 / 2	105.00 / 2.00	1,460.81 / 60.09	$1,400.72
03/11/00	SOTE006	16 / 2	1.20	23,400	70%	0.0017	48 / 2	105.00 / 2.00	1,460.81 / 60.09	$1,400.72
09/28/00	SOFU015	24 / 2	1.20	12,817	70%	0.0079	50 / 2	105.00 / 2.00	3,677.70 / 146.16	$3,531.54
Total Annual Cost Avoidance Through Thermography										$15,711.84

Figure 6.13: Actual Report on Annual Cost Avoidance Through Thermography

One of the best KPIs for Predictive Maintenance is to determine its Return on Investment (ROI) since this requires investment for industries in terms of the instrument and the user. ROI will directly measure the amount of return for a given investment, relative to the investment's cost. In calculating the Return on Investment (ROI), the benefit or return gained from an investment is divided by the cost of the investment. The result is either expressed as a percentage or a ratio.

ROI= Current Value of Investment − Cost of Investment / Cost of Investment

Chapter 7

MRO Spare Parts and Storeroom Indicators

> *Like any other continuous improvement effort, improving the storeroom is a continuous process. It should not only be done inside the storeroom. Improving the storeroom should be done both inside and outside. It means that if maintenance is doing its job of preserving the equipment, then there will be lesser failures which means lesser parts needed from the storeroom.*

7.1: Who Manage Your MRO Storeroom?

Before we discuss the different measurements in the storeroom, some questions need to be raised. First, who manages your industry's MRO storeroom? Although I might be a bit biased that the maintenance function should be in the best position to manage the storeroom because they are the users and they know the parts better than any other functions in the organization. Yet in other industries, perhaps it is the warehouse, supply chain, purchasing, administration, or even the finance people that manage the MRO Storeroom. The question is how can we manage the storeroom if other functions are controlling it. Well, the only answer I got is that whoever is managing or controlling the storeroom should be in constant and direct communication and collaboration with the maintenance or engineering function of the plant, since these people are the users of the storeroom. There are many problems the industry face in their storeroom which I wrote in my fifth book on Problems and Solutions on MRO Spare Parts and Storeroom, and 99.99% of them are man-made problem. Therefore, if man created the problem, then man should find ways to solve their problems.

First and foremost, the storeroom people should be included in the communication loop within the maintenance function. In the majority of cases, the problem will start with a lack of communication. Management together with operations and maintenance meet to decide on retiring some equipment since the product will be phased out in the next 3 months. CMMS has been installed in their storeroom and some EOQ calculation for reordering fast-moving parts are in place. The machine is finally decommissioned. After a couple of months, the Boss gets mad because of the high cost of inventory and when asked about the cause of the problem, they simply forgot to include the people from the storeroom and so the lists of obsolete parts grow. In a reactive industry, maintenance will often experience that those parts that are not needed are available in plenty but those parts they needed are not available when maintenance needs them most. And by definition, a Spare Part is defined as a part of a machine ready to replace an identical part if it becomes faulty. It is also defined as those parts of the machine which are kept on standby to be substituted when a part of equipment fails, a repair is required, or the part simply becomes worn-out and needs to be replaced during Preventive Maintenance. Hence, a good Spare Parts Management system ensures the right parts get to the right place at the right time. When a part is not around when a piece of equipment is in a failed state, downtime increases, operations are pissed off while maintenance lost their trust in their storeroom.

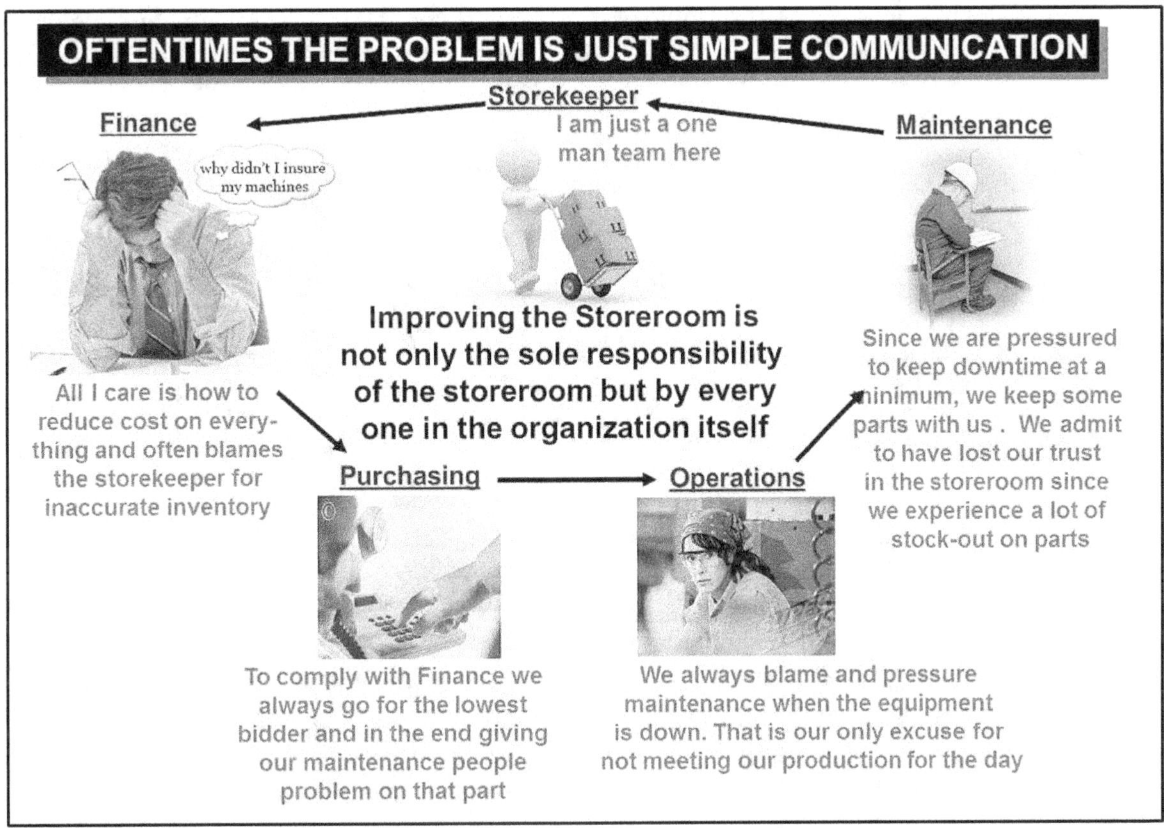

Figure 7.1: Oftentimes the Problem is Simple Communication

One problem in the storeroom can lead to a domino of other problems and as the problem multiplies, then we now find it even more difficult to control and manage the storeroom and the problem becomes worst over time. So the question is, if maintenance or engineering do not manage the storeroom, how can we control it. The only logical answer I can provide is that whoever manages the storeroom should allow the storeroom people to report indirectly to the maintenance function. This is very important so that maintenance can help in controlling the storeroom. Does maintenance squirrel parts in your plant? If you ask, what is squirreling in spare parts. Squirrelling is the act of keeping spare parts to themselves. A squirrel store is often said as maintenance secret hiding place for spare parts and they are scattered mostly in the plant, and maintenance are the only ones who know where they are kept. Charlie might be keeping bearings in his locker, Peter might be keeping belts somewhere, and so on. This has become a common practice in industries and the reason is quite simple. Why? Because maintenance has lost their trust in the storeroom, and whenever a part is unavailable inside the storeroom, maintenance gets the blame even if maintenance did not cause the equipment to fail.

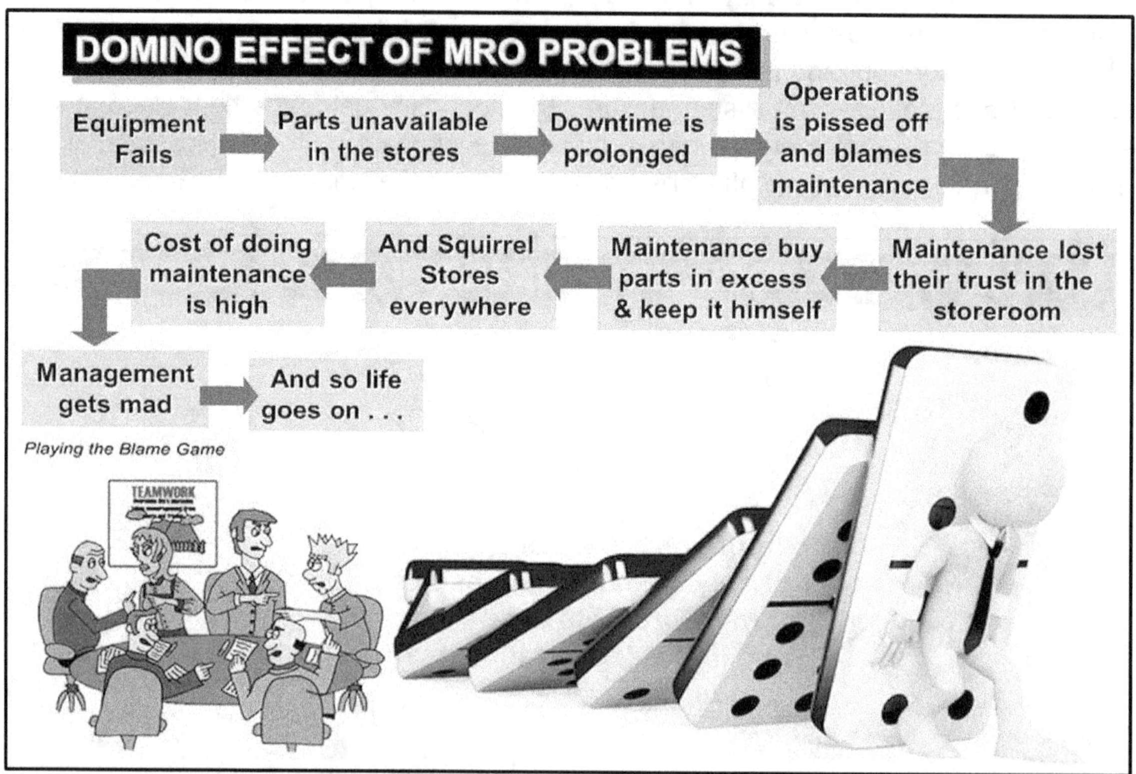

Figure 7.2: Domino Effect of MRO Problems

Maintenance people are the real customers of the MRO storeroom; therefore, they should be given a chance to manage the storeroom. MRO Spare Parts and storeroom are part of the overall Maintenance Management Strategy. The largest and the most common de-motivator for maintenance is waiting for the parts if they are unavailable and worse if the parts are not stored in the storeroom when they needed them most even if the system indicates that the part is available. Usually, if a part is not around when maintenance needs them most, expect maintenance to make a McGuyver move, which is not the appropriate thing to do. They do this because there is always pressure from operations people to keep the equipment running at all costs even if everyone knows that the part is not available or in stock in the storeroom. In short, maintenance people will resort to cannibalism to get the parts from idle equipment just to make the equipment run, which has been part of the maintenance culture since the beginning of time. Maintenance people must finally convince top management and C-level people that they are the best people to manage and control the storeroom. Real savings can be generated from the MRO spare parts if the right people are the ones managing the storeroom.

MRO Spare parts management is one of the strategies needed to sustain the equipment since parts do not last forever. When the plant finally decides to embark on the journey of sustaining the reliability of the equipment, then the storeroom should be part of the strategy, which means that the parts should be well managed, maintained, and controlled. If the lifespan of a spare part has been lengthened, then the quantity to be ordered decreases, and the interval of placing the call to make an order will be lengthened, which is good. When Predictive Maintenance is in place, then some parts can only be reordered once a potential failure can be detected on the equipment, since they have some knowledge that the delivery time of the part is shorter than the failure development period. Obsolete parts are finally disposed of and decided on what is the best course of action to make. ABC analysis for those non-moving parts is analyzed. Their physical and systems inventory is high at 98% or more all the time. MRO Spare Parts are now properly cataloged for easy reference. Parts with a different part number are consolidated and reflected in the system.

The MRO storeroom is now operated 24 hours because they know that operation also runs for 24 hours. Spare parts inventory is now controlled. This time maintenance is no longer squirreling because they know that parts they need are now available at the storeroom. The goal and direction of maintenance and the storeroom are consolidated and headed forth in the same direction. Remember that maintenance people are the humble users of the MRO storeroom. They know what parts work best with their equipment and not. They know which parts will keep on failing as well as the lifecycle of parts that they replaced frequently. MRO Spare Parts Management can only be managed if the right people are the ones managing the storeroom.

The role and responsibility go beyond just issuing and receiving parts. People in charge of the Storeroom should also have some background on how to maintain the spares. For example, For electronic parts with semiconductor parts, the storekeeper must have an understanding of two important factors that can damage electronic parts, ESD, and moisture. ESD stands for Electro Static Discharge. This means that our human body contains static electricity. We cannot touch electronic parts directly unless we are properly grounded with a wire strap and protective clothing, as these ESD parts are sensitive to ESD damage. Surface Mounted Devices (SMDs) will absorb moisture, especially during solder reflow operations. The rapid increase in temperature can cause moisture to expand and delaminate the internal package, which can result in an early or premature circuit board failure. All PCBs (Printed Wiring Boards) stored inside the storeroom should be stored in a moisture barrier bag (MBB) that is completely vacuum-sealed. In addition to the bags, desiccant bags and humidity indicator cards must be used for proper moisture protection. Motors that have been stocked for more than 6 months or more should be checked if they needed to be regreased. Rubber belts should be stored at room temperature as the rubber can lose its resiliency or elasticity when stored in a cold environment and more. These are just a few of the things that storekeepers should understand since if the parts become defective, they might contribute to infant mortality failures when used on the equipment.

Another important role of the storekeeper is to perform some analysis such as FSNO (Fast, Slow, Non-Moving and Obsolete part), ABC, VED, or HML. It is difficult to manage something you cannot control, and you cannot control something you cannot manage. MRO storeroom and spare parts management are some of the major strategies that can help industries reduce their maintenance cost big time if the right people manage them. I believe that this is just one of the missing links in any reliability and maintenance strategy. For the record, the best people in a position to control and reduce the cost of maintenance are the users of the parts themselves, and I think that is all I have to say about that.

7.2: Common Measurements and Indices for the Storeroom

The total parts and items inside the storeroom can range from around 2000 to 200,000 parts and, in some cases, even more. The bigger the industry, the more parts are stocked inside the storeroom. As we have said, the goal of MRO spare parts and storeroom is to create a balance on minimizing the cost of spares inventory as well as providing all the required and necessary materials needed to keep the plant operating, which is quite contradicting. Practically we can achieve one of the goals of the storeroom by keeping every single part of the equipment inside the storeroom, that is if your industry still wants to remain in business. If we do this, then we sacrifice the other goal of minimizing the cost of inventory, yet still, we can provide the required parts the user needs. To do this, the MRO storeroom should have several important indices, measurements, and KPIs that they need to monitor to keep on track for them to know whether they are serving their purpose well or not. Like maintenance, there is no single KPI that will tell us the whole story in the storeroom; we need to at least have a minimum of five KPIs for the storeroom for a start.

```
• Total Inventory Last Week          = 145,895
• Items Withdrawn                    = (-1,436)
• Items Received                     = 856
=================================================
• Total Inventory Status             = 145,315 items
```

In figure 7.3, the total line inventory is decreasing, which satisfies the goal of reducing the inventory in the storeroom. Still, maintenance has an appalling encounter on lots of stock-outs experienced with their MRO storeroom. In this case, the goal of the storeroom was achieved, but creating excessive downtime from operation had not been satisfied. There are several KPIs and indices for the MRO storeroom to choose to determine the success or failure of the MRO Spare Parts and Storeroom Management effort that can be tracked and monitored by the storekeeper. These KPIs and measurements are one of the methods at the heart of challenging an organization towards continuous improvement. This generally takes the form of performance indicators, KPIs, measurements designed to focus the attention on various areas of performance that deserve our utmost attention. Some of the common and important measurements for the storeroom are available if CMMS or EMS is in place in the plant, and the correct data is populated in the system.

The only concern I have mentioned is if the storekeeper does not report directly or indirectly under the control of the maintenance function. Their goals and KPIs might not be exactly aligned with the maintenance function. For example, the number of stock-out is a very important KPI for the storeroom function as it indicates the number of times that the part is not available and had reached a zero level that prolongs the downtime of the equipment. Perhaps if the storeroom reports to purchasing or finance, their goals may have something to do about reducing the overall inventory of the storeroom. If the storekeeper reports directly to the warehouse, then the warehouse may be interested in maximizing the space inside the storeroom and may be measuring the storeroom's space utilization, which is equal to the actual space consumed by the storage and spares divided by the total space of the storeroom. What is important that makes sense is that the KPI of the storeroom must be aligned with maintenance. This is where the real value of the storeroom is. The revenue of the plant is dependent on the outcome of the equipment to produce the products or services the industry provides to their clients, and I think that is the bottom-line of it all.

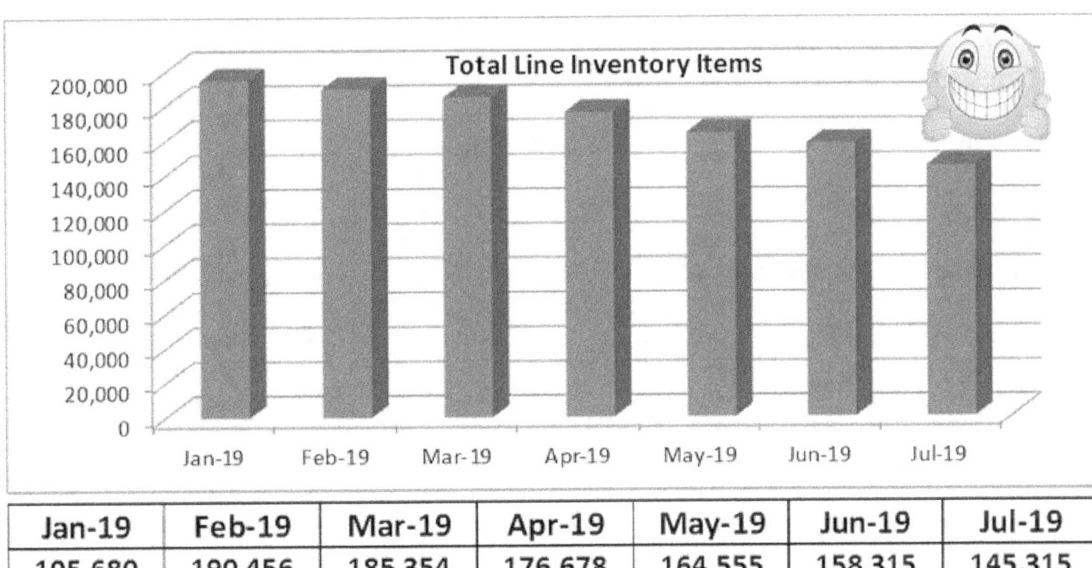

Jan-19	Feb-19	Mar-19	Apr-19	May-19	Jun-19	Jul-19
195,680	190,456	185,354	176,678	164,555	158,315	145,315

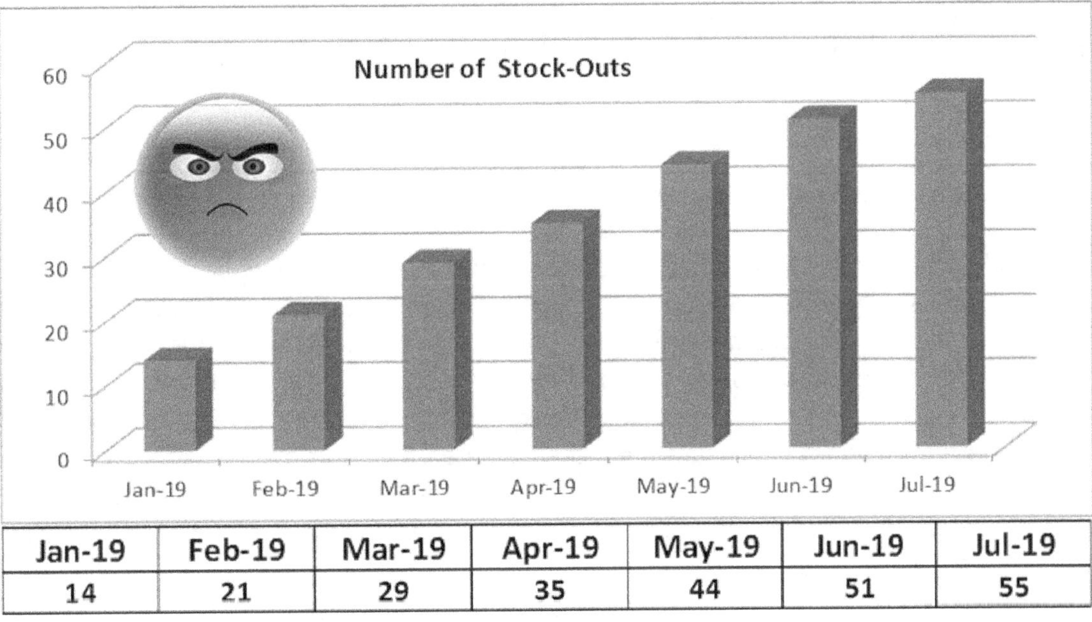

Jan-19	Feb-19	Mar-19	Apr-19	May-19	Jun-19	Jul-19
14	21	29	35	44	51	55

Figure 7.3: Sample MRO KPI on Line Inventory versus Stock-outs

The KPIs for the MRO Storeroom should not be in conflict with other departments, especially for both operations and maintenance. Since most inventory management, goals will be to reduce the cost of inventory and line items inside the storeroom. Still, on the other hand, this can lead to a large amount of downtime and stock-outs, which results in the unavailability of the parts when it is needed by both operations and maintenance. So the question is, is it really possible to satisfy both requirements? The answer is definitely yes but this can only be made possible if both the storeroom people and the maintenance function are both cooperating and collaborating frequently. These two functions cannot exist without each other.

STOREROOM MEASUREMENTS		
Recommended Measurements For Your Storeroom		
Measurements	**Frequency**	**Trend**
Total Line Items and Cost of Inventory	Weekly	The lower the better
Number of Stock Outs	Monthly	The lower the better
Inventory Accuracy in Percentage	Twice/Year	The higher the better
List of Obsolete Parts and Cost	Weekly	The lower the better
List of Non-Moving Parts and Cost	Weekly	The lower the better
List of Fast-Moving Parts and Cost	Weekly	The lower the better
Lead Time for Critical Parts	Weekly	The lower the better
Average Waiting Time for Spares	Weekly	The lower the better
Carrying and Holding Cost of Spares	Yearly	Should be equal
Costs of Emergency Buying	Yearly	The lower the better

Supply Chain   Warehouse   Logistics   Finance   Purchasing   Maintenance

Figure 7.4: MRO Spare Parts most Common Indices and KPIs

On the other side of the building, the storeroom and Purchasing Department are happy and celebrating because they have reduced the cost of their spare parts inventory by localizing some critical parts and changing vendors by going to the lowest bidder with a lower quality, which in the end create more problems than they realized for the users. This situation posed a greater risk due to the quality of parts being delivered to the storeroom. The bottom line is that whatever KPIs and measurements are monitored by the storeroom people, it should not just be measured to satisfy the storeroom. The important thing is, it should not pose problems with the other functions of the organization, especially maintenance. These KPIs must always be aligned to the company goals at all times.

As the saying goes, we can only measure what we can control. The success or failure of the things that we control can only be known by these KPIs and measurements that we measure. One of the critical factors for a successful MRO Spare Parts Management is achieving a high level of inventory accuracy. An accurate inventory is defined as the correct parts and quantity physically in the physical location inside the storeroom and in the system. A loss of confidence from the maintenance on the storeroom people can lead maintenance to stock the parts on their own if these indices are not satisfied, which will be the beginning of the problem. Other books indicate that the inventory accuracy should be at 95 %, in my book Problems and Solutions on MRO Storeroom and Spare Parts, I indicate that the inventory accuracy should be 98 % and beyond at all times. What I do believe is that inventory accuracy is the most important KPI for the storeroom people.

Total Inventory costs: This is the overall cost of parts and items inside the storeroom. The direction should be to lower the cost of the inventory while still providing the parts needed by the users. This can be made possible if we can control and manage our spare parts more intelligently. Disposal of obsolete parts inside the storeroom is one of the priorities that can help us reduce the overall inventory of the storeroom. Having an MRO Decision Diagram or Algorithm which I provided in my 5th book on Problems and Solutions of MRO Spare Parts and Storeroom will allow us to understand and determine what parts need to be stock and not.

7.3: Overall Lists of Obsolete Parts and Costs

Frequently, there are many parts held in the storeroom that do not belong to any equipment in the plant since the equipment had either been removed, decommissioned, or retired. The equipment in the plant may be retired and no longer on site. The problem, in this case, is that since the parts of the equipment are still inside the storeroom, there is a cost of holding the part in the storeroom that is called the carrying or holding cost. If the cost of an obsolete motor in the storeroom is 100,000.00 USD, the cost of holding this part can range from 10 to 30 percent of the cost of the motor per year, which can range from 10,000 to 30,000 USD per year. In this case, we are only speaking here about one item in the storeroom. Just imagine if you have 100,000 to 200,000 items inside the storeroom, everything will have a holding cost just to place the item in the storeroom. Remember that storing obsolete parts in the storeroom costs money and space. A plant can utilize the space consumed by these obsolete parts. The worst case is when the storekeeper doesn't know that the equipment was already retired and the possibility of still reordering fast-moving parts for this equipment.

When the equipment is retired or removed from operations, maintenance tends to forget that the parts needed for repairing the equipment also become obsolete. The important thing is for maintenance to communicate this to the storeroom people so that all parts that are affected can be identified. Remember that these parts will no longer be used unless there are still the same equipment types operating in the plant. If a plant has an automated CMMS or EAM generating system for spare parts, the fast-moving parts will be reordered continuously if they will not be removed from the system. This is like having a patient die in the hospital, yet the bill keeps on adding up. The worst part is that some hospitals do not release the bodies of the dead to the family if the entire bill has not been fully paid especially in third world countries. I am beginning to think about whether hospitals are here to save lives or just prolong the agony of the family affected. Instead of spending the time to heal, additional problems come in. Same with industries, things get a bit more complicated because there are just too many red tapes, bureaucracy, politics, signatories, or even corruption needed to remove a specific obsolete item from the list or perhaps a procedure on what to do after the equipment is decommissioned does not exist at all. The thing is, if the machine is no longer around, the best thing to do is to identify all the parts that come with it. Not only do we need to identify the parts in the system, but we also need to locate all the parts inside the storeroom that belong to this equipment, isolate them, and finally remove them from the storeroom, as they will no longer be used permanently.

Another reason why the part becomes obsolete is that the shelf life of the part has already been reached. There are also cases where an engineer modified or redesign a part that has an inherent design weakness, and this part is stocked in the storeroom. This means that if a part had been modified, the original part inside the storeroom automatically becomes obsolete. Therefore, for parts subject to modification or redesign, this must be thoroughly coordinated with the storekeepers and purchasing to control its ordering especially if this is a fast-moving item. A decision must also be made if these original parts will be consumed before using the modified part. If the decision is to immediately use these modified parts then the original parts held inside the storeroom will no longer be used and will become obsolete. When the equipment is retired, the spares associated with it must be identified so that it can be freed up from the storeroom. This will free up space, which can be used to store other items. Remember that there is a cost of storing or holding obsolete parts inside the storeroom. The problem is that not all storekeepers provide a report on this to management, and maintenance might be thinking that there is actually no cost of holding these parts inside the storeroom. For the record, nothing is free anymore on this planet.

[7]The very first thing that MRO Storekeepers should do is to create an FSNO Analysis on all their spares held inside the storeroom. F for fast-moving which is used monthly, S for slow-moving which is used 5 times or more in a year, N for non-moving which can start with those items that have not moved in a year, two years, and so on. The last will be O for Obsolete parts. Figure 7.5 is a sample of an FSNO Analysis.

There are three main reasons why a part becomes obsolete. First, if the equipment is retired, all the parts that are stored in the storeroom should be identified, that is if the equipment has no similar type of equipment operating in the plant. If the equipment is retired, and we still have similar types of equipment operating, the spares are not yet classified as obsolete. Maintenance must provide the storeroom with a list of equipment that has already been decommissioned so that the parts held on stock can be identified if they can still be used on other equipment or permanently disposed of. Another reason why spare parts become obsolete is that someone modified an existing part due to some weaknesses in the design or perhaps it has a short lifespan. This means, if the modified design will now be used on the equipment, then the original part becomes obsolete unless the original existing parts are consumed first to zero out the stock. If this will not be the case and the original parts will not be used, then the parts become obsolete. The third reason why spare parts become obsolete is that some spares have an expiration date, as in the case of consumables such as chemicals, lubricants, grease whatnot. According to leading belt manufacturers, belts have shelf lives of two years. This means that if spare parts are not maintained well and cared for when placed in the storage, expect the life of these parts to be shortened.

There is also the possibility of human error that the part was purchased by mistake, and the vendor had a no return policy. In this case, the company shoulders the burden for the mistake. The worst thing that can ever happen to obsolete parts is that if these parts are considered as

---

[7] Angeles, Rolly, **Problems and Solutions on MRO Spare Parts and Storeroom**, Centralbooks , Pages 238 to 239

fast-moving items and Purchasing might still be replenishing these parts since they do not know that the equipment was already retired, especially if the replenishment is fully automated and have not been adjusted. Unlike non-moving parts, there are many options on what to do with these obsolete parts. Listed here are just some of them.

Mining Industry Spare Parts 2020

Item Description	Part Number	Location	FSNO Analysis	Consumed last year	UOM	Unit Cost in PHP	Total Costs
Roller Bearing	03-54392405	2FB26	F	15	PC	950.00	14,250.00
Roller Bearing	03-54392604	2FB27	F	20	PC	1,500.00	30,000.00
O-Ring	29-54059687	2FB28	F	40	PC	460.00	18,400.00
Hyrdraulic Oil Tellus 68	26-48349032	4FTEMP	F	15,283	Liters	57.14	873,270.62
Transmission Oil Rimula X30	26-48349650	4FTEMP	F	20,872	Liters	52.19	1,089,309.68
Engine Oil Shell 15W-40	26-48349651	4FTEMP	F	2,317	Liters	52.84	122,422.88
Drilling Oil, Torcula 150	26-48349652	4FTEMP	F	6,916	Liters	49.14	339,852.24
Multi-Purpose Grease	22-48349654	4FTEMP	F	150	Pale	4,600.00	690,000.00
Face Dusk Masks	33-49568798	2FB29	F	243	PC	756.00	183,708.00
Mechanical Seal 4506	27-56768799	2FB30	F	26	PC	2,457.00	63,882.00
Mechanical Seal 4507	27-56768800	2FB31	F	54	PC	2,950.00	159,300.00
Mechanical Seal 4508	27-56768801	2FB32	F	156	PC	3,500.00	546,000.00
Tires Cat 777g	29-54059898	5FTEMP	F	54	PC	112,200.00	6,058,800.00
Tires Cat 740g	29-54059867	5FTEMP	F	45	PC	98,000.00	4,410,000.00
Tires Cat 750g	29-54059868	5FTEMP	F	40	PC	85,000.00	3,400,000.00
Ford Pick-up Tire Standard	29-45030453	5FTEMP	F	85	PC	6,600.00	561,000.00
Safety Apparel	34-48349653	2FB30	F	500	Set	2,400.00	1,200,000.00
Safety Shoes	33-48349645	2FB31	F	500	Pair	1,250.00	625,000.00
Working Gloves	33-48349653	2FB30	F	500	PC	750.00	375,000.00
Overall Costs for Fast Moving Items				**47,816**		**323,584.31**	**20,760,195.42**
Motor, Hydraulic	53-38101278	1FB14	N	1	PC	15,701.30	15,701.30
Motor, Hydrostatic	53-33881759	1FB9	N	3	PC	188,687.40	566,052.20
Motor, Wagner	53-40125638	1FB10	N	1	PC	139,428.87	139,428.87
Pump, Hydrostatic	38-43092582	1FB9	N	1	PC	198,554.67	198,554.67
Pump, Wagner	38-35567418	2FTEMP	N	2	PC	149,264.51	298,529.02
Pump Charge Asssembly	38-10723745	2FB27	N	2	PC	26,674.04	53,348.08
Gear Pump	29-45789849	2FB28	N	1	UNT	35,514.48	35,514.48
Pump Hydraulic Rextroth	38-45623745	2FTEMP	N	1	PC	86,522.01	86,522.01
Pump Torque Converter	38-34623746	2FTEMP	N	2	PC	50,805.74	101,611.48
Pump, Fuel	38-22395505	2FB16	N	1	PC	5,042.71	5,042.71
Pump, Steering	38-64118134	2FTEMP	N	1	PC	36,039.00	36,039.00
Pump, Charging	38-44696665	2FB27	N	3	PC	52,787.88	158,363.64
Pump Assembly	38-55365965	3FTEMP	N	2	PC	81,593.07	163,186.14
Overall Costs for Non Moving Items				**21**		**1,066,615.68**	**1,857,903.60**
Pump Hydraulic	38-66555691	1FB17	O	2	PC	156,287.78	312,575.56
Pump Assembly	38-10775546	2FTEMP	O	2	PC	40,229.96	80,459.92
Overall Costs for Obsolete Items				**4**		**196,517.74**	**393,035.48**
Pump, Hydraulic	38-43083169	2FB26	S	6	PC	31,672.70	190,036.20
Pump, Supply	38-01176335	2FTEMP	S	5	PC	4,295.45	21,477.25
Pump Assembly	38-55365962	2FTEMP	S	6	PC	76,510.90	459,065.40
Overall Costs for Slow Moving Items				**17**		**112,479.05**	**670,578.85**
Overall Spare Parts Costs				**95,677.72**		**2,935,737.76**	**45,534,523.60**

Figure 7.5: Overall Cost derived from FSNO Analysis

• List the obsolete items accordingly and place them in the newspaper for bidding

• Ask Purchasing to find a vendor where these obsolete parts can be sold as scrap

• Some of these obsolete items can be used for training purposes for the operators and maintenance. For example, have the cross-section of a motor to be cut and use it as a teaching aid for operators.

- If the plant has other business units, they can sell these obsolete parts to them.
- Try to seek the help of Purchasing to locate similar industries that are still using the equipment you have retired, and sell it to them directly at a lower cost.
- Sell them back to the OEM at a depreciated cost.
- Join online maintenance forums or bulletin boards and post about the equipment you retired and spare parts that you will no longer be using. Once you found someone, get his contact details, and let your Purchasing handle this case.
- Trade them back for something that you can use.
- Nowadays, the internet is the greatest source of information. During my time, we got the Britannica Encyclopedia. You just need to type the correct keyword on what you are looking for, such as vendors or traders who buy this sort of stuff. Some vendors buy scrap and obsolete items.
- Sell them at e-bay, sulit.com, Amazon, and other online platforms.
- You can post the items for sale or auction on maintenance forums or LinkedIn.
- Seek vendors in the same category and sell them to them at a depreciated cost.
- If none of these items work, then they can be totally scrapped or just sell them in a junk shop.

Remember that obsolete parts now become garbage to your industry. Still, there is a saying that your garbage can be someone else treasure. It is also important that at the very beginning of automating the tasks for the MRO Storeroom, the software should include what equipment do the spare parts belong to. If a part belongs to several machines, and one of the equipment is retired, the parts that belong to this equipment are still active and not yet considered obsolete. The spare part only becomes obsolete if all the equipment that utilized this part is finally retired from the industry. For obsolete parts, all I can say is remember the theme song from Frozen, which is Let it Go, and not the Beatles song which is Let it Be.

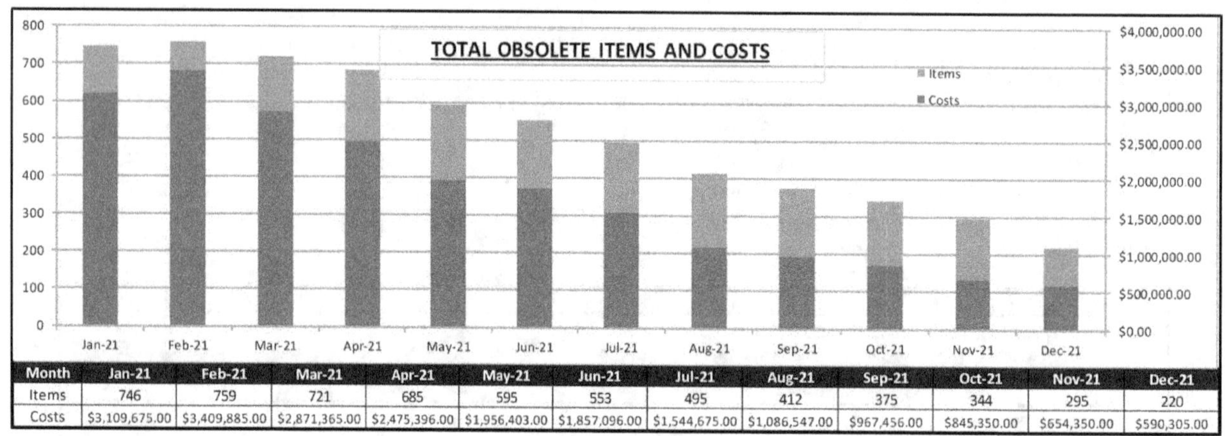

Month	Jan-21	Feb-21	Mar-21	Apr-21	May-21	Jun-21	Jul-21	Aug-21	Sep-21	Oct-21	Nov-21	Dec-21
Items	746	759	721	685	595	553	495	412	375	344	295	220
Costs	$3,109,675.00	$3,409,885.00	$2,871,365.00	$2,475,396.00	$1,956,403.00	$1,857,096.00	$1,544,675.00	$1,086,547.00	$967,456.00	$845,350.00	$654,350.00	$590,305.00

Figure 7.6: Tracking for Obsolete Parts Inside MRO Storeroom

7.4: Overall Lists of Non-Moving Parts and Costs

Non-moving parts will be more challenging than obsolete parts and these parts pose a real big problem inside the storeroom as almost all of these parts are big in size as well as big in cost. If we explain non-moving parts in the simplest way I can. You think you might need them but in reality, you don't want to use them. Just like a spare tire in your car, we keep them but we do not want to use them in reality. In fact, one of the main reasons why they become non-moving is because the OEM recommended keeping a spare of this part due to its criticality and lead-time. They told us ten years ago to keep this part, and we agree to their demands. Now the part is still in the storeroom and remained untouched. Although there are only a limited number of options on what we can do with non-moving parts. One of them is to sell them back to the OEM.

Non-moving parts are also termed as rarely used inventory. These items are difficult to obtain, or their lead-time to acquire usually takes several months, others even years because most of these parts are unavailable or imported from other countries. The difference between non-moving and obsolete parts is that the machine or asset is still used and operating in the plant. For obsolete parts, the equipment or asset has already been retired or decommissioned. An item or part becomes non-moving because the OEM recommended most of them to be stocked inside the storeroom. They claimed that the part is critical. Today, they sit idle inside the storeroom shelves for years and remain untouched. All parts inside the storeroom have a cost for storing them. They are called carrying costs or holding costs. Most maintenance thinks that there is no cost to hold or keep something in the storeroom. They do not want to get rid of the part by hanging on to them on the assumption that these parts might be needed some day; besides, their thinking is that it does not cost anything to keep them. This is where they got it all wrong because all parts inside the storeroom have a cost on holding them.

An analysis of 100 MRO stores tells us that 60% or more of the total cost of inventory has no usage for the past couple of years. Yet, most of these items should be on hand when needed because the impact and consequences of the failure are merely unacceptable. It is recommended that the industry should have a list of what items have not been used or withdrawn for one year, two years, three years, and so on. It is best to do this with a team, involving the maintenance, storekeepers, Purchasing, Finance, and Top Management with the Decision-Making process on what to do with these non–moving parts. Start with the oldest spare part. Know the consequences of the failure if it will occur. Will the equipment be running for the next 3 to 5 more years? I would also recommend that spare parts that have not moved for a year or more must be inspected and thoroughly checked for any traces of deterioration, rust, or corrosion as they have remained idle for a long time. The most logical and practical way is to use them, starting with the oldest spare part. The original part that had been removed is still functional, so we preserve it or even refurbish it, and this now becomes our spare. This way, the part had been withdrawn from the storeroom and will definitely lower your inventory cost.

My point is, why not start using these non-moving parts, starting with the old one. You might be having second thoughts regarding the consequences since the lead-time for these items usually take six months, and others may take years. My point is we will be replacing a part that is still working, and we still have a spare, which was the original part that had been replaced in the equipment. For example, I have an induction motor with specification Siemens 1LE1 Reversible Induction AC Motor, 5.5 kW, IE2, 3 Phase, 2 Pole, 400 V, 690 V, foot mounting horizontal motor in the storeroom. The cost of this motor is $ 2,500.00, and it had not moved for five consecutive years inside the storeroom. Your current setup of the cooling system is that it has redundancy or backup in place; hence why not replace the backup motor with the one in the storeroom. The old motor will be inspected, cleaned, and refurbished when necessary. We now keep the old motor in a safe place or return it back to the storeroom for safekeeping. Note that there will be no more holding or carrying cost on this motor since this is not considered as a spare part since it had already been withdrawn a long time ago. The motor now becomes our spare. Although I contradict myself in my writings on my other books that we need to replace parts that are actually on the verge of failing and not on parts that are still working. This will be an exception to the case since our goal is to reduce the Non-Moving parts of the storeroom. My point, in this case, is that if the part that will be replaced by the non-moving part can be preserved or refurbished by maintenance, then we can use them as a spare. I believe that this is the only option we got for non-moving parts, and that is to use them.

This should likewise be one of the measurements that must be tracked by the storekeepers. This is an important measurement and KPI not only for the storeroom but also for the maintenance function because so far, this is the most difficult to address. For non-moving parts, make a decision on whether they will still be kept in stock inside the storeroom or finally be used on the equipment. This is one of the most difficult decisions to make, as there are only limited options on what to do with non-moving parts, unlike obsolete parts. Remember that there is a cost to hold these parts inside the storeroom.

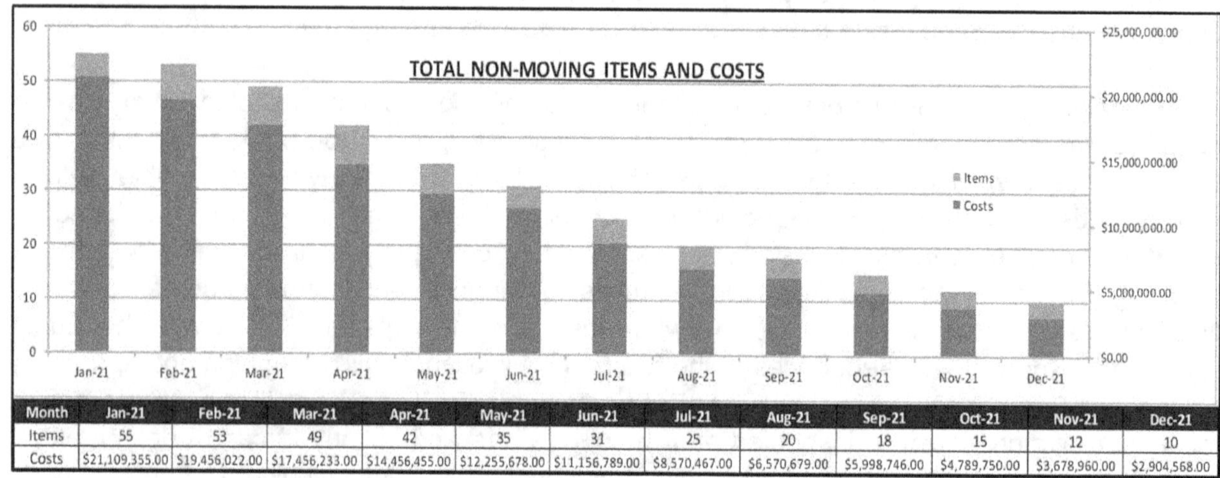

Month	Jan-21	Feb-21	Mar-21	Apr-21	May-21	Jun-21	Jul-21	Aug-21	Sep-21	Oct-21	Nov-21	Dec-21
Items	55	53	49	42	35	31	25	20	18	15	12	10
Costs	$21,109,355.00	$19,456,022.00	$17,456,233.00	$14,456,455.00	$12,255,678.00	$11,156,789.00	$8,570,467.00	$6,570,679.00	$5,998,746.00	$4,789,750.00	$3,678,960.00	$2,904,568.00

Figure 7.7: Tracking for Non-Moving Parts Inside MRO Storeroom

This should likewise be one of the measurements that must be tracked by the storekeepers. This is an important measurement and KPI not only for the storeroom but also for the maintenance function. For non-moving parts, make a decision on whether they will still be kept in stock inside the storeroom or finally be used on the equipment. This is one of the most difficult decisions to make, as there are only limited options on what to do with non-moving parts. Remember that there is a cost to hold these parts inside the storeroom.

7.5: Inventory Turnover for the Storeroom

In many industries, the MRO parts represent a significant portion of overall inventories and incur an amount of holding cost on all items stored due to its typically slow turnover level. The inventory turnover is a ratio that measures the number of times an inventory is withdrawn or consumed in a given time by the end-user. Inventory turnover is sometimes called stock turns or stock turnover. The formula for Inventory Turnover is,

$$\text{Inventory Turnover (months)} = \frac{\text{Total Value of the Storeroom}}{\text{Total Value Withdrawn in a Year}} \times 12 \text{ months}$$

Higher inventory turnover will indicate better performance in the storeroom. This is because a high inventory turnover will indicate that the storeroom is purchasing too much, and the holding or carrying cost of the item is high. This may be good for the storeroom as the items are easily withdrawn, which lowers the inventory of the store, but on the other end will indicate that the plant is reactive and have many repairs to perform. On the other side, this can make the parts idle for a long time, which is not a good sign since the item can deteriorate much rapidly as they sit idle inside the storeroom while incurring holding costs.

Let us assume that the total value of the MRO Storeroom is at $4,000,000.00 and the total amount of withdrawals for the past year is $ 1,500,000.00, the number of Inventory Turnover will be given as

$$\text{Inventory Turnover (months)} = \frac{\$ 4,000,000.00}{\$ 1,500,000.00} \times 12 \text{ months}$$

$$\text{Inventory Turnover (months)} = 32 \text{ months}$$

The reciprocal of Inventory Turnover is called the number of turns

$$\text{Number of Turns} = \frac{\text{Total Value Withdrawn in a Year}}{\text{Total Value of the Storeroom}} = \frac{\$ 1,500,000}{\$ 4,000,000} = 0.375$$

[8]The figures from the inventory turnover can be improved by either reducing the dollar value held in the storeroom or by increasing the number of withdrawals per year in which the storeroom has no control whatsoever. Although our goal is to increase the number of turns, this can also have a negative effect as this can imply a large number of individual purchases, which eventually can increase the overall cost of purchasing. Let us state this example on how to improve the number of turns. The storekeeper performed an FSNO Analysis and submitted it to purchasing who, in turn, source vendors who are willing to provide consignment. Because of these efforts, the total value of the storeroom was reduced from a total value of $ 4,000,000.00 to $ 3,200,000.00. In this case, the number of turns will be;

$$\text{Number of Turns} = \frac{\text{Total Value Withdrawn in a Year}}{\text{Total Value of the Storeroom}} = \frac{\$ 1,500,000}{\$ 3,200,000} = 0.46875$$

$$\text{Inventory Turnover (months)} = \frac{\$ 3,200,000.00}{\$ 1,500,000.00} \times 12 \text{ months}$$

$$\text{Inventory Turnover (months)} = 25.6 \text{ months}$$

In this case, both the number turns and the inventory turnover have improved so far in the storeroom. Most storekeepers would be satisfied with an inventory turnover of 12 months. We

---

[8] Michael Brown, ***Managing Maintenance Storerooms***, Wiley Publishing Inc., 2004, Page 138 to 140

also need to take into account that in most MRO storerooms, those items that have not moved so far are considered non-moving parts, which are big in size and cost as well. Their withdrawals or consumption is extensively low, but their dollar value is extremely high.

7.6: Inventory Accuracy

One of the essential factors in our MRO Spare Parts Management quest is having control over the inventory of the parts inside our storeroom. To have an accurate inventory means that the number of parts physically present inside the storeroom should match the number of parts indicated in the system or computer. Some books say that the benchmark is at 95%. My standards would be to have an inventory accuracy of 98% or more. This means that if I have 100,000 parts inside the storeroom, 98,000 of these parts or higher should reflect the same numbers physically in the actual location in the storeroom and in the system. When the system reflects a higher quantity than what is physically present, then there is an excellent possibility that the storeroom can experience a stock out. At the same time, if too much inventory is present than what is reflected in the system, then that industry remains costly. Either way, the company loses. If maintenance checks the system to determine if a particular belt is in stock, and the system indicates that the belt is there, but when you go to the storeroom and withdraw the part where the storekeeper says that they no longer have stock of the belt, how would you feel? You will lose your trust and confidence in the storekeeper and start squirreling your way on the belts for a start, and it goes on and on. It is like going to a restaurant known for its specialty menu. When you place your order, the waiter tells you that they run out of stock and advise you to order other food or return tomorrow. When maintenance starts to lose their confidence with the storeroom people, there is a great tendency to hoard some of these parts in their cabinets, lockers, and perhaps even in their homes. Once hoarding begins, then there is no more control over the storeroom. We just do not know the number of spares held in stock since there are also stocks or unofficial stores everywhere in the plant.

If the actual inventory is lower than the system recorded, then the risk is high that an out-of-stock condition will occur because parts will not be reordered. At the same time, if the actual physical inventory is higher than the system recorded, then parts will be reordered by the system, even if it is not needed creating excess inventories and higher carrying costs. This is true mostly for fast and even some slow-moving items. It is critically important that not only the storeroom people should understand the importance of inventory accuracy, but also everyone in the entire organization. If the inventory does not match the system, then the users will lose their confidence in the storekeeper. Why do you think maintenance has squirrel stores, which will be the next domino in the entire lineup of the problem?

One of the main reasons why system inventory and its physical inventory do not match is that everyone has access to the storeroom especially during late night and peak hours when the storekeeper is no longer around, which often results in stock-out of parts when the users need them most. Why does maintenance do that, because the storekeeper is just from 8:00 to 5:00 pm? If you ask why management refuses to add additional storekeepers even if their operation is 24 hours, then everything is written in the bible, for the love of money is the root of all evil. Management people might be thinking that they are saving money by just hiring one shift from the storeroom.

A function cannot be managed unless it can be measured. The storeroom has its own function, which is to provide the needed spare parts and keep its inventory to a minimum. If you are the storeroom manager, the storeroom must be well managed and controlled. We cannot let the problems of the storeroom overwhelm and pressure us. For as long as the parts are available when needed, then everyone is happy. Basic storeroom management consists of two areas, managing the inventory and the daily storeroom operations. I cannot say for the rest of the experts, but what I believe is that inventory accuracy is the most important measurement for the storeroom people.

$$\text{Inventory Accuracy} = \frac{\text{Actual Physical Count}}{\text{Inventory in the System}} \times 100\%$$

Here is the catch, the only way to determine the accuracy of your inventory is if all the stocks and items inside the storeroom are counted, and compared to the system. It is very unlikely that a report on inventory accuracy can be provided every week unless you have around 1000 or more workforce inside the storeroom whose job is just to count. If an industry has an RFID system, then counting the inventory will be easy. Pragmatically, the inventory accuracy report can be provided twice a year at the most or once a year. However, there will be signs or symptoms indicating that the storeroom's inventory accuracy is low. First, if the number of stock-outs experienced by maintenance is high. Second, if the number of cases of emergency buying is high. These two indicators will guarantee that your inventory accuracy is low.

The storekeepers should seek the help of other functions in their organization on counting all the parts inside the storeroom. Once the count is completed, it will be tagged and checked on the system. If there is a discrepancy, then the actual count will be recorded in the system. After the count had been completed, the storekeeper can invite other functions of the organization every 2 months or every quarter to make a random audit. The storeroom can invite different functions of the organization, and audit the storeroom at random for 1 to a couple of hours. They will select at their own discretion, what bins, shelves, or drawers they want to open and count the items respectively. Once they complete the counting, they can record it on the sheet together with the part number of the item and check it on the system. Doing this consistently will assure a high inventory accuracy for the storeroom. This will ensure that the inventory accuracy will always be high. Another important thing is that the storeroom is always for the storeroom people and only authorized people should access the storeroom except for special reasons such as audits, or when the storekeeper needs the help of maintenance to identify alternative or substitute parts consolidate them in the system. The accuracy of inventory levels physically and in the system can be measured and reported as part of the spare parts management indices. Having a high accuracy of inventory levels can once again build confidence and trust with the maintenance people. They know that they can trust their storeroom on having the parts they need to perform their regular PM and repairs because it is what is reflected from the system.

7.7: Other MRO Storeroom and Spare Parts KPIs

Carrying or Holding Costs: Carrying Costs (C), which is also called the holding cost, is the cost that is accumulated on the parts held in stock in the storeroom for some time, usually per year. This is the cost of keeping the item inside the storeroom. The storeroom is like a condominium in which each of the items placed in the storeroom has a specific location or space they occupied. Each of these items pays rent yearly to their landlord. The carrying or holding cost usually includes the electricity, salary of the storekeepers, telephone, lighting, depreciation, air conditioning, and so on. Usually, this will be around 20 to 30% of the item's value, also depending on the size of the industry. The carrying percentage is constant for industries at 20 to 30 %. In most cases, the typical carrying percentage would be at 20%. This means that the parts are sitting inside the storeroom have to pay rent for being stored inside the storeroom. The trend we want for carrying costs will be the lower, the better for the storeroom.

$$\text{Carrying Cost} = \frac{L \times U \times I}{2}$$

Where:
- U = Unit Cost
- I = Carrying Percentage
- L = Ordered Quantity

Case 1: If the overall inventory items stocked inside the MRO Storeroom is 85,890 items and the Overall Inventory Cost is at $ 35,709,454..80, what will be the Carrying Cost of all these inventories in the storeroom if the Carrying Percentage (I) used is at 20%.

$$\text{Carrying Cost} = \frac{L \times U \times I}{2} = \frac{\$\,35,709,454.80 \times 85,890 \times 0.20}{2} = \$\,613,417,014,554.4 \ / \ year$$

Carrying Costs = $ 613,417,014,554.40 / 12
Carrying Costs = $ 51,118,084,546.20 per month

This means that the cost of storing all these 85,890 items inside the storeroom will cost around $ 613,417,014,554.40 per anum or per year. This will be a whopping $ 51. 118,084,546.20 per month. Perhaps that is the reason why some Storeroom does not want to report the carrying costs.

The number of Stock-out: The storeroom people should keep logs and records to determine the number of times an item was requested where the inventory had been depleted and reached zero. This means that the part has run out of stock. A good target value for stockout items should be less than 0.5% for each month, but the ultimate goal is to have a zero stock-out, as this is our main goal and objective. A stock-out condition occurs when the item needed by maintenance is not available in the storeroom when it is requested to be withdrawn. This means that the inventory had reached zero before the replenishment stock arrives. A safety stock level, as well as its lead-time, are important factors to reduce or zero out the number of stock-outs experienced by the user. Remember that frequent stock-outs can make maintenance people lose their confidence in the storekeeper and start to keep the parts to themselves.

Stock-outs (%) = (Number of Withdrawals with Stock Out / Total Number of Withdrawals) x 100%

The target for stock out will be around 0.5 % per month, and for emergency buying, it will be less than 2% per month. Inventory Accuracy is often the cause of why maintenance people lose their trust in the storeroom people and squirrel parts on their own. The logic here is simple if the parts are available when needed, then there is no need to squirrel the parts.

Emergency Purchases: This is the overall cost of purchases that resulted when a part is needed in the storeroom and was not available. In this case, a stock out occurred, or either the part is not held inside the storeroom. Usually, emergency buying is the result of massive failures and breakdowns experienced in operations. Maintenance is usually reactive in this case. The only maintenance strategy in the plant is playing the hero stuff that saves the day syndrome. If a failure occurred in operations, maintenance will try to look for similar equipment that is idle or not currently used and start cannibalizing the parts that they need. If there is no equipment to be cannibalized, then emergency buying will be the ultimate resort. The formula for emergency buying is equal to the total costs of items bought on an emergency basis divided by the total costs of spare parts purchased for a particular period say monthly.

Emergency Buying (%) = (Total Costs of Emergency Purchases / Total Purchased Costs) x 100%

Although this KPI is directly connected with the total number of stock-outs, it is important to measure the overall cost in dollars or whatever currency is used versus the total costs of spare parts that are purchased based on an emergency basis. There are cases where making an emergency purchase will be more expensive as the order will be rushed by the Purchasing and the vendor might add additional expenses on the freight and delivery. The higher the stock-out, the higher will be the number of emergency purchases made. Our goal will be to zero out emergency purchases once we understand what to stock and not to stock inside our storeroom. Emergency Purchases can be computed as the Total Emergency Purchases in dollars divided by the Total Costs of Inventory purchases in dollars. An acceptable range will be less than 2%, and these indices can be calculated monthly. The trend we definitely want will be the lower, the better.

Cost for Fast and Slow Moving: Knowing our consumption and usage of these parts can help us manage and improve our spare parts better. For fast-moving items, maintenance must challenge themselves if something can be done to lengthen the life span of these parts. If they agree, form a team and challenge themselves to increase the MTTF or MTBF of these items. It would be a good idea if the storekeeper can provide maintenance an updated list of the top 10 to 20 items consistently withdrawn from the storeroom with their corresponding cost. These parts can be the subject of continuous improvement efforts by maintenance.

Storeroom Response Time for Request: I think that this is one important measurement for the storekeeper, where this indicates the accuracy and speed the storeroom can provide the parts requested from the storeroom to the user. Remember that if the equipment is in a failed state, the response time by the storekeeper will always be included as part of the overall downtime and repair time. Fast-moving and slow-moving parts must be located inside the storeroom where it is easy to retrieve. At the same time, non-moving and big items can be stored at the back of the storeroom as this is seldom requested unless a non-moving part will finally be used or parts are being withdrawn before a Preventive Maintenance shutdown or schedule. Although we will limit these indices on occasions when there is an actual repair, and a part needs to be withdrawn from the storeroom. We will not include the storekeeper's response time in acquiring the parts needed for a Preventive Maintenance schedule. The trend we want will be the lower, the better.

Whatever the MRO Storeroom's KPI, the storeroom is currently using, it is important is not only the results but likewise the trend. The trend will indicate if we are making progress in our efforts to improve the storeroom. Improving the storeroom cannot be done overnight or even a week, just like playing the guitar the first time around. It will take time. This means that if the stock out been dropping for the past 6 months, then the storeroom is improving its performance. These KPIs and storeroom indices will spell the success or failure of how the storeroom is definitely being managed. They provide us to have an understanding of the performance by comparing their actual performance from one period to another.

Chapter 8

Indicators for Autonomous Maintenance Operators

> *Reliability is always a shared responsibility for operators and maintenance. Most failures start from small things that have been left unattended and neglected in our equipment. Autonomous Maintenance emphasizes the importance of establishing the basics. Once these simple basic equipment conditions have been established well, their impact will be felt immediately.*

8.1: Why Operators are Important in the Reliability Strategy?

Operators are important because they are the ones responsible for providing the revenue of the plant. However, for the majority of industries they provide a restriction on operators since according to Human Resources, their main responsibility is to operate the equipment, and because of this operators in industries remain just switch flickers and operate their equipment without any understanding of their equipment intimately. Let me cite a simple example, if you want to hire a driver for your car or you plan to have a Grab Business and want someone to do this, which one would you prefer to be a driver?

a) A driver with a valid professional license with 5 years of experience in driving
b) A driver who not only knows how to drive but has a background in basic engine knowledge and how to repair minor problems.

If we discuss the two drivers, if the driver has some background and knowledge about the engine, any unusual noise from the car will be observed by the driver and advise the owner to have the car checked or if the driver is capable will address the problem themselves. Now going back to our equipment, which one do we prefer to be our operators, those who just operate or those who know the equipment as well and can perform repairs and corrections on minor problems on the equipment? I believe that this is a No Brainer Question. Having this level of operator in industries will not happen overnight or not even in a couple of months. An average Autonomous Maintenance team doing the 7 steps of Autonomous Maintenance will complete all steps in 5 years or perhaps, even more, depending on the pace of the team. But as the operators complete each step, the more they become knowledgeable and intimate with their equipment. The key is, to improve the people first before they can improve their assets.

Although RCM and TPM come from different origins, there is one thing that both of them agree on. Both unanimously agree that operators should be involved in any reliability strategy on improving the equipment. As I keep on saying, the equipment is always a shared responsibility for both operators and maintenance. TPM has a thorough and detailed plan for operators. The journey on empowering operators is done through the Seven Steps of Autonomous Maintenance, in which the goal is to develop empowered operators that can decide what is best for their equipment and assets. On the other hand, operators are strongly recommended to be part of the team conducting the RCM Analysis because both strategies believe in the importance of operators even if both these strategies come from different origins.

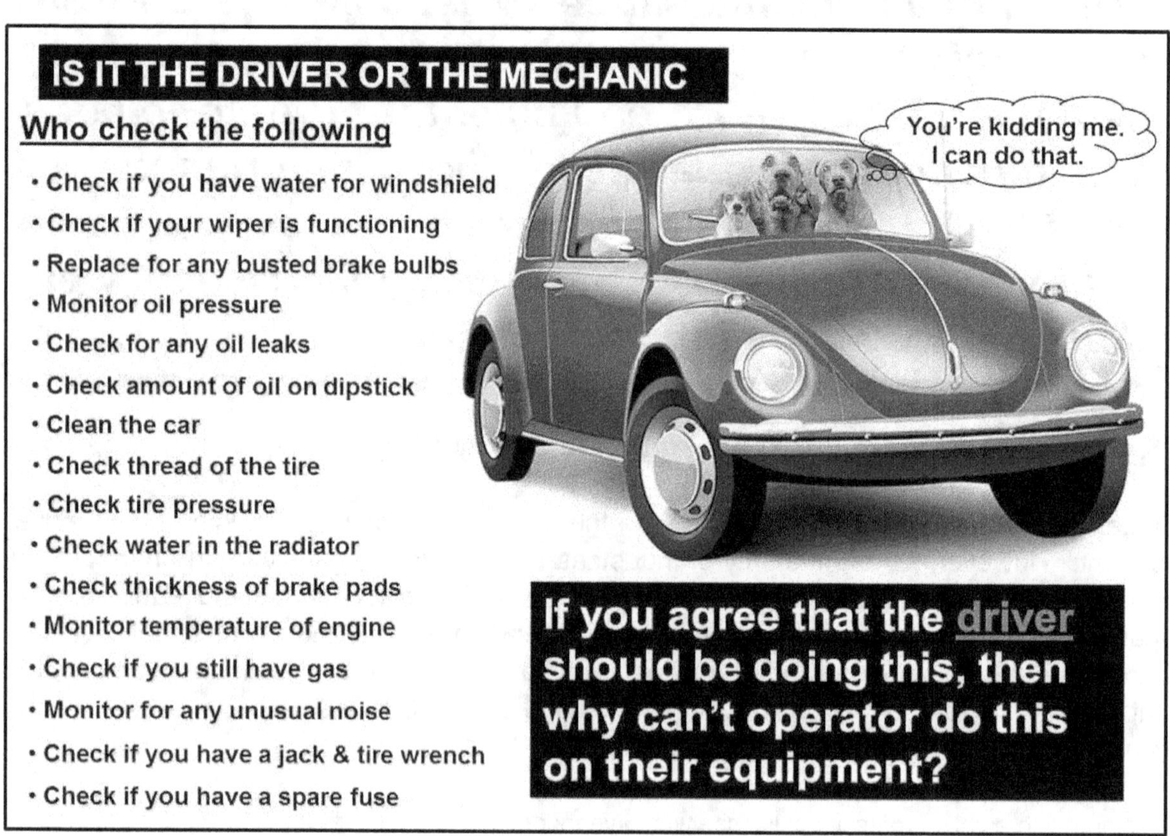

Figure 8.1: Who Usually Does This In Your Car, Mechanic or Driver?

In TPM, Operators Learn from Maintenance: From the traditional mindset I operate, which is the operator, and you fix which is the maintenance, to a paradigm that the equipment is a shared responsibility for both operators and maintenance. They both must take care of their equipment. Operators must understand their role on the equipment by helping maintenance establish the very basic equipment condition on the asset. But to make this happen, maintenance must take the time to teach operators about their equipment both formally and informally. Just like in a game of chess, the white pieces move first, which means that maintenance should start by teaching operators about the equipment they are operating. Here are some of the things that maintenance must teach operators:

• Basic Machine Function
• Safety on the Equipment

- Importance of Establishing the Basic Equipment Condition
- Preparing Inspection Standards
- Basic Lubrication Standards and Techniques
- Correct Tightening of Bolts
- Minor Repairs and Troubleshooting
- How to Use their Senses to Detect Early Equipment Problems
- Importance of Addressing Sources of Contamination

Autonomous Maintenance is a slow process. Do not expect results in a month. It takes several years to transcend and empower operators in this journey, but it will be very much worth doing it. On average, it will take 5 years or more for operators to complete the 7 steps. Autonomous Maintenance is not just a mere transfer of maintenance responsibilities to operators. It is changing the mindset of operators and their bosses which are the operations managers that they too had a role to spare by helping maintenance establish the very basic equipment condition in their equipment by keeping the equipment clean, having the correct amount of lubrication, and completing all bolts and nuts in their equipment with the correct torque. These small things are often the cause of sporadic and catastrophic problems in the equipment. While in most industries, they don't allow their operators to think but just to operate their equipment till their shift is done. Those industries that are now making their journey to World Class Maintenance Management are finally learning to understand that maintenance cannot do this alone. They need to be partners with operators if they want to move forward. This is I think the biggest missing link ingredient in any maintenance and reliability improvement strategy. If the operator changes, expect that they will do the same on their equipment and this is one of the benefits of having an operator and maintenance partnership and alliance in place in the plant.

In RCM, Maintenance Learn from Operators: RCM strongly recommends that an operator should be part of the team to compose the RCM analysis. When equipment fails, the first line of defense will always be the operator. Kindly take note that when something fails on the equipment, it is not the maintenance that is in direct contact with the failure but rather it is the operators. Therefore, operators must have experienced or witnessed something on the equipment before it fails. This information is vital and critical in the RCM Analysis. By allowing operators to share their experiences when encountering failures, maintenance can learn a great deal about the failure itself. Maintenance usually goes to the equipment after the failure occurred because the operator called them. But if there was a change in sound, a change in vibration, some changes in the gauges reading, it is the operators who will experience them first and not the maintenance. One problem with most industries is that there is little or no communication between operators and maintenance. When the equipment fails, and the maintenance is around, the operators take a break and leave the maintenance alone to fix the equipment. The operator only comes back to the equipment when maintenance calls them to tell them that their equipment has been repaired. When both operators and maintenance communicate, maintenance can have a deeper understanding of the failure that took place. Operators might not be technical people like the maintenance but they knew something for sure about the failure before it took place in the equipment. RCM believes that the best source of failure modes is the operators themselves because they are simply the first line of defense on failures.

8.2: What is Autonomous Maintenance?

Autonomous Maintenance is one of the key major pillars of Total Productive Maintenance (TPM) which involves the operators of the equipment. It means that operators should not only be confined in just operating their equipment but should also understand their equipment intimately by creating a solid partnership with Planned Maintenance in establishing equipment basic condition. The basic equipment condition means that the equipment should be clean with no form of leaks, with the correct lubrication, and all bolts completely tightened on the equipment. These neglected functions on the equipment lead to further problems. Just by performing these basics, expect equipment breakdowns and minor stoppages to be reduced. Autonomous Maintenance believes that operators will be the first line of defense on any equipment problems since they are the people closest to the asset when equipment fails.

The goal of Autonomous Maintenance is to enhance the operator's senses to spot problems and deviations on the equipment at its earliest possible stage. The role of Autonomous Maintenance is to detect problems, irregularities, and abnormalities in their equipment by addressing the basics and accepting these basic maintenance responsibilities which include keeping the equipment clean, completing the bolts with the correct torque, applying the correct lubrication, and equipment free from any form of leaks. This means that we take care of the problems when it is still small, as these small problems will lead to bigger problems. Autonomous Maintenance is also the activity in which each operator performs their daily inspection, lubrication, cleaning, minor repairs, and accuracy checks on one's own equipment which aims in keeping the equipment in healthy and good operating condition. Training and nurturing operators with strong equipment skills can develop operators to detect early signs of problems before they can become serious enough to cause catastrophic failures and even disasters. One unique feature of Autonomous Maintenance is to develop a checklist for operators which will include cleaning, lubrication, and inspection to be done on the equipment. This pillar is done on a standard seven-step approach which will begin with Step 0 which is the preparatory step for Autonomous Maintenance. The goal of Autonomous Maintenance is to develop a self-directed and empowered operator who aims to continuously improve the performance of their equipment making their job no longer routine for them.

Cleaning: Cleaning is the process of removing any form of an unwanted object from the equipment. It consists of removing dirt, contaminants, grime, stain, rust, corrosion, excess oil, grease, and other foreign objects that can affect equipment parts and components as well as its operations. Cleaning is performed not just to satisfy the cosmetic looks of the equipment but because this dirt and contaminants shorten the life of certain parts and components of the equipment. Detailed cleaning of components and equipment is often a no man's island because while everybody agrees that it is important, nobody wants to do it actually. World-Class Companies perform these activities. Components and equipment are kept clean in every detail. Such an organization realizes that inspections cannot be done without this level of cleaning. These industries understand that cleaning definitely extends the life of parts and components. Through cleaning, we can expose problems on the equipment that has been left ignored for a very long time. When operators clean their equipment, they touch parts, and by touching parts, they are actually inspecting them. Leaks, cracks, fractures that have been in the equipment for a very long time can be detected and exposed when operators touch and clean their equipment.

Dirty and untidy equipment will tend to deteriorate more rapidly than the equipment that is well maintained and cleaned at all times.

Lubrication: All machines contain lubricants in one form or another. The purpose of lubrication is to reduce friction for mechanical parts moving inside our equipment. It is not only important to lubricate the equipment but also to use the correct lubricant and the correct amount. Equipment that lacks lubrication or is over-lubricated can induce problems. Operators should be taught about the importance of checking the lubrication in their equipment and maintaining the correct amount of lubricant. So many failures can be attributed to lubrication that can be avoided if only these basic equipment conditions can be well established. Maintenance must teach operators not only the necessary points in their equipment to lubricate but also what proper and improper lubrication can do with their equipment. Just like humans, lubrication is the life-blood of the equipment, and it should be maintained clean and adequate all the time. Many failures are attributed due to lack or inadequate lubrication. These things can be avoided and controlled in our equipment if we understand and learn the role of what lubrication plays in our equipment.

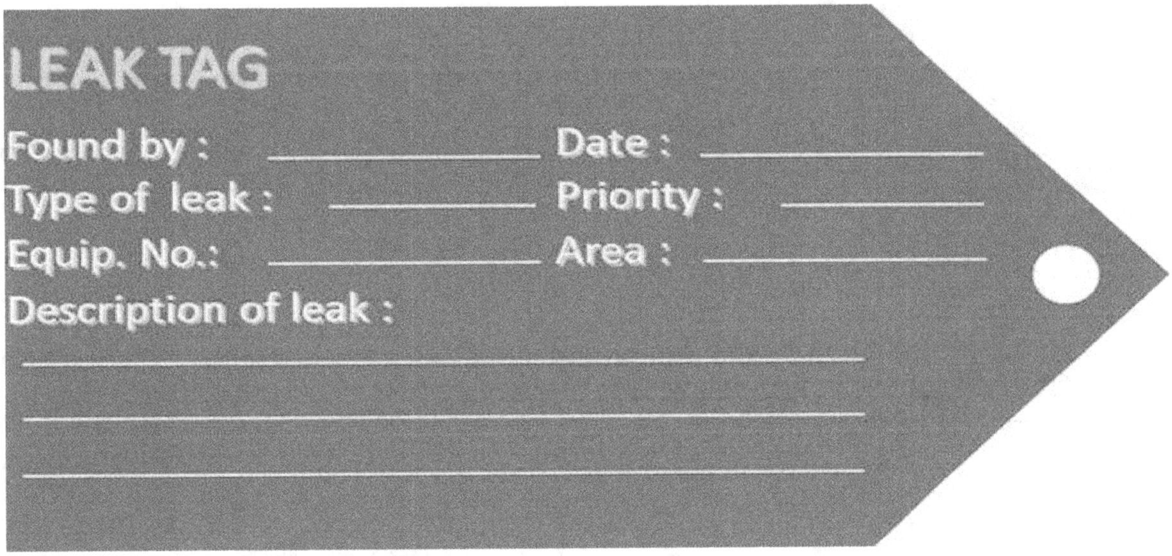

Figure 8.2: Sample Tags for Leak (Red Color)

Addressing Leaks: Equipment must have no amount of any leaks. Leaks may be in the form of an oil leak, air leak, vacuum leak, or any other form. Once the leaks are detected, they will be tagged, easy to correct leaks will be addressed by the operators, while others will require a job order for the maintenance function to address during their regular scheduled Preventive Maintenance shutdown activities. As a rule of thumb, tags will only be removed once the leak has been completely restored.

Bolting: Machinery contains bolts, nuts, and fasteners as part of their construction, and they serve a particular purpose. The equipment functions properly only if all bolts and fasteners are securely tightened. All equipment vibrates, but excessive vibration can be destructive as this can induce secondary damages to other parts and components affected by the vibration. When we touch the equipment, we can feel its vibration. What we feel is simply the sum of all vibrating forces moving inside the equipment. Bolts, screws, and fasteners are placed to minimize the vibration in the equipment. When vibration increases, it can result in cracks and fractures on mechanical parts, and these fractures can propagate to the point of rupture as a result of excessive vibration

7 Types Of Equipment Abnormalities

		Minor Flaws	Examples
1		- Contamination - Damage - Play - Slackness - Abnormal Condition	- Dust, dirt, powder, oil, grease, rust, corroded parts - Cracking, crushing, deforming, chipping, bending - Shaking, falling-out, eccentricity, backlash - Belts, chains - Excessive noise, overheating, vibration, strange smell - Over or under pressure,

		Unfulfilled Basic Condition	Examples
2		- Lubrication - Lubricant Supply - Oil Level Gauges - Tightening	- Insufficient Oil Level, dirty, unidentified, unsuitable or leaking oil - Dirty, damaged or deformed lubricant inlets, faulty lubricant pipes - No minimum and maximum points for lubricants - Loose threads, slackness, missing bolts and nuts, worn out heads, incorrect bolts and missing washers

		Inaccessible Places	Examples
3		- Cleaning - Checking - Lubricating - Tightening - Operation - Adjustment	- Machine construction, covers, layout, foothold, space - Covers, construction, layout, instrument position & orientation - Position of lubricant inlet, construction, height, lubricant outlet - Covers, construction, layout, size, foothold, space - Machine layout, position of valves, switches & levers, foothold - Position of pressure gauges, thermometers, flowmeter, moisture gauges

		Contamination Sources	Examples
4		- Product - Raw Material - Lubricants - Gases - Liquids - Scrap - Others	- Leaks, spills, spurt, scatter, overflow - Leaks, spills, spurt, scatter, overflow - Leaking, split, seeping lubricating oil, hydraulic fluid, fuel oil, etc., - Leaking compressed air gases, steam, vapors, exhaust fumes, etc., - Leaking spill cold & hot water, leaks on cooling water, heat exchanger - Flashes, packaging materials, non conforming products - Contamination induces by people, dusty places, leaks on roofs etc

		Quality Defect Sources	Examples
5		- Foreign matter - Shock - Moisture - Grain Size - Concentration - Viscosity	- Inclusion and entrainment of rust, chips, wire scraps, insects, etc - Dropping, jolting, collision, vibration - Ingression of moisture, defective elimination of moisture - Abnormality on screens, centrifugal separators, compressed air etc - Inadequate warming, heating, compounding, mixing, etc - Inadequate warming, heating, compounding, mixing, etc

		Unnecessary / Non urgent item	Examples
6		- Machinery - Piping equipment - Measuring Instruments - Electrical Equipment - Jigs and Tools - Spare parts - Makeshift repairs	- Pumps, fans, compressors, auxiliary machines, columns, tanks - Pipes, hoses, ducts, valves, dampers - Temperature and pressure gauges, vacuum gauges, ammeters, etc., - Wiring, harnessing of wires, piping, switches, plugs, octopus wiring - General tools, cutting tools, jigs, molds. Dies, frames, etc., - Standby equipment, spares, permanent stocks, auxiliary machines - Tape, strings, wire, metal plates etc.,

		Unsafe Places	Examples
7		- Floors - Steps - Lights - Rotating Machinery - Lifting Gears - Others	- Unevenness ramps, cracks, leaks floors, wet floors, peeling, wear - Too steep, missing handrails, wet and leaky steps, dirt - Dim, dirty, busted lights, explosion proofed - Displaced, broken off or broken covers, no emergency stopping device - Unsafe wires, hooks, brakes, and other part of hoisting mechanisms - Special substance, solvents, toxic gases, chemicals etc.,

Figure 8.3: Lists of Abnormalities for Autonomous Maintenance

Remember that it only takes one loose bolt to create a chain reaction of destruction and havoc in our equipment. As a result, other bolts become loose, and vibration increases. Match-marks are placed on critical bolts and nuts so that operator can easily detect if bolts have been loosening due to excessive vibrations for some time. Imagine driving your car, and each of your wheels has 3 instead of 5 stud bolts on each of your tires. Would this be all fine with you? I guess not. Then why don't we treat our equipment in the same way? We can see many bolts missing or loose as a result of many activities performed by the equipment. For a start, why don't we complete them?

8.3: Most Common Operator Indices

Operators involved in Autonomous Maintenance also need to tract indices and measures as they go deeper into the 7 steps of Autonomous Maintenance. Although their indices might not be as tangible compared to the maintenance indices. They are likewise important as they can be included as leading indicators that can improve the lagging indicators. One important role of Autonomous Maintenance is to work hand in hand together with the Planned Maintenance pillar and establish the basic equipment condition. In implementing Autonomous Maintenance, here are some indicators that the operators need to track and monitor as they journey themselves into the 7 Steps of Autonomous Maintenance.

Number of Suggestions: Autonomous Maintenance will be done in small group activities, perhaps with a membership between 4 to 8 members. As they complete their initial requirements for Step 0, the team proceeds to Step 1 of Autonomous Maintenance, which is the Initial Cleaning process. Operators will be exposing problems and abnormalities on their equipment. Their suggestions will be mainly focused on how to address these abnormalities, address sources of contamination, improve their workplace, minimize their cleaning, lubrication, and inspection time, making hard to clean and hard to reach areas in their equipment easily accessible. But the real challenge here is not only about making a suggestion but rather if the suggestion will be implemented successfully or not. This will be the start and will be one of the main activities that will boost the team in pursuing the higher steps of their Autonomous Maintenance activities. I have seen operators that reached step 6 and even step 7 of Autonomous Maintenance with my very own eyes. Operators that reached these steps have changed dramatically in how they deal with their work. It is no longer a routine for them. Believe me when I tell you that they can decide on their own. Operators know their equipment better than anyone else in the organization, even the maintenance because they are the people day in and day out that come face to face with their equipment all the time. They know the problems, and they can provide improvements together with the maintenance function. The problem with most industries is that they have not been allowed to voice out their suggestions on how to improve their equipment because they are only limited to operating the equipment from the start till the end of their shift, making the operator's work routine.

Number of Question Lists Generated and Answered: As the teams start their Autonomous Maintenance activities, they raised questions about their equipment. Usually, the team will answer these questions raised by their fellow operators. If no one can answer the question raised by a team member, they will refer to their mentors, who are the Planned Maintenance people. This is part of the learning curve for operators on doing Autonomous Maintenance.

The goal is for the operators to know their equipment intimately. If they think that this question is important, either the operators or maintenance will generate a One Point Lesson to provide a clear answer to their questions. As they complete each step of the 7 steps of Autonomous Maintenance, every single question generated by operators and those that were answered will be recorded and placed on their activity board. No matter how hard or easy the question is, it will be listed and answered by the team leaders or solicited from the maintenance function. The purpose of generating questions by the operator is to generate awareness. For example, asking a name of the part will not be part of the question lists, but if I have a pressure gauge and the pressure goes beyond its limit, an operator may ask maintenance what will be the effect of going beyond this limit will now be included in the question lists generated.

Percentage of Abnormalities Detected and Corrected: As Autonomous Maintenance Step 1 activities are initiated, operators will tag their equipment for any abnormalities, slight or major deviation that they can see during the initial cleaning process. Maintenance will also have its own tags when they performed the Planned Maintenance Phase 1 restoration stage. The goal of both Phase 1 and Step 1 of both Planned and Autonomous Maintenance is to bring the equipment back to its original basic equipment conditions. The distinction between Autonomous Maintenance and Planned Maintenance tags is that maintenance will be focused more on the interior part of the equipment, including deteriorated parts that need to be restored, while the Autonomous Maintenance team will focus more on the exterior part of the equipment. As much as possible, it is best for the operators who locate the abnormality to correct it themselves if they can. Abnormalities that the operators cannot correct themselves will be passed on to the Planned Maintenance team. It is highly recommended that when Step 1 Initial Cleaning is performed on the equipment, both operators and maintenance should perform their initial cleaning and restoration activities on the same day so that operators tagging abnormalities can be assisted by the Planned Maintenance teams. Those abnormalities found by operators that cannot be corrected themselves can be passed on to the Planned Maintenance team. One thing important to note is never ever passed all abnormalities to maintenance. There will be abnormalities that can easily be corrected by operators. The Autonomous Maintenance team should provide a before and after picture of the abnormalities, they have corrected. Hence, when doing step 1, let us say that 180 abnormalities were detected, and 90 of them have been corrected so far, which means that their accomplishment rate will be at 50%. This percentage is what the members of Autonomous Maintenance will be tracking. It is also highly recommended for both Autonomous Maintenance and Planned Maintenance to have the same pilot equipment at the start so that Planned Maintenance can easily focus on supporting the needs of the operators instead of having different pilot equipment for operators and maintenance. One more important point is that Planned Maintenance activities should be ahead before Autonomous Maintenance is implemented. This means that Phase 0 or the Preparatory Stage of Planned Maintenance should already be completed since Planned Maintenance will be performing the machine ranking on all equipment in the plant. The operator's pilot equipment should belong to the Rank A (worst) category with the possibility of large replication or fan-out.

The number of One Point Lessons Generated: One-point lesson also called a single-point lesson provides us specific information on a single topic. Usually, this will contain just one topic. This will include a picture, chart, graph, or diagram, and a one-point explanation. Usually, it is best to place these One Point Lessons on Flip Charts. Maintenance or the operators themselves will discuss these One Point Lessons with the Autonomous Maintenance team, where they spend around 10 to 15 minutes for each One Point Lesson generated. Once they understand the message, then all team members will sign on to the One Point Lesson. Suggestions, questions generated, improvement, and Kaizen activities can be included in the One Point Lesson. The important thing to consider is to discuss just one point at a time. A One-Point Lesson is a learning tool for communicating standards, problems, and improvements in the work processes and equipment. Workers and supervisors use One-Point Lessons to provide key information about their everyday work and improvement opportunities. Thus, a one-point lesson may contain information on a wide range of topics. In other words, whenever a worker needs key information to perform their jobs, One-Point Lessons can be an effective tool for delivering this information. OPL is a good tool for sharing knowledge among the team. This translates knowledge into practical information that the Autonomous Maintenance team can use to effectively perform their jobs correctly with confidence. This will be part of the operator's continuous learning process.

The number of Kaizen Activities: The team will also be tracking the number of improvements or Kaizen activities they have generated as a way to motivate the team and inspire other operators to participate in the Autonomous Maintenance implementation in the plant. These kaizen activities will be focused on how to address sources of contamination, hard-to-reach areas, hard-to-clean areas, how to reduce their cleaning, lubrication, inspection time, and how to improve the workflow process. The team will generate kaizen activities starting on Step 2 onwards when the operators address the sources of contamination and hard-to-reach or clean areas on the equipment. These indices for operators will be placed on their activity boards and will be updated by the operators themselves. The team will be monitoring the number of improvements generated versus those that have successfully been implemented in their machine and workplace.

WHAT AUTONOMOUS MAINTENANCE TEAMS SHOULD TRACK

WHAT TO TRACK	STEP 1 :	STEP 2 :	TOTAL	PERCENTAGE
1. Abnormalities Tagged	145	+ 25	170	Percent = 90 %
2. Team Suggestions	25	+ 20	45	Percent = 67.0 %
3. One Point Lesson	20	+ 25	45	Percent = 100.0 %
4. Questions Generated	20	+ 25	45	Percent = 100.0 %
5. Difficult to Clean Areas	5	+ 6	11	Percent = 100.0 %
6. Contamination Sources	4	+ 0	4	Percent = 75.0 %
7. Kaizen Activities	0	+ 20	20	Percent = 75.0 %

Figure 8.4: What Autonomous Maintenance Should be tracking

8.4: Tracking Small things Matter Most

Big problems start from small things. In fact, the majority if not all of the failures and breakdowns we experience in the equipment are simply an accumulation of small problems that have been left neglected in our equipment. It will just take one piece of loose bolt to create a chain of destruction on our equipment as vibration increase incrementally. Cracks and fractures can propagate as a result of this. As the machine vibrates, other bolts start to loosen.

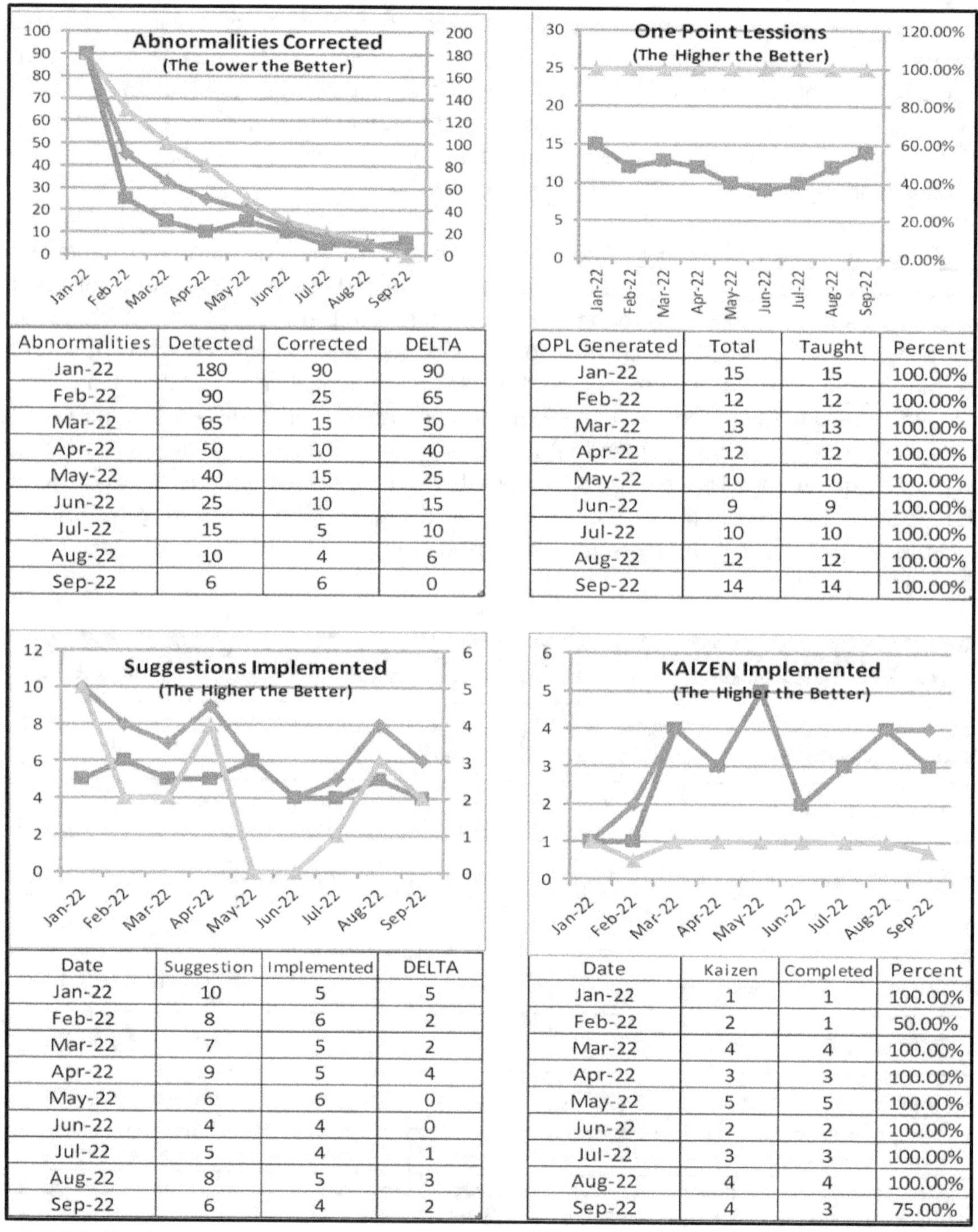

Abnormalities	Detected	Corrected	DELTA
Jan-22	180	90	90
Feb-22	90	25	65
Mar-22	65	15	50
Apr-22	50	10	40
May-22	40	15	25
Jun-22	25	10	15
Jul-22	15	5	10
Aug-22	10	4	6
Sep-22	6	6	0

OPL Generated	Total	Taught	Percent
Jan-22	15	15	100.00%
Feb-22	12	12	100.00%
Mar-22	13	13	100.00%
Apr-22	12	12	100.00%
May-22	10	10	100.00%
Jun-22	9	9	100.00%
Jul-22	10	10	100.00%
Aug-22	12	12	100.00%
Sep-22	14	14	100.00%

Date	Suggestion	Implemented	DELTA
Jan-22	10	5	5
Feb-22	8	6	2
Mar-22	7	5	2
Apr-22	9	5	4
May-22	6	6	0
Jun-22	4	4	0
Jul-22	5	4	1
Aug-22	8	5	3
Sep-22	6	4	2

Date	Kaizen	Completed	Percent
Jan-22	1	1	100.00%
Feb-22	2	1	50.00%
Mar-22	4	4	100.00%
Apr-22	3	3	100.00%
May-22	5	5	100.00%
Jun-22	2	2	100.00%
Jul-22	3	3	100.00%
Aug-22	4	4	100.00%
Sep-22	4	3	75.00%

Figure 8.5: Tracking Autonomous Maintenance Indices

Dirt and foreign matter penetrates rotating parts, sliding parts, pneumatic and hydraulic systems, electrical control systems, and sensors which can cause loss of precision, malfunction, short stoppages, and breakdown as a result of early wear, blockage, frictional resistance, and other problems ending up in catastrophic problem failures. Compressed air leaks can contribute to operational problems which included fluctuating pressure causing air tools, and other air-operated equipment not to function properly.

Establishing these basic equipment conditions simply means eliminating the causes of accelerated deterioration. This is when a certain part or item of its component in our equipment does not reach its useful life. It means cleaning to remove dirt and sources of contamination, proper lubrication to prevent early wear, and understanding that bolts need to be complete and secure. Parts do not achieve their desired lifespan due to dirt and contamination. The wrong lubricant is poured into the equipment because the guy is new. The machine fails simply because several bolts are missing. The maintenance uses the wrong tools that later damaged the equipment and so on.

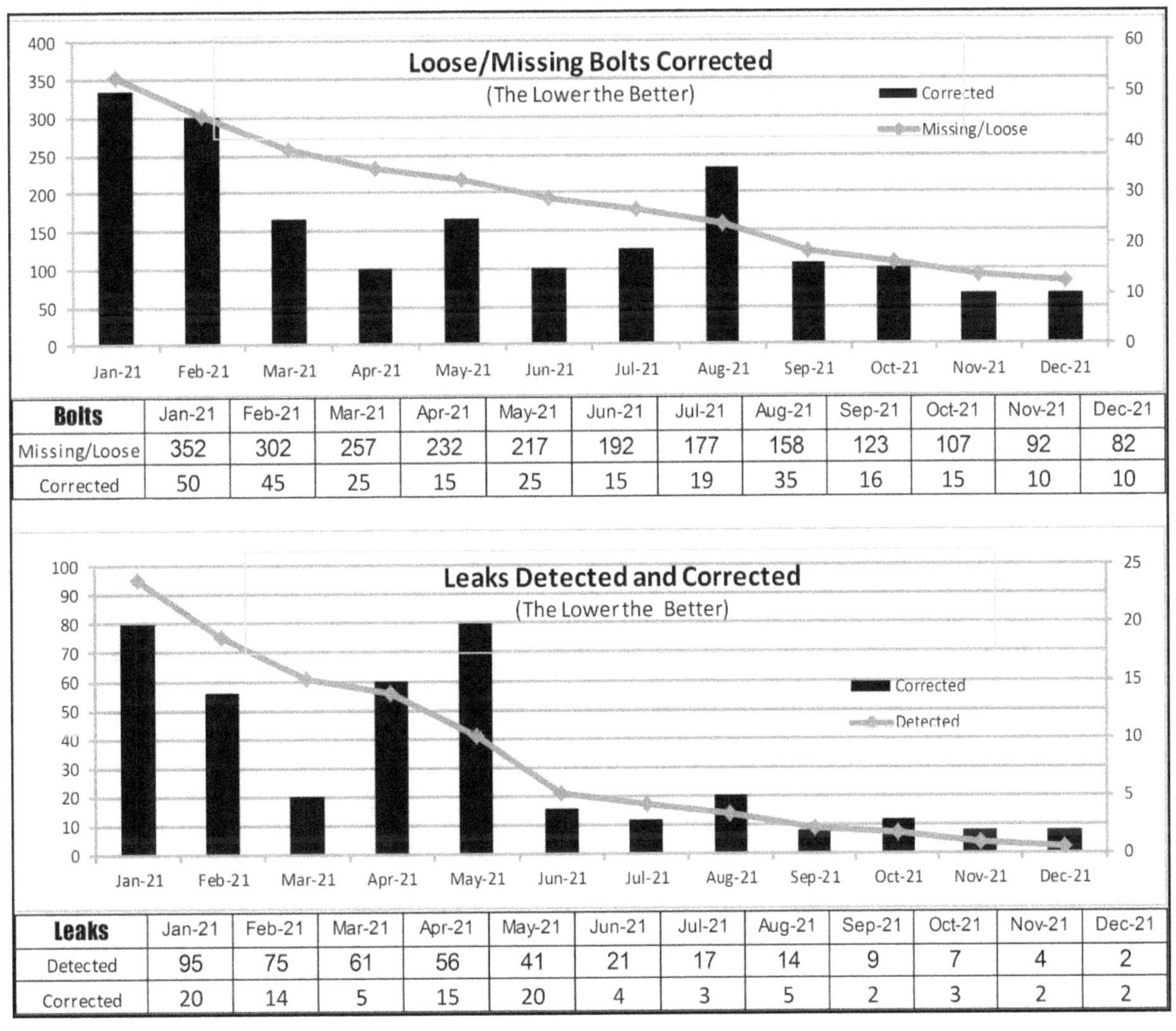

Bolts	Jan-21	Feb-21	Mar-21	Apr-21	May-21	Jun-21	Jul-21	Aug-21	Sep-21	Oct-21	Nov-21	Dec-21
Missing/Loose	352	302	257	232	217	192	177	158	123	107	92	82
Corrected	50	45	25	15	25	15	19	35	16	15	10	10

Leaks	Jan-21	Feb-21	Mar-21	Apr-21	May-21	Jun-21	Jul-21	Aug-21	Sep-21	Oct-21	Nov-21	Dec-21
Detected	95	75	61	56	41	21	17	14	9	7	4	2
Corrected	20	14	5	15	20	4	3	5	2	3	2	2

Figure 8.6: Tracking Leaks, and Missing or Loose Bolts

As a starting point in monitoring the basic equipment condition, we can assign a group of people from maintenance to track and tag all equipment with leaks, missing, and loose bolts for all equipment and machines in the plant. As a general rule, the tags will be located exactly where the leak or missing/lose bolt is detected and will only be removed once it is restored and corrected, but not on moving parts. This can be measured monthly and results are updated not only on the system but can also be placed on bulletin or activity boards.

Tracking the percentage of abnormalities corrected versus detected, the number of leaks, missing or loose bolts perhaps is a unique indicator in the sense that these are the only indicators I know that measure the failure of secondary functions or function-reduction breakdowns of the equipment. What TPM taught me which I believed with all my heart and soul is that small problems matter most since neglecting them will just result in a bigger problem on our equipment and assets.

Chapter 9

Having a Crystal Clear Direction on Maintenance

> *Whatever KPIs, indices, goals and measures we choose should be aligned with the Overall Corporate objectives of the industry. If this will not be agreed upon, then whatever strategies we implement will just be treated as a separate program and will just fade away one day.*

9.1: Let's Not Commit the Same Mistake as the Top 500 Fortune Companies

Although these companies were considered giants during their time, today, only around 52 US companies have made it to the Fortune 500 companies since 1955. Although these companies may have different reasons why they did not last, survival is still the name of the game for the majority of industries This means that only 10.4% of these companies have remained until today after 64 years since 2019. More than 89% of these companies have either gone bankrupt, merged, or acquired by other industries or have fallen out from the original Top 500 Fortune Companies in the United States. Many of these companies were big shots during their time in 1955. Still, today, they are now unrecognizable, forgotten, merged with other industries, or merely gone for the record. In Chapter 1, Section 1.1, for those industries who want to improve their maintenance function, we can borrow the concept from Stephen Covey.

1) Determine where we are?
2) Where do we want to go?
3) How will we get there?
4) How will we know we have arrived?

The first and last step in implementing a World Class Maintenance Management Structure is defining and measuring performance indices that tell us that we have achieved our goals and whether we are headed in the right direction or not. This will answer the question of how we would know we have arrived at our destination. These measurements and indicators should be defined at the very beginning of implementation and must be reviewed regularly. The success or failure of all our efforts depends entirely on these measurements. Measurement has always played an important role in our day-to-day lives. Still, the most important reason for measuring KPIs is to improve the organization's performance. By knowing our performance, we ask ourselves what we should do differently to improve our goal? Generally, there is no single measurement, indicator, or KPI for maintenance to tell us the whole story. We need at least a minimum of five measurements, and this should be defined well at the beginning of the journey on World Class Maintenance Management. While it is vitally important to measure the results, we should also look into the process of how the results were achieved. This creates better communication, more participation, more involvement, high morale, and motivation among the team by focusing on the process. When managers focus on both processes and results, the team knows the management cares about how the results were obtained. When people focus on the process, the results will follow in place. Therefore, if we want to improve the results, we also need to improve the process. This is what separates a good manager from a better manager. A good manager always looks at the results, while a better manager will look at how the results were achieved.

The journey to World Class Maintenance involves the commitment of each and everyone in the maintenance organization. I have been asked several times how long shall it take the industry to achieve this stage, and I provided them with figures only to learn that they abandoned the process completely. Today, when people ask me the same question, I revert the question back to them and tell them that the time it takes an industry to achieve a World Class Maintenance stage depends on how serious they are about it. If you ask me, who is the greatest NBA basketball player of all time? My answer is Michael Jordan. The reason why he is the GOAT (Greatest of All-Time) and the best is that he does not want to be compared to anyone. Even if he is already a world-class basketball player, he never stopped there and aimed for more. When he did not make it in a bid to join his high school basketball team, he cried so hard that his mother told him that if you want to be the best, then he should practice really hard. Michael replied and said to his mom that he practiced hard, just like any other one else. His mother replied that if you want to be the best, you need to practice harder than the rest, which is what Michael Jordan did exactly in his basketball career. Even at his peak, he still continues to improve himself because he has a passion for winning.

9.2: The Importance of having a Strategic Planning for Maintenance

Just imagine a game of basketball where the scores are evenly close and with just a couple of seconds left in the game, the coach signals a time-out to the referee. For the defensive team, the coach creates a plan informing his players of the best defense so the opponent would not score, while the coach of the offensive team, will generate his last-second plan on how to score and win the game. Hence, just like in a game of basketball, the maintenance team needs to have a time-out which is in the form of strategic planning. This can be done before the year ends or in the very first month of the Year, which is January. The objective of this strategic planning for maintenance is to evaluate the past year's performance and provide a clear direction and strategy in improving our performance so we can achieve a close to perfect high-reliability equipment and safe workplace. This is done by identifying all hindrances, obstacles, and pitfalls in achieving a World Class Maintenance Management level. Maintenance needs to understand why they need some time to reflect on their previous activities on things they need to change for the better.

Know where we are and where maintenance is going. Without Strategic Planning is like sailing to the sea on a sailboat without a rudder but rather at the mercy of the wind or driving a car in which the windshield is painted black. Maintenance needs to look back and reflect on the past. This is where they ask themselves, have the goals been met? What are the things that maintenance should improve upon? What is important for maintenance is developing a short-term and long-term strategy for the year by asking, am I satisfied with our current maintenance system? Ask ourselves, are we ready for the change, or can we just work the way we were last year? Is my current maintenance organization equipped with the right tools and tactics to carry out these changes? Do we need to change now? Do we need to change things here? Can we better our performance last year?

Figure 9.1: Actual TPM Planned Maintenance Strategic Planning 1999

Figure 9.2: Sample of Maintenance Vision and Mission Statement

Every industry has its own unique culture, and even more to that, each function in the organization may have a different culture. Just like our 10 fingers in which there is no finger with the same fingerprint. Culture is the way of doing things around here. It is about people's common values, shared beliefs, shared things, shared sayings, shared doings, and shared feelings. It is the pattern of basic assumptions that a given group has invented, discovered, or developed in learning to cope with its problems of external adaptation and internal integration that has worked well enough to be considered valid and therefore taught to new members as the correct way to perceive, think and feel about these problems. From Schein 1983. A Culture Change is possible if we can alter or change the way we think and do things around here. People are willing to make changes in the organization for the better. What is important is for people to provide ownership of the initiative for change. Members should know their role in the change initiative by making amendments to previous rules, policies, and procedures for the better. It is important to understand the roadblocks and hindrances in these initiatives and recognize people for their efforts. Cultural change is possible, but don't expect it to happen overnight. Changing culture can happen if we think and do things outside the box, and move beyond our comfort zone. A strong culture is embedded by having a common vision and mission, a sense of inclusion among its members, and pride in the workplace. People need to have a sense of belonging to their organization. The main difference between reactive and pro-active industries is the way they think about what they believe.

Although a company has its own Vision and Mission statement, maintenance can also have its own Vision and Mission statement, but what is important is that this must be aligned with the Overall Corporate Vision and Mission. Although most industries, I visited and trained have their Vision, Mission, Core Values framed on the wall for the visitors to read. What is important is that each employee from whatever function must perform their day-to-day duties in achieving its Vision and Mission. In thinking about your maintenance Vision and Mission, think about Mount Everest. Placing a flag on the tip of Mount Everest will be the vision, while the Mission represents the path to reach the tip. There are 6 camps on the route, and the higher you reach the tip, the more difficult it is due to the lack of oxygen. Vision must be time-bounded, simply stating that your industry wants to be the best is not enough. A vision statement is a document that states the current and future objectives of an organization.

<u>Why do Improvement Programs fail In A Company?</u>

- Failure to create a powerful mandate for change. In our experience, the biggest barrier to change is that the organization is not ready for it.
- Failure to deliver early, tangible results. If we are not showing tangible benefits in six months, expect support to cut in half, barriers to double, and an increase in people's resistance to change. Most managers always look at the results, but the better manager should not only look at the results but at the process of how the results were achieved. This creates better communication between their subordinates and people.
- Measurements and indices are not aligned with the company's corporate goals. If our improvement goals and measures are different from the company's goals and objectives, we can be sure that the program will be treated separately, and one day will die a natural death.
- What's in it for me is unclear? Although this is a management prerogative, management must seek ways to recognize the team; rewards must be in the form of recognition. A guaranteed job, career growth, and other financial rewards. The old thinking simply states that everyone is waiting for a miracle man who will change and do everything; it won't happen this way.

• Another reason I experienced why improvement programs fail is due to frequent reorganization by the company. I have seen improvement projects that have been abandoned because the person in charge was transferred to another responsibility.

Step	What to Do	Key Questions to Raise
Step 0	Define the Maintenance Strategy.	• What is the Vision and Mission of maintenance?
		• What are the short and long-term goals for maintenance?
		• What maintenance strategies will be adopted?
		• Will everyone be involved in the journey?
Step 1	Establish the need for the change.	• Where are we now?
		• Why are we always reactive?
		• Are we doing the right things in maintenance?
Step 2	Define the objective of the change.	• Why do we need to make this change in maintenance?
		• Where do we want to go?
		• What kind of culture do we need to adopt in maintenance?
Step 3	Plan the means to achieve the objective.	• How is maintenance going to get there?
		• Does maintenance have a roadmap?
		• How do we bridge the gap between the current and future?
Step 4	Implement the change plans	• How are we going to manage the implementation?
		• What interventions do we need to bridge the gap?
Step 5	Evaluate the changes.	• How do we know we have arrived at our destination?
		• What can be done further to improve maintenance?

Figure 9.3: Maintenance Strategy Planning

It is important to clearly state the reason for the change and do it right the first time. Otherwise, any initiative for change will be treated as "Another Program of the Month." Some factors are essentially and equally important; if one is missing, changing the maintenance culture will be impossible. This means having the right tools and a heart to make the change having a system in place and the right attitude to do the job. Many people do not want to improve since their thinking is that improvement programs are fads, and they aim to reduce the workforce in the long run. Programs fail because they are poorly planned, developed, and implemented. Most improvement strategies and programs look good on paper and PowerPoint slides. People view these programs as the latest in a series of fads that come and go. When a program takes too long to implement, management will abandon it, and they're back to their old ways of doing things in the plant.

9.3: Measuring Maintenance Effectiveness on Training

One thing important for any industry includes skilled and competent people who understand their equipment intimately and know how to perform their jobs correctly. Training is always the backbone of any cultural change. It is mainly the missing link ingredient in any change or continuous improvement effort. Skill is the ability to do one's job, to apply knowledge and their experience correctly in all kinds of events over an extended period. It is the product of personal motivation and thorough training. The end result is mastery. Mastery is when we are capable of transferring our knowledge and skills to others. And to enable to achieve this, industries must develop the most effective training methods. The first step in any training program is to identify the level of knowledge, technology, skill, and competency of their people. The second is to assess their skills from time to time. Training should not be the focus of any cost-cutting initiative and management must understand that it will play a vital role in improving the skills of their people. The bottom line is that the skill will come from knowledge.

Training will always be an investment for any industry whether this will be done by the training people in your plant, SMEs (Subject Matter Experts) who are regular employees, or by independent third-party training contractors just like myself if there are no people in your plant capable of conducting the training. In fact, in my first book on World-Class Maintenance Management – The 12 Disciplines, I indicated that training and education are the foundation and cornerstone of any cultural change. This is where we gain the knowledge needed for our maintenance people to build their skills to perform their jobs correctly and right the first time around.

But here's the irony, although I believe that training is one of the important functions of any organization, it is also one of the top candidates for cost-cutting schemes in any industry. The first thing to do is establish key performance indicators (KPIs) for training and development. These indicators tell you how well your training program is working and show executives and boards that your work is worth investing in. However, one question that is often asked by managers, decision-makers, or those who will approve the budget on training will be, what evidence can you provide me that training is worth the investment? We have been doing things this way, why should it be different then? How will it impact the bottom line results?. Or, how can we convince these people that this training will help improve the KPIs we measure. If we hire Rolly and pay him this much for a 2-day training on lubrication, when can we expect the results or the ROI (Return on Investment)?

Although I was already conducting training when I was still employed, however, there is a big difference today working as an independent body on training. When I was still employed, once I conducted training and completed it, you can expect me to make a consistent follow-up to the students I taught on when they are going to implement what they have learned from my training since it is part of a bigger program since we were implementing TPM during those times. You will be pissed since I will stalk you everywhere you go. But I can no longer do that because I am not employed by any of these industries, and when my training ends, it will be up to them. I received emails from previous students thanking me for the learning and telling me that they have benefited from them, while others will just upgrade their CVs and resumes so it looks cool to the Human Resources. I usually go to India 2 or 3 times a year to conduct training on reliability and maintenance, and one of my students' names was Mr. Rajendran. He was easy to spot among the class since he was the tallest guy in the class. He also heads the overall maintenance function in their organization. After a few months, this guy was again in my class and during the break time, he approached me shook my hand, and gave me a pack of peanuts. He was very thankful and told me that they applied what they learned from the module on lubrication and purchase some off-line filtration carts which they use regularly on their hydraulic equipment. To their surprise, their hydraulic failures were reduced as well as their maintenance costs. It really humbled me that I made some difference to the working lives of these good maintenance people.

There may or may not be an answer to all of this but let me start off that training consists of two types, we have the hard skills and soft skills. Hard skills are related to specific technical knowledge and training while soft skills are personality traits such as leadership, communication, or time management. Hard skills are technical knowledge gained from training needed to develop the skills to perform the people's job correctly. Soft skills are training that improves our personal well-being that shapes how we work and interact with others. An example of these training includes supervisory skills, Leadership, Team-Building, Work Ethics, Effective Communication, Teamwork, Time-Management and the like. This type of training may not be easy to quantify as to how it impacts the bottom-line indicators. Although we can still generate KPIs for this training. Here are some indicators on training that we can measure

Pre-Post Quiz: One measure of the training I conduct is to include a pre and post-quiz, which is given usually before the start of the training and before the training ends. The question for this pre-post quiz is usually the same. The format of the quiz can be a multiple-choice, true or false question, or other means. The logic of this is quite simple, their score on the post should be higher than their score on the pre-test or quiz. If this is reversed, then something is wrong in how we taught the subject.

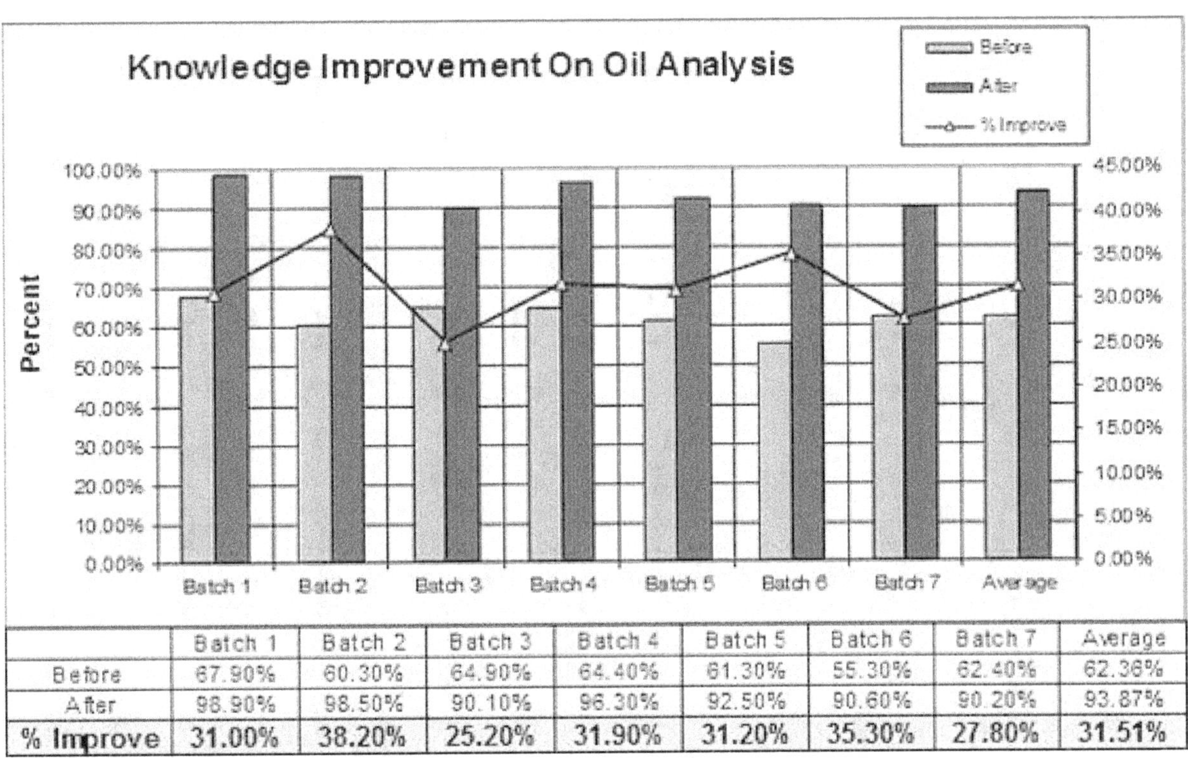

	Batch 1	Batch 2	Batch 3	Batch 4	Batch 5	Batch 6	Batch 7	Average
Before	67.90%	60.30%	64.90%	64.40%	61.30%	55.30%	62.40%	62.36%
After	98.90%	98.50%	90.10%	96.30%	92.50%	90.60%	90.20%	93.87%
% Improve	31.00%	38.20%	25.20%	31.90%	31.20%	35.30%	27.80%	31.51%

Figure 9.4: Actual Result of Pre-Post Test on Oil Analysis

Figure 9.4 graph represents the knowledge gained by the participants as reflected on their pre and post-test results where a quiz is provided at the start and end of each training to determine the amount of knowledge they have absorbed from the training itself. Although this is a measurement to gauge the effectiveness of the training. It does not mean that it contributed to the bottom line. This can be a good measurement if the students have learned from the training class. However, the downfall is that the retention decreased as time passed by on what the students has learned from the training.

Training Needs Analysis: Another important role of the training function, while others call this Learning and Development is to provide a Training Needs Analysis to their people. This is used to assess their people's training needs from time to time. To begin with, a skills gap is necessary to conduct, to assess the difference between the skills required to perform a specified task and the skills maintenance currently possesses. With these scores, a sensible skills training plan can be developed to transform the maintenance workforce into a highly skilled team that is fully trained and qualified to perform these tasks in a precise state every time maintenance tasks will be performed on the equipment.

Planned Maintenance Skills Evaluation

Gearing Towards A Pro-Active Maintenance System

Division : Central Equipment   Team Name: The Untouchables   Equipment type handled : All Types
Station : PLCC Department   Leader: Karl Richards

Legend :

	LEVEL 1	LEVEL 2	LEVEL 3	LEVEL 4
	Knowledge & Skill not Satisfactory	Knowledge Satisfactory	Skill Satisfactory	Knowledge and Skill both Satisfactory
	(0 points)	(0.5 points)	(0.75 points)	(1 Point)

Classification	No.	Knowledge / Skill Item	KARL	BOB	CHARLIE	RACQUEL	CAS	JOHN	BUDDY	NENA	FRANZIN	JB
BASIC MACHINE FUNCTION	1	Basic Machine Function										
	2	Machine Specs, Parts and Function										
	3	Knowledge in Actual Set-up and Conversion										
	4	Basic Lubrication Knowledge										
	5	Bolts, Screws and Fasteners										
	6	Basic Repair and Troubleshooting										
ANALYTICAL SKILLS ENHANCEMENT	7	Failure Mode and Effect Analysis										
	8	Root Cause Failure Analysis										
	9	P-M Analysis										
	10	MTBA Snapshot and Analysis										
	11	Single Minute Exchange of Dies (SMED)										
	12	Statistical Process Control										
PNEUMATICS & HYDRAULICS	13	Knowledge and use on FRL's										
	14	Knowledge and use on Pipings and Connectors										
	15	Knowledge and use of Cylinders										
	16	Knowledge and use on Filtration										
	17	Knowledge and use on Speed Controllers										
	18	Knowledge on Leaks and Seals										
MAINTENANCE TECHNOLOGY & OTHERS	19	Bearing Failures and Causes										
	20	Sensors Technology										
	21	Motors and Pumps										
	22	Total Productive Maintenance										
	23	MRO Spare Parts Management										
	24	Reliability-Centered Maintenance										
	25	Maintenance KPI and Indices										
PREDICTIVE MAINTENANCE (Specialization)	26	Basic Knowledge on Vibration Monitoring										
	27	Basic Knowledge on Heat and Thermography										
	28	Oil Analysis and Tribology										
	29	Condition-Based Maintenance Tactics										
	30	CMMS Structure and System										
S5-03			30.00	22.50	12.50	12.50	13.50	13.50	14.50	14.50	13.50	16.25

Figure 9.5: Knowledge and Skills Assessment Gap for Maintenance

Training Needs Analysis (TNA) is the process in which the company identifies the training and development needs of its employees so that they can do their job effectively, depending on the specific needs of that function in the organization. Purchasing people may require skill in negotiating with its vendors, while a maintenance function will be focusing on training that can sustain and improve the reliability of their equipment and assets. Typically, the manager or supervisor will identify the needed skill that is required to perform the job correctly. This means that the job should be completed both effectively and efficiently. This should be a continuous process since technology changes. As automation changes and more systems begin to rely on process indicators, their need to produce accurate data is growing exponentially, especially in this era where maintenance is in the digitalization age. Training gap is defined as the difference between the skills required to complete the job and the existing skillset of any particular team member. Focusing all our training based on the needs may or may not still contribute to the bottom line results. For example, if we train our people on the correct practices on repairing hydraulics by hiring a hydraulic consultant and expert will allow our people to know how to repair yet leave us a lot of repair works on our hydraulic system since many are still failing.

Training Hours Per Employee: Another indicator I see often industries used in training is the training hours completed by the employee per year. This means that if a plant has 1000 employees, and the plant operates 365 days or 8760 hours per year. Each employee may be required to complete several hours before the year ends as a requirement from the training department. This can be measured either in hours or in percentage. Assuming that each person is required to complete 240 hours of training and you only consumed 200 hours, then your Training Hours Completion is at 83.3 %. The backlog of 40 hours can be added the following year.

Return on Investment on Training: I believe that this is the best indicator for training but this can only happen if the training is implemented and results were achieved. Management people should talk to their people and set expectations before sending them to training to ensure that what has been learned will be implemented by the plant. Once a success story is generated, then this must be looped back to the training department.

9.4: Maintenance is as Strong as its Weakest Link

There are several functions on maintenance in a mature organization, and each of these functions is as good as its weakest link, just like a chain. Each of these functions can never be an isolated group, but it will also depend on the other functions of the maintenance organization. A football team has 11 players. Each of these players will have their own designation and role in the game. The positions include 1 goalkeeper, 4 defenders, 3 midfielders, and three forward. The goal of the goalkeeper is to block the ball from the opponent, the midfielders are the bridge between the defenders and forwards usually responsible for carrying the ball, while the forward's role is to score a goal for the team. Just like in a game of football, the different functions of the maintenance organization have a specific role to play and will be needing each other, where the maintenance manager acts as the coach. Each of these people involved in the different functions relies on other functions so that together, they can achieve their goals on maintenance. What I am trying to say is that every function on maintenance is connected, and the strength of your maintenance organization depends on its weakest link.

• People involved in executing PM depend on the plan of the maintenance planner.

- The planner is also dependent on the feedback of the Preventive Maintenance team.
- The planner is also dependent on the availability of the parts, items, tools, and consumables needed for maintenance.
- The Predictive Maintenance must be in constant communication with the Preventive Maintenance so that only parts with problems will be subject to replacement, or overhauls
- For manufacturing plants, the entire operations and maintenance will depend on the Facilities / Utilities of the plant. These are the people who provide the power, air, energy, and other resources for production equipment and machines to run.
- Instrumentation and calibration people ensure that test instruments, gauges, protective devices are calibrated and functioning without any deviations from the norm.
- CMMS depends on the output of the entire function on what to automate and streamline. The outcome entirely depends on what we want to populate.
- But all of these functions depend on the very foundation which is training as training will provide us the knowledge on how to perform and execute our work correctly. The main role of each of these functions includes:

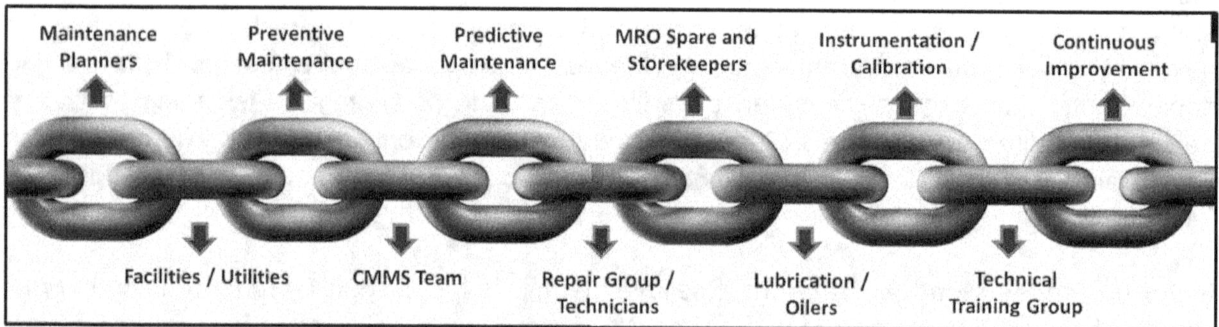

Figure 9.6: Maintenance is a Good as its Weakest Link

Maintenance Planners: Provide the Planned Activities for the Preventive, Predictive, and Corrective Maintenance. The planner provides the details of the maintenance tasks to be performed, labor estimates, duration, bill of materials, consumables, MRO spares needed for a scheduled Preventive Maintenance and Corrective Maintenance from the PdM group.

Preventive Maintenance Group:: People involved in these activities will be responsible for the execution of tasks dictated by the maintenance planner based on the schedule provided. Common activities will include cleaning, routine lubrication, inspection, conducting failure-finding tasks, calibration, scheduled replacements, and overhauls.

Facilities / Utilities Group: The role of the maintenance people from the facilities is responsible for sustaining and maintaining the industry's buildings, facilities equipment, and assets to ensure the production operates without any problems. The role of the people involved in the facilities is to ensure that their facilities are operating daily by completing daily inspections and conducting repairs, contracting works, and performing scheduled Preventive Maintenance, and Predictive Maintenance as well

Predictive or Condition-Based Maintenance Group: The main role of this group is to anticipate or Predict failures through the use of their Predictive Maintenance instruments. Their role is to monitor and analyze critical assets in the plant for any signs of Potential Failure. Once a Potential Failure is detected, these people will recommend corrective maintenance before a functional failure is likely to occur on their equipment. Predictive Maintenance can be performed online, offline, or both.

CMMS or EAM Group: Usually the people involved in this case will be the IT people or those with knowledge of computerization. The main difference between a CMMS and EAM will be the scope. CMMS or Computerized Maintenance Management Software is designed for the maintenance department whose function is to streamline and automate the maintenance process. An effective CMMS should support these needs by capturing and automating administrative tasks for maintenance that usually eats up time in doing it manually, and minimize the chances of human error as well. CMMS is a software design to automate and streamline the process of maintenance, while an EAM or Enterprise Asset Management Software will not only include automating maintenance but other functions of the organization.

MRO Storekeepers: The role of the MRO Spare Parts and Storekeepers is to ensure the right parts get to the right place at the right time when operation and maintenance needed them most. People in charge of the storeroom should provide the item or part needed by maintenance, supply them as quickly as possible, yet at the same time control the overall cost of spares. The storeroom should support the following, items, parts, consumables needed for a scheduled Preventive Maintenance, corrective maintenance performed as a result of Predictive Maintenance, and parts needed for repairs. Although in other industries, other functions besides maintenance control and manage the storeroom, these people must be in constant communication and collaboration with the maintenance function as these are their primary customers.

Lubrication Group: These people are responsible for all lubrication activities in the plant. Although routine lubrication can be performed by Operators, especially those involved in TPM's Autonomous Maintenance activities. People involved should be responsible for generating a plant-wide Oil Contamination Control Awareness Campaign, and oil analysis activities. Note that Oil Analysis can also be included in the Predictive Maintenance or CBM group.

Instrumentation and Calibration Group: Calibration is the process in which instrumentation and equipment used in industries are monitored and maintained to ensure they continue to give accurate and reliable results and ensure that deviations are corrected. The role of these people is to inspect their test instruments on a scheduled basis to ensure that the reading does not deviate and meet expected results. If there is a deviation found, that piece of equipment or instrument will be further analyzed, adjusted, calibrated, or repaired as necessary.

Technical Training Group: Although in other industries, these functions may be absorbed by the Training Department, Human Resources or Learning and Development function, the role of these people is to conduct a Training Needs Analysis for Operators and Maintenance and provide the training needed so that both operators and maintenance can perform their job correctly right the first time around. Training should be based on the needs so that both operators and maintenance can build their skills in performing their jobs correctly. Training is the venue where we absorbed knowledge. Knowledge will provide us the means to develop our skills. Mastery is the point in time where we can transfer this knowledge to others.

Repair Group: These are people responsible for repairing equipment when it failed or breakdown in operations. These people can also be involved in performing corrective maintenance based on the recommendation of the Predictive or CBM group. Their role is to restore the equipment back when it fails.

Continuous Improvement Group: Although in some industries a full-time person will be in charge of this group which monitors all the improvement activities of the plant. People involved in the maintenance improvements are usually a cross-selection of people from the maintenance function addressing a particular problem using different analytical problem-solving tools such as FMEA, RCM, RCFA, basic problem-solving analytical tools. Once the improvement project has been completed, the team disbands and goes back to their usual roles and responsibilities.

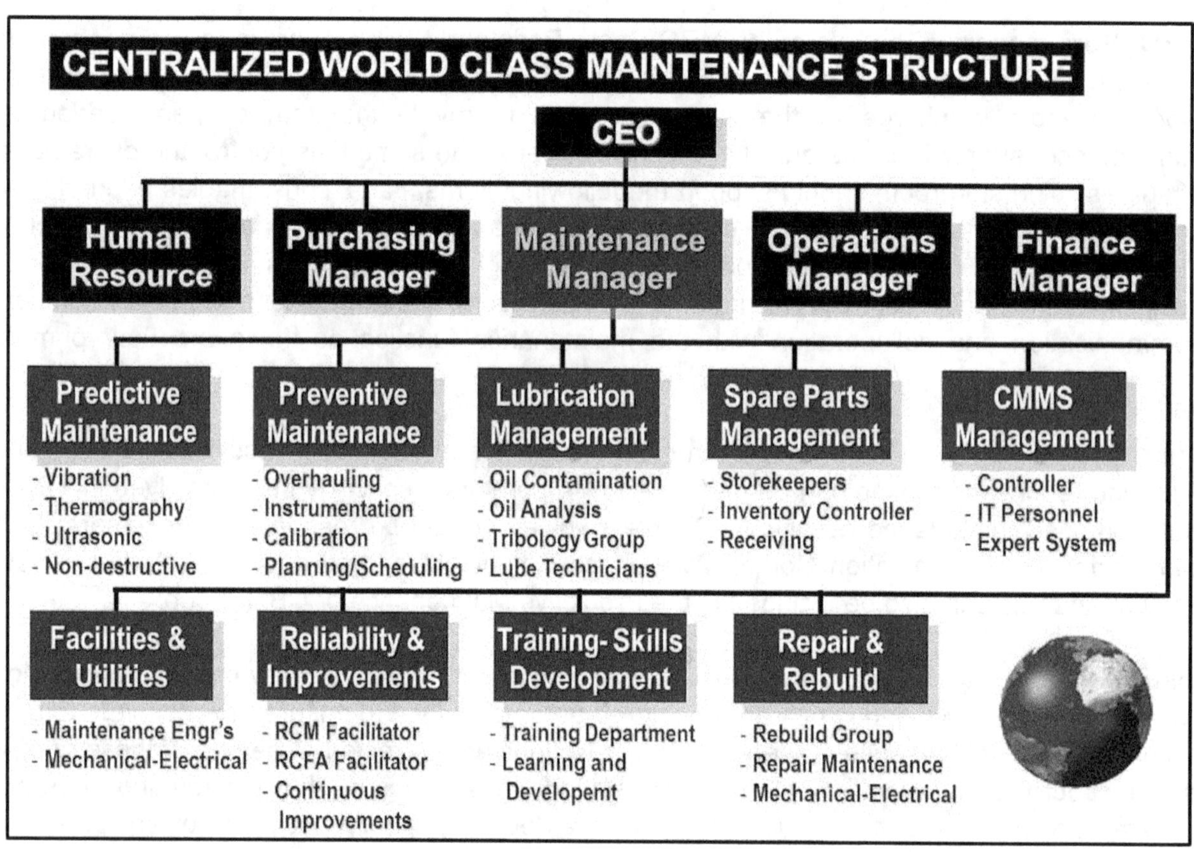

Figure 9.7: World Class Maintenance Organization

9.5: Achieving World Class Maintenance Level Can Now Be Measured

Although I have already included, this in my other books on Book 2, Maintenance – Roadmap to Reliability (MRR), Book 3, Reliability, A Shared Responsibility for Operators and Maintenance (OAM), and Book 4, Cutting Edge Maintenance Management Strategies (CEMMS), let me again include it in case the reader does not have a copy. All these books I have written and will be writing are in every way connected to my first book which is World

Class Maintenance Management – The 12 Disciplines. These 12 disciplines are categorized into 3 groups, the basics, the strategies, and the advanced disciplines.

Basic Discipline: The basic discipline will start with its very basic foundation which is training, as knowledge will serve as the key to performing their jobs correctly. Training and Education will be a continuous process as technology travels fast. Once the training needs are identified, they will be provided to both operators and maintenance. All the different functions of the maintenance organization should sit down and discuss what do they need to measure. They need to determine which measures, indices, and KPIs are important to them so that they can start measuring their performance. These indices will dictate if they are on the right track or not. Maintenance also needs to create a lifetime partnership with operators that cannot be divorced just like the Rolling Stones so that they can both address the basic equipment condition of their equipment. Maintenance will try to review their lists of tasks performed on their Preventive Maintenance so that it can better serve its purpose. These are the basic disciplines of maintenance.

Intermediate Discipline: These disciplines refer to the different reliability and maintenance strategies that can be adopted once the basic equipment condition has been well established. These refer to strategies such as conducting Root Cause Failure Analysis, Reliability-Centered Maintenance or streamlined RCM process, Total Productive Maintenance, Life Cycle Management, Lubrication Management, and MRO Spare Parts Management. Once the basics have been well established, we can now develop these strategies. Have we performed a root cause analysis investigation just to find out that the cause of the problem is a wrong tool used by the maintenance technician? Or equipment that lacks some bolts and nuts due to some previous overhauls done, causing excessive amplitude from a vibration reading. Or perhaps lapses on the maintenance to bring back all the bolts together. These simple things, when neglected, cause big problems in our equipment, and I cannot overemphasize the importance of addressing the basics first before proceeding with the reliability and maintenance strategies.

Advance Discipline: One of the requirements in pursuing this discipline will be an investment in both the people, the instruments, and the software needed. It is very difficult to advance to this discipline if the basic condition of the equipment has not been well established. Savings generated through the application of basic and intermediate discipline can be well spent on the acquisition of these technologies. Most industries try to resort to these technologies first only to be abandoned in a couple of years. Do not be misled by those sweet-talking vendors who promise a silver bullet solution to every single maintenance problem. There is no single strategy or rocket science solution that will address every single maintenance problem at hand. Knowledge of what maintenance management strategy to use and when to use it will be the key to improving reliability.

Implementing the Twelve Maintenance disciplines can be done by conducting a maintenance assessment, which can be done during the strategic planning for key people in the Maintenance Department. I would recommend assigning an outside and experienced Facilitator to guide the members in the planning session. Start by defining your current maintenance situation. Set this as your focal baseline on where the maintenance organization structure currently is at this point. Make a selection of people and start by assessing your current status using the questionnaires on the assessment. Next, organize a group of cross selections of members and have them conduct an internal assessment based on the 12 disciplines of maintenance. Assessment should include some form of a survey, line audit, or perhaps having a standard list of questionnaires for interviewing both operations and maintenance people about their work. Knowing where we are will be the starting point on any reliability and maintenance improvement drive and effort. Start with the Basic Discipline first and assess how we are doing based on this point.

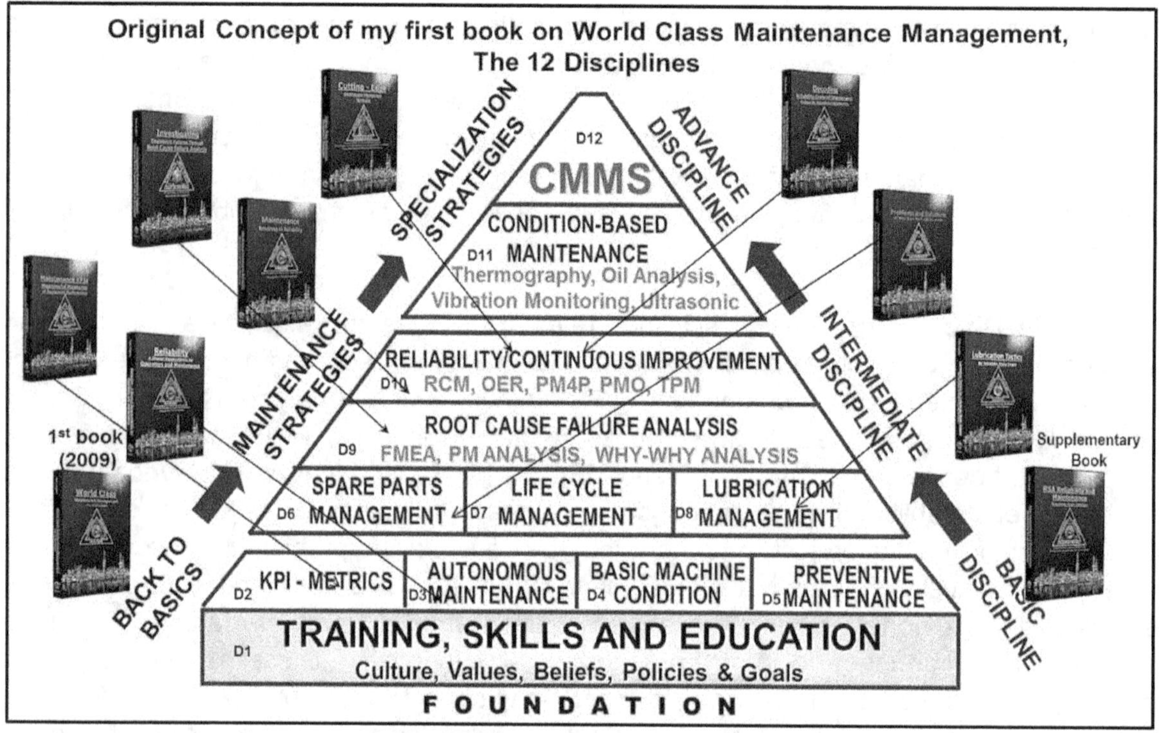

Figure 9.8: The Original World Class Maintenance Management – The 12 Disciplines

At this stage, maintenance must agree on what initial maintenance indices to measure to create a benchmark or starting point on this drive. These indices identified should be closely monitored to determine the success or failure of this initiative. The goal is to improve these maintenance indices as we journey along with the implementation of each of these disciplines. The journey to World Class Maintenance should be done on a plant-wide scale, which means that everyone must be part of the process. Getting everyone on board requires a great amount of initiative and passion as we will be changing the culture of the entire maintenance organization. To be honest, this will be the role of the facilitator. It is a very frustrating job, but believe me when I tell you that there is no better reward once everything falls in place and the benefits each of these disciplines will reap. The facilitator must have a strong passion, commitment, patience, and empathy in driving this initiative. He must bear with him a lot of patience, as there will be many resistance and obstacles at the beginning, which is quite normal. What is not normal is if nobody is resisting. Start by identifying how many departments are involved in the maintenance and assign a leader representing every department in the plant. At this stage, it is also important to organize a Maintenance Core Group or Maintenance Steering Group composed mostly of people from management from both Maintenance and Operations. If we talk about the Mafia Mob, these people will serve as the God Fathers.

<u>The objective of this Assessment:</u> The objective of this world-class maintenance assessment is to determine where your maintenance organization's strengths and determine its areas and opportunities for the industry to grow and improve. This maintenance assessment will allow the organization to view its structures, relationships, processes, and the people involved in its day-to-day maintenance activities. This is the first step in the overall maintenance system improvement and development. There will be 129 survey questions that need to be answered, which are divided into the Twelve Disciplines of Maintenance Management. The more people that will take this survey, the more accurate the results will be. Try to answer the best you can on the question at the bottom of the assessment; your perspective definitely counts on this survey. If 20 people took the survey, the overall score would be divided by 20 to get the average score. The average score will be compared to the ratings below to determine where your industry currently ranks.

<u>Please Read and Follow the Instructions Carefully:</u> We would like each of you to respond and answer the question of how you see your organization is currently doing it now and not what it needs to be done in the future. You do not need to write your name to create an unbiased result at the end of each survey. Kindly select only one of the letters you think you and your organization are currently doing by checking the box. If, after some thought, you want to change your answer, kindly place an X on the checked portion letter and check the letter for your final answer. Note that this is not a quiz and the overall output of this assessment will determine where your industry stands from the worse to a World-Class Maintenance level.

- A) 1 point - I do not know the answer to this question.
- B) 2 points - No, means you know the value, but this simply does not exist in our plant.
- C) 3 points - Sometimes, this is being used to some extent but not regularly.
- D) 4 points - Yes, our people know its value, and we use them to a good extent.
- E) 5 points - Always, this question is very important in my organization, and we comply with it.

<u>Instructions:</u> • If there are any survey or assessment questions that you don't understand, you may call the facilitator to clarify the question. Answer the question on the bottom part that you feel is relevant and important to your plant. If space is not sufficient, you may write at the back. We appreciate it if you can provide us your thoughts regarding this. Take your time to answer this survey; you will be given 2 hours for the assessment duration. Do not rush in answering each question, as the output of the overall survey is important to your organization. We would like to repeat the instruction you need to answer based on how you are doing them now and not what you think your industry should be doing. Our point is to determine how you compare with the best in class industries. If you believe that the question is not applicable or relevant, just check letter (A) 1 point. The score below will determine where your industry currently is. If the current question does not exist in your plant, check A on the box, meaning you will get 1-point credit for this question. It is important to complete this survey and do not leave any blank.

World Class Maintenance Scoring Ranges (Revised as of March 25, 2020)
- 600 to 645 points: (93 to 100%) World Class Maintenance – Best in Practice
- 516 to 599 points: (80 to 92%) Very Good Performer
- 387 to 515 points: (60 to 79%) Average Performer
- 258 to 386 points: (40 to 59%) Below Performer, many opportunities for improvement
- 257 and below   (39% and below) Poor and Worst Industry Performer

No.	Discipline 1: Training, Skills, and Knowledge Assessment	A 1 point	B 2 points	C 3 points	D 4 points	E 5 points
1.	Does your company allocate a budget for training based on the training needs of your people?					
2.	Is your training based on the needs of both your maintenance and operations people?					
3.	Does your management provide some expectations as to why you need to attend this training and the plan after completing the training?					
4.	Is there a regular yearly calendar set of training schedules provided with your maintenance and operations people?					
5.	Is there a training department in your plant? If yes, do they have a clear goal, plan, and objective to foster this knowledge and skills to their people?					
6.	Do other departments in your industry support the existence of your training department as a vital tool in improving the plant's productivity?					
7.	Does your training make a follow-up to the people they teach if they have been using the knowledge gained in their daily work activities?					
8.	Does management allocate time for their people to attend training and not be pulled back to work in the plant?					
9.	Are the skills and competency of the people updated, reviewed, and regularly evaluated by the training department?					
10.	Do the people know what training courses they need to perform their jobs and develop their skills?					
11.	Are the people equipped with the knowledge, both practical and theoretical, to perform their jobs in the plant?					
12.	Do your management leaders, superintendents, and supervisors also attend training together with their people?					
Please write your feedback or recommendation on how to improve this discipline in your industry.						

Figure 9.9: Discipline 1: Training, Skills and Knowledge Assessment

No.	Discipline 2: Maintenance Indices and KPI Assessment	A 1 point	B 2 points	C 3 points	D 4 points	E 5 points
1.	Are there KPI's (Key Performance Indicators) and measures tracked and reviewed by your plant's organization regularly?					
2.	Are maintenance craftspeople involved in setting as well as meeting the goals and objectives with the maintenance department?					
3.	Are any of the following KPIs measures reviewed regularly by the maintenance group, such as PM Compliance, Ratio of PM hours versus Emergency Repairs hours, MTBF, MTTR, maintenance costs, Breakdown Occurrence, and so on?					
4.	In line with item 3, are there clear goals and targets on these measurements and KPIs and known to all people from the organizations?					
5.	Are your people familiar with what constitutes a breakdown or when to say a breakdown occurred?					
6.	Does your management team, together with your people, regularly review the maintenance cost? Are there any strategies or targets to reduce your maintenance costs?					
7.	Are maintenance people motivated to challenge the goals? When the goals are exceeded, is the maintenance rewarded or recognized in any form?					
8.	Are the whole maintenance people involved in the KPI, and are you familiar with the plant's goals and targets for the year? Are goals clear to everyone in the organization down to the shop floor people?					
9.	Are the KPIs and measures that are important currently being tracked by the maintenance organization? Are they relevant, updated, or do you think that they need to be changed or retained?					
10.	Are data such as downtime, breakdown captured regularly and can easily be known or retrieved? For example, the total downtime last month can be retrieved easily, and the main contributor to the downtime is easily known.					
Please write your feedback or recommendation on how to improve this discipline in your industry.						

Figure 9.10: Discipline 2: Maintenance Indices and KPI Assessment

No.	Discipline 3: Operator and Maintenance Partnership Assessment	A 1 point	B 2 points	C 3 points	D 4 points	E 5 points
1.	Does maintenance teach or coach operators about their equipment? Do operators know how to perform minor repairs on their equipment?					
2.	Is there an operator's checklist done on their equipment? Does the operator perform these checks correctly and completely?					
3.	Is there a program to enhance the operator's skills on their assets and machines to understand their equipment intimately?					
4.	Does the operator assist the maintenance during the repair period when the equipment is down? Does maintenance take this opportunity to teach operators about the failure?					
5.	Can the operator spot problems easily, such as noise, excessive vibration, or others, before a breakdown happen? Do they communicate the problem with the maintenance immediately?					
6.	Are your operators confident of performing minor repairs themselves instead of maintenance? Do they have the basic tools to perform these repairs?					
7.	Are PM schedules on time and not deferred by operations? Does the operation understand why PM needs to be done on the equipment?					
8.	Is there some form of training for operators to familiarize themselves with the functions of each sub-assembly of their assets and equipment?					
9.	Does the operator perform the basics on their equipment, such as cleaning, basic lubrication, and other things? Do they know the importance of doing these basics on their equipment?					
10.	Does the operator understand the importance of cleanliness on their equipment and maintaining it clean always?					
11.	Does your operator understand the importance of lubrication and performing it regularly on the equipment?					
12.	Does the operator understand the use of visual control to simplify inspection and detect abnormalities on their equipment?					
Please write your feedback or recommendation on how to improve this discipline in your industry.						

Figure 9.11: Discipline 3: Operators and Maintenance Partnership Assessment

No.	Discipline 4: Basic Equipment Condition Assessment	A 1 point	B 2 points	C 3 points	D 4 points	E 5 points
1.	Do the operator and maintenance know what basic equipment condition is all about and who is responsible for doing them?					
2.	Are leaks exposed and address immediately? Do operators and maintenance know its impact on the equipment besides safety issues?					
3.	Are sources of contamination such as dirt and leaks addressed on the equipment? Are there any permanent countermeasures deployed to prevent them from recurring?					
4.	Are all the gauges, charts, sensors, and protective devices in the equipment working and readable? Does the operator know the limits of these gauges in their asset and equipment?					
5.	Can operators spot problems on their equipment easily, such as abnormal noise, excessive vibration, unwanted smells such as burnt wires and motors?					
6.	Do operators perform lubrication in the equipment themselves? Do they know the lubricating points, min-max, quantity needed, and type of lubricant to be used?					
7.	Does the equipment have complete bolts, fasteners, and screws to keep vibration minimum and controlled?					
8.	Does the operator know how to interpret mechanical drawings and schematic diagrams of the asset or system?					
9.	Is the equipment maintained clean by the operator? Are operators provided sufficient time to clean their equipment either before or after their shift?					
10.	Are operators trained on lubrication? Are they knowledgeable about lubricants and how they are being contaminated?					
11.	Are deteriorations, abnormalities, defects, and deviations that are found addressed immediately on the asset or equipment?					
Please write your feedback or recommendation on how to improve this discipline in your industry.						

Figure 9.12: Discipline 4: Basic Equipment Condition Assessment

No.	Discipline 5: Preventive Maintenance Assessment	A 1 point	B 2 points	C 3 points	D 4 points	E 5 points
1.	Is the frequency of Infant Mortality Failure or start-up failures minimal, which means that equipment does not fail after performing Preventive Maintenance on the equipment?					
2.	Do the new and old equipment of the same type receive only the required PM needed during the planning stages?					
3.	Do both Preventive Maintenance and planners review their PM tasks for accuracy, revisions, additions, deletions, and feedback before executing these PM tasks on the equipment?					
4.	Is the amount of emergency work and repairs reduced as a result of the Preventive Maintenance activities done? Is there any KPI or evidence on this?					
5.	Are spares used for replacing the equipment for a PM schedule well communicated ahead of time with the Storeroom and Purchasing people?					
6.	Do operations allow maintenance to access the equipment for a PM scheduled, which means that Preventive Maintenance activities are not being waived by operations or others 95% of the time?					
7.	If contractors are used for PM, are they provided with a detailed list of tasks on what to do on the equipment? Do they comply 95 % of the time?					
8.	If Predictive Maintenance or CBM exists in your plant, do they communicate with PM before the scheduled PM regarding their assessment on the equipment using their non-destructive Predictive Maintenance instruments?					
9.	Do the maintenance planners have a job plan, detailed procedures, and Bill of Materials for a major PM Shutdown on the equipment?					
10.	If a spare part is still good and in running condition, but it is subject to a PM replacement as indicated in the PM tasks, will the spare parts be replaced with a new one?					
11.	Are equipment undergoing PM prioritized based on its criticality and severity? Does maintenance have a guideline or matrix on how to determine critical equipment in the plant?					

Figure 9.13: Discipline 5: Preventive Maintenance Assessment

No.	Discipline 6: MRO Spare Parts Management Assessment	A 1 point	B 2 points	C 3 points	D 4 points	E 5 points
1.	Are obsolete MRO spare parts resulting in decommissioned machines well communicated with the MRO storeroom people and removed from the storeroom immediately?					
2.	Are the people from the MRO storeroom for spare parts managed by the Maintenance and/or Engineering Department?					
3.	Is the number of stock-out low? A stockout is defined as the number of times your people request a part in the storeroom and the part has zero stocks.					
4.	Is your spare parts inventory automated and included in your CMMS (computerization software) or EAM (Enterprise Asset Management) software? Are the actual inventory and systems inventory (computer) accuracy is 95% or more accurate?					
5.	Are the parts under warranty, shelf-life, and expiration dates known to the storeroom and the Maintenance Department?					
6.	If your plant is 24 hours operating, does the storeroom also run for 24 hours? For example, is your storeroom manned for 3 shifts if you are operating for 24 hours a day?					
7.	Do storeroom people provide a card catalog for all the spare parts? Are the spare parts catalogs regularly updated for additional parts and those that are obsolete?					
8.	Is the maintenance knowledgeable in identifying the correct parts code or part number of a spare when a breakdown happens where a part needs to be replaced?					
9.	Is the amount of emergency buying and stock out by maintenance low and excess parts returned to the storeroom by maintenance?					
10.	Does your storeroom use barcoding or any other form of automation in issuing spare parts or transactions with the storeroom?					
11.	Can maintenance decide whether to stock or not to stock a spare part and no longer rely on their Vendor or OEM recommendations?					
12.	Does maintenance review the cost of non-moving and obsolete parts currently in your storeroom? Do they have plans on what to do with them?					
Please write your feedback or recommendation on how to improve this discipline in your industry.						

Figure 9.14: Discipline 6: MRO Spare Parts Management Assessment

No.	Discipline 7: Life Cycle Management Assessment	A 1 point	B 2 points	C 3 points	D 4 points	E 5 points
1.	When awarding a part to a vendor, do maintenance and purchasing agree to base their decision based on the Life Cycle of the part even if the cost is a bit more expensive than their competitors?					
2.	When deciding to purchase new equipment, does your organization also consider the cost to maintain the equipment before making any further decision to purchase?					
3.	When deciding to localize some spares, does your team study the metallurgical aspect or strength of the part's material and evaluate its life span?					
4.	When commissioning new equipment in the plant, does maintenance find ways to reduce the time to commission based on previous purchasing equipment and assets?					
5.	Do purchasing or people involved in requisitioning spares aware and understand what Life Cycle Management is all about? This means that the decision to purchase is based on the life span of the part and not on the costs.					
6.	When your plant purchases new equipment, does your organization compare and study other costs in maintaining the equipment such as spares, maintenance costs, commissioning costs, cost of consumables, energy or utility costs, and others?					
7.	When the same spare are obtained from two different vendors, does your team provide feedback to purchasing regarding which part actually lasts longer and provide recommendations to whom to award it?					
8.	When trying to modify parts with a short life span, does your team evaluate the life span of the modified part compared to the original part?					
9.	Does maintenance keep track of how much it costs to maintain their equipment and make studies to reduce the costs of doing maintenance?					
10.	Is your maintenance team part of the decision-making process on which vendor or suppliers to award a certain spare part or when purchasing new equipment in the future?					
Please write your feedback or recommendation on how to improve this discipline in your industry.						

Figure 9.15: Table on Discipline 7: Life Cycle Management Assessment

No.	Discipline 8: Lubrication Management Assessment	A 1 point	B 2 points	C 3 points	D 4 points	E 5 points
1.	Does your maintenance team understand the effects of oil contamination, how your oil is being contaminated? How to control oil contamination?					
2.	Is your lubricant storage facility centralized and kept in an area with a controlled temperature, humidity, lighting, and minimum dust?					
3.	Is your maintenance team aware of the operating temperature of your equipment? Do they know how to select the type of grease to be used and up to what temperature it can be used? For example, Aluminum base grease is not recommended for temperatures above 175°F or 79.4 °C.					
4.	Is your maintenance team aware of different types of grease and why it should not be mixed? The reason for not mixing is that they have different additives and can cause incompatibility issues.					
5.	Does your maintenance team sample their oil regularly to determine the condition and health of the oil?					
6.	Are your people aware of the correct practices regarding lubricating grease and oil and how it affects the performance of their equipment and assets?					
7.	Does maintenance sample their oil regularly? Do they monitor and track their oil analysis report and understand how to interpret them?					
8.	Does maintenance understand the correct interval on when to re-grease their rotating equipment and assets in the plant?					
9.	Do the people performing greasing understand the correct quantity to apply and know when to stop greasing? Do they have any form of measuring tools such as meters or ultrasonic monitoring devices on their grease gun to determine the precise amount of grease dispense?					
10.	Does your maintenance team use closed-type containers when tapping oil in their equipment, as this directly affects oil contamination?					
11.	Does maintenance use any means of filtration on equipment when transferring new oil or changing the oil on the equipment?					
Please write your feedback or recommendation on how to improve this discipline in your industry.						

Figure 9.16: Discipline 8: Lubrication Management Assessment

No.	Discipline 9: Root Cause Failure Analysis Assessment	A 1 point	B 2 points	C 3 points	D 4 points	E 5 points
1.	Does your maintenance team perform Root Cause Failure Analysis on recurring failures, especially those with high impact and consequences?					
2.	Is your team familiar with the most basic types of wear, as well as how to distinguish them, such as abrasion, erosion, adhesion, fatigue, and how to control them?					
3.	Is your maintenance team equipped with the knowledge and technique on how to investigate equipment breakdowns and failures?					
4.	Is your management team aware that the reasons for conducting RCFA are not to blame or punish anyone but to learn from the things that go wrong with our equipment in the plant?					
5.	Do your people understand that Root Cause Failure Analysis can only be done when the evidence is preserved, meaning that it is not technically possible to perform a root cause when the equipment is already repaired.					
6.	Is your maintenance team practicing any of these analytical problem-solving tools such as 8 Disciplines, brainstorming, Pareto Analysis, FMEA or FMECA, why-why analysis, fault tree, and fishbone diagram to find the possible cause of the problem?					
7.	Do your people understand that performing root cause failure analysis is not only about performing countermeasures and corrective actions but also addressing the latent cause of the problem?					
8.	Is your organization no longer on the blame culture that someone will always be blamed for the failure when something goes wrong in the plant?					
9.	Are your people in your industry well trained in conducting a proper Root Cause Failure Analysis investigation on how to probe in dealing with equipment failures and breakdowns?					
10.	Is there any strategy or training program in your plant design to deal with human errors and reduce or manage them for both operations and maintenance?					
Please write your feedback or recommendation on how to improve this discipline in your industry.						

Figure 9.17: Discipline 9: Root Cause Failure Analysis Assessment

No.	Discipline 10: Reliability and Continuous Improvement Assessment	A 1 point	B 2 points	C 3 points	D 4 points	E 5 points
1.	Is there any strategy that currently exists in your plant besides PM that is designed to improve or sustain the reliability of your equipment and assets?					
2.	Does your industry perform Reliability-Centered Maintenance, Total Productive Maintenance, or any form of reliability and continuous improvement strategy for maintenance?					
3.	Does your industry provide benchmarking activities on similar industries to determine how well or not you are currently doing maintenance?					
4.	Are your maintenance strategies aligned with your company's overall corporate goals, targets, mission, and vision?					
5.	Are your maintenance people highly motivated and involved in any initiative to improve their skills and knowledge on maintaining their equipment and assets in the plant?					
6.	Are your current Preventive Maintenance activities effective in reducing the amount of crisis or emergencies, and are the activities on your current PM regularly reviewed before executing them?					
7.	Are your safety and Quality people also part of the reliability initiative of the plant, or do these people work hand in hand to improve or sustain the quality, safety, and reliability of the equipment?					
8.	Are people involved in continuous improvements on maintenance recognized or rewarded for their efforts by the company?					
9.	Are your maintenance people familiar with protective devices in their equipment? Are these protective devices inspected regularly for functionality tests to check if they are still working or not on a regular interval? (Examples of protective devices such as alarms, led, emergency stops, lighting arrester, and others.)					
10.	Does your maintenance understand the failure consequences that can lead to any environmental or safety consequences, and is there any strategy on your maintenance to address them?					
Please write your feedback or recommendation on how to improve this discipline in your industry.						

Figure 9.18: Discipline 10: Reliability and Continuous Improvement Assessment

No.	Discipline 11: Condition-Based Maintenance Assessment	A 1 point	B 2 points	C 3 points	D 4 points	E 5 points
1.	Does predictive maintenance exist in your plant? Are there full-time users and dedicated people that use these non-destructive instruments to check the actual condition of their equipment?					
2.	Are your Predictive Maintenance activities done in-house by regular employees? Are the users of these instruments qualified and certified?					
3.	Does your Predictive Maintenance team provide reports regarding cost avoidance or savings generated by predicting failures before they happen?					
4.	Do your Predictive and Preventive Maintenance people communicate regularly and consistently, especially on what to focus on during a Preventive Maintenance shutdown?					
5.	Does your maintenance and management team find value in doing any Predictive Maintenance such as vibration, thermography, ultrasonic monitoring, oil analysis, or other PdM techniques?					
6.	Are the users of these instruments in your plant knowledgeable not only on operating the instruments but also on its principles? Is your management supportive of having the users certified as an initial requirement?					
7.	Do your people and management have knowledge or understanding of Predictive Maintenance and the role they play in your maintenance and reliability strategies?					
8.	Can your Predictive Maintenance users decide to stop the equipment if a functional failure is on the verge or process of happening in monitoring their equipment?					
9.	Does your Predictive Maintenance people conduct a regular monitoring schedule, and are plotting the results of these date on every piece of equipment that is monitored, and tracked?					
10.	Is the amount of corrective emergency work and repairs reduced because of your Predictive and PM activities in your plant?					

Please write your feedback or recommendation on how to improve this discipline in your industry.

Figure 9.19: Discipline 11: Condition-Based Maintenance Assessment

No.	Discipline 12: Computerized Maintenance Management Assessment	A 1 point	B 2 points	C 3 points	D 4 points	E 5 points
1.	Does your maintenance organization use any CMMS (Computerized Maintenance Management Software) or EAM (Enterprise Asset Management) software for maintenance in your industry?					
2.	Is your maintenance storeroom fully automated and linked to your CMMS or EAM software? Are reports from the storeroom easily seen and retrieved?					
3.	Is your CMMS or EAM software regularly updated? Are maintenance craftspeople knowledgeable about using them, such as inquiring inventory of a spare, updating work orders, what equipment is scheduled for PM, KPI reports, and others?					
4.	Is your CMMS or EAM software capable of providing important technical information such as KPIs, total maintenance costs, maintenance budget, downtime, and other information maintenance needs?					
5.	Is your CMMS or EAM software capable of communicating with other software on maintenance such as vibration, thermography, oil analysis, and other PdM software?					
6.	Does your management find value in using their CMMS and that it has benefited the industry in reducing the amount of manual work in your plant?					
7.	Are management decisions, goals, targets, and strategies based on the outcome of your CMMS or EAM data and reports?					
8.	Are there clear guidelines and procedures on encoding downtime and other failures from the equipment to the CMMS or EAM software, and are the data encoded updated and reliable?					
9.	Are the features of your CMMS or EAM software used and maximized by the users? This means that at least 75% of your CMMS or EAM features are utilized by the maintenance and other functions in the organization.					
10.	Have your CMMS or EAM software reduced the amount of paperwork done on maintenance? Is your maintenance benefitting from your software?					
Please write your feedback or recommendation on how to improve this discipline in your industry.						

Figure 9.20: Discipline 12: Computerized Maintenance Management Assessment

COMPANY ABC OVERALL MAINTENANCE RESULT

Training & Education — Rank: 9th

Disciplines	Mtce 1	Mtce 2	Mtce 3	Mtce 4	Mtce 5	Mtce 6	Mtce 7	Mtce 8	Mtce 9	Mtce 10	Mtce 11	Mtce 12	Mtce 13	Mtce 14	Mtce 15	Mtce 16	Mtce 17	Mtce 18	Mtce 19	Total	Ave
1	1	5	4.5	3	5	3	5	3	5	4	3	3	3	3	2	2.5	3	3	3	64	3.37
2	1	1	4.5	4	4	3	1	4	1	4	5	4	4	4	2	2.5	4	4	4	61	3.21
3	1	5	4	1	4	3	5	4	5	4	1.5	4	4	4	2	4	4	4	4	67.5	3.55
4	1	5	3	1	3	4	5	2	5	3	1	2	3	3	2	2	3	3	3	54	2.84
5	3	1	3.5	4	5	1	1	2	1	4	2	3	2	2	2	3.5	2	2	2	46	2.42
6	1	2	4	3	4	1	2	2	2	3	1	1	1	1	2	3	1	1	1	36	1.89
7	1	3	3.67	3	1	1	3	3	3	3	2	1	4	3	2	3	3	3	3	48.67	2.56
8	1	5	5	1	4	1	5	3	5	4	3	4	3	4	2	4	4	4	4	66	3.47
9	1	5	4	3	1	1	4	3	4	4	1	3	4	4	3	2.5	4	4	4	59.5	3.13
10	1	4	3	5	1	1	4	3	4	3	1	3	3	4	5	3	4	4	4	60	3.16
11	5	5	4	4	4	1	5	4	5	5	5	4	4	3	4	3	3	3	3	74	3.89
12	1	3	4	1	4	1	3	3	3	5	1	4	3	4	3	4	4	4	4	59	3.11
Total	18	44	47.2	33	40	21	43	36	43	46	26.5	36	38	39	31	37	39	39	39	695.7	36.61

Perfect	Average	Percent
60	36.61	61.02%

Maintenance Indices & KPI — Rank: 1st

Disciplines	Mtce 1	Mtce 2	Mtce 3	Mtce 4	Mtce 5	Mtce 6	Mtce 7	Mtce 8	Mtce 9	Mtce 10	Mtce 11	Mtce 12	Mtce 13	Mtce 14	Mtce 15	Mtce 16	Mtce 17	Mtce 18	Mtce 19	Total	Ave
1	5	5	5	4	4	5	5	4	5	4	5	4	4	4	3	4	4	4	4	82	4.32
2	5	1	3	4	4	5	1	5	1	3	5	4	4	4	3	4	4	4	4	68	3.58
3	5	1	4	4	4	4	1	5	1	5	4	4	4	4	3	4	4	4	4	69	3.63
4	5	4	3.5	1	1	3	4	4	5	4	1	4	4	4	3	4	4	4	4	66.5	3.50
5	5	5	4	4	3	2	5	4	5	3	4	3	4	1	2	1	1	1	1	58	3.05
6	5	3.5	4	1	1.5	3.5	3.5	4	3.5	5	4	3	4	4	3	4	4	4	4	68.5	3.61
7	5	3	4	4	1	3.5	2.5	4	3	4	3.5	4	3	3.5	2.5	4	4	4	3.5	66	3.47
8	5	1	4	4	2.5	4	1	4	1	3	2.5	4	4	4	2	4	4	4	4	62	3.26
9	5	5	4	4	2	4	5	4	5	4	4	4	4	4	2	4	4	4	4	76	4.00
10	5	1	3.67	4	4	3.5	1	4	1	5	4	4	4	3	2.5	4	4	4	3	64.67	3.40
Total	50	29.5	39.2	34	27	37.5	29	42	30.5	40	37	38	39	35.5	26	37	37	37	35.5	680.7	35.82

Perfect	Average	Percent
50	35.82	71.65%

Autonomous Maintenance — Rank: 7th

Disciplines	Mtce 1	Mtce 2	Mtce 3	Mtce 4	Mtce 5	Mtce 6	Mtce 7	Mtce 8	Mtce 9	Mtce 10	Mtce 11	Mtce 12	Mtce 13	Mtce 14	Mtce 15	Mtce 16	Mtce 17	Mtce 18	Mtce 19	Total	Ave
1	5	1	3	5	1	2.5	1	3.5	1	3	4	2	3	2.5	3	2.5	2.5	2.5	2.5	51.5	2.71
2	5	1	3.5	4	4	4.5	1	4	1	3	4	4	4	2.5	3	2.5	2.5	2.5	2.5	60.5	3.18
3	1	1	3	4	1	2	1	3	1	4	3	1	4	4	3	4	4	5	4	56	2.95
4	5	1	3	4	4	2	1	2	1	2	2	3	2	2	2	2	2	1	2	47	2.47
5	5	1	3.67	4	1	3.5	1	2.5	1	4	3.5	3	3	3.5	3	3.5	3.5	3.5	3.5	61.67	3.25
6	3	1	3	4	4	2.5	1	2	1	3	2.5	1	3	3	2	3	3	3	3	54	2.84
7	3	1	3	4	2	3	1	4	1	3	3	4	3	3	3	3	3	3	3	61	3.21
8	1	1	4	4	2	2	1	3	1	4	4	2	1	4	3	4	4	4	4	61	3.21
9	1	1	3.5	4	2	2.5	1	2.5	1	3	3.5	1	4	2.5	2.5	2.5	2.5	2.5	2.5	54	2.84
10	1	1	4	4	2.5	2	4	2	5	3	1	4	3	3	2	3	3	3	3	63.5	3.34
11	1	5	3.5	4	3	2.5	4	2.5	5	3	2	1	3	3	2	3	3	3	3	67.5	3.55
12	5	1	3	4	3	3	1	3	1	4	4	1	1	4	3	4	4	4	4	69	3.63
Total	36	16	40.2	49	29.5	32	18	34	20	39	36.5	27	35	37	31.5	37	37	37	37	706.7	37.19

Perfect	Average	Percent
60	37.19	61.99%

Basic Equipment Condition — Rank: 8th

Disciplines	Mtce 1	Mtce 2	Mtce 3	Mtce 4	Mtce 5	Mtce 6	Mtce 7	Mtce 8	Mtce 9	Mtce 10	Mtce 11	Mtce 12	Mtce 13	Mtce 14	Mtce 15	Mtce 16	Mtce 17	Mtce 18	Mtce 19	Total	Ave
1	5	4	4	4	3	2	4	3.5	4	4	4	1	4	4	2.5	4	4	4	4	70	3.68
2	5	1	4	4	3	2.5	1	4	1	3	4	4	3	3	3	3	3	3	3	59.5	3.13
3	5	4	4	4	3	2	4	4	5	3	4	4	3.5	3.5	3.5	3.5	3.5	3.5	3.5	73	3.84
4	5	1	3.5	4	1.5	2	1	3.5	1	4	3.5	2	4	3.5	3	3.5	3.5	3.5	3.5	60.5	3.18
5	1	1	4	4	3	3	1	2	1	3	4	1	4	3	3	3	3	3	3	55	2.89
6	1	1	3	4	2	2	1	2	1	2	1.5	2	2	2	2	2	2	2	2	42.5	2.24
7	5	4	4	4	1	3	4	4	4	5	4	4	3	3	3	3	3	3	3	74	3.89
8	5	1	3	4	2	2	1	3	1	4	4	3	1	1	2	1	1	1	1	49	2.58
9	3	1	3	4	3	3.5	1	2	1	3	3	1	3	3	3	3	3	3	3	58.5	3.08
10	1	1	3	4	1	2	1	2	1	1	2	1	1	1	2	1	1	1	1	38	2.00
11	3	1	3	4	1	2	1	4	1	1	4	3	3	4	2	4	4	4	4	64	3.37
Total	39	20	38.5	44	23.5	26	20	34	21	33	38	26	31	31	29	31	31	31	31	644	33.89

Perfect	Average	Percent
55	33.89	61.63%

Preventive Maintenance — Rank: 4th

Disciplines	Mtce 1	Mtce 2	Mtce 3	Mtce 4	Mtce 5	Mtce 6	Mtce 7	Mtce 8	Mtce 9	Mtce 10	Mtce 11	Mtce 12	Mtce 13	Mtce 14	Mtce 15	Mtce 16	Mtce 17	Mtce 18	Mtce 19	Total	Ave
1	1	1	3	5	2.5	1.5	1	2	1	3	2.5	4	4	4	4	4	4	4	4	55.5	2.92
2	1	1	3	4	2	3	1	4	1	4	1	4	4	4	4	4	4	4	4	57	3.00
3	5	4	3	4	4	4	4	4	5	4	3	4	4	4	4	4	4	4	4	76	4.00
4	5	1	4	4	3	3	1	4	1	5	4	4	4	4	4	4	4	4	4	67	3.53
5	5	5	4	4	4	4	5	4	5	5	4	1	1	1	3	3	3	3	3	67	3.53
6	3	1	3	4	4	3	1	4	1	4	3	4	4	4	4	4	4	4	4	63	3.32
7	5	1	4	4	4	4	4	4	1	4	4	4	4	4	4	4	4	4	4	71	3.74
8	1	1	3.5	4	4	4	1	3	1	3	4	4	4	4	4	4	4	4	4	61.5	3.24
9	5	1	3	4	1	4	1	3	1	4	4	2	2	2	2	2	2	2	2	47	2.47
10	5	1	4	4	2	3	1	3	1	4	5	4	4	4	4	4	4	4	4	65	3.42
11	5	5	4	4	4	3	5	4	5	4	4	4	4	4	4	4	4	4	4	79	4.16
Total	41	22	38.5	45	34.5	36.5	25	39	23	44	38.5	39	39	39	41	41	41	41	41	709	37.32

Perfect	Average	Percent
55	37.32	67.85%

MRO Spare Parts — Rank: 10th

Disciplines	Mtce 1	Mtce 2	Mtce 3	Mtce 4	Mtce 5	Mtce 6	Mtce 7	Mtce 8	Mtce 9	Mtce 10	Mtce 11	Mtce 12	Mtce 13	Mtce 14	Mtce 15	Mtce 16	Mtce 17	Mtce 18	Mtce 19	Total	Ave
1	5	3	3	4	1	1	3	4	3	4	4	3	1	1	1	1	1	1	1	45	2.37
2	5	5	2	4	2	2	5	3	5	4	2	3	4	4	2	4	4	4	4	68	3.58
3	1	4	4	4	1	1	4	2	4	3	2	1	3	3	3	3	3	3	3	52	2.74
4	5	5	4.5	4	1	3	5	3.5	5	1	3.5	3.5	1	4	4	4	4	4	4	69	3.63
5	5	3	3	4	1	1	3	3	4	4	2	1	3	3	1	3	3	3	3	53	2.79
6	5	4	3	4	2	1	5	3	5	4	3.5	3	3	2	2	2	2	2	2	57.5	3.03
7	5	5	4	1	1	1	5	3	5	4	1	4	4	1	1	1	1	1	1	49	2.58
8	5	3	4	1	4	1	3	4	3	4	4	4	4	4	4	4	4	4	4	68	3.58
9	1	1	1	1	2	1	1	2.5	1	2	2	2	3	2	2	2	2	2	2	32.5	1.71
10	5	1	2	1	1	1	1	4	1	1	1	4	2	2	2	2	2	2	2	37	1.95
11	5	5	4	1	1	1	5	4	5	3	4	1	3	4	4	4	4	4	4	66	3.47
12	5	1	4	1	1	1	1	2	1	3	4	1	4	2	2	2	2	2	2	41	2.16
Total	52	40	38.5	30	18	15	41	38	42	37	33	30.5	35	32	28	32	32	32	32	638	33.58

Perfect	Average	Percent
60	33.58	55.96%

Figure 9.21: Summary of Results on Assessment Sample Part 1

Life Cycle Management — Rank: 12th

Item	R1	R2	R3	R4	R5	R6	R7	R8	R9	R10	R11	R12	R13	R14	R15	R16	R17	R18	R19	Total	Average
1	1	1	3	1	1	1	1	3	1	1	3	1	4	3	3	3	3	3	3	40	2.11
2	5	3	3	4	4	1	3	3	3	1	3	4	4	4	4	4	4	4	4	65	3.42
3	5	5	4	1	4	1	5	3	5	1	3	4	1	1	4	1	1	1	1	51	2.68
4	1	1	4	4	1	1	1	3	1	2	2	4	4	2	2	2	2	2	2	41	2.16
5	5	1	3	1	1	1	1	4	1	1	3	3	1	1	1	1	1	1	1	32	1.68
6	5	1	4	4	2	1	1	4	1	1	4	4	4	4	4	4	4	4	4	60	3.16
7	5	1	3	1	1	1	1	4	1	2	2	4	3	3	1	3	3	3	3	45	2.37
8	5	5	4	4	4	1	5	3.5	5	4	3	4	4	4	4	4	4	4	4	75.5	3.97
9	5	1	3	4	1	1	1	3	1	4	3.5	4	4	4	4	4	4	4	4	59.5	3.13
10	1	1	3	1	1	1	1	3	1	1	4	1	2	1	2	2	2	2	2	32	1.68
Total	38	20	34	25	20	10	20	33.5	20	18	30.5	33	31	27	29	28	28	28	28	501	26.37

Perfect	Average	Percent
50	26.37	52.74%

Lubrication Management — Rank: 11th

Item	R1	R2	R3	R4	R5	R6	R7	R8	R9	R10	R11	R12	R13	R14	R15	R16	R17	R18	R19	Total	Average
1	5	3	4	4	1	1	5	3	5	3	4	3.5	4	4	4	4	4	4	4	69.5	3.66
2	5	1	4	4	1	1	5	2	5	4	3	2	3	1	4	4	4	4	1	58	3.05
3	5	5	4	4	1	1	1	3	1	3	3	1	1	1	4	4	4	4	1	51	2.68
4	5	1	3	4	1	1	5	4	5	4	3	4	4	4	4	4	4	4	4	68	3.58
5	1	1	3	1	1	1	1	3	1	4	3	2	2	2	2	2	2	2	2	36	1.89
6	5	1	4	4	1	1	1	3	5	4	4	4	4	4	4	4	4	4	4	65	3.42
7	5	1	4	4	1	1	1	2	1	4	4	4	3	1	1	1	1	1	1	41	2.16
8	1	1	1	4	1	1	1	2	1	4	4	4	2	2	2	2	2	2	2	39	2.05
9	5	1	4	4	1	1	1	3.5	1	5	3.5	2.5	3	4	4	4	4	4	4	59.5	3.13
10	1	1	4	4	1	1	1	4	1	5	4	4	4	4	4	4	4	4	4	59	3.11
11	1	1	3	4	1	1	1	3	1	4	2	1	2	1	2	2	2	2	2	36	1.89
Total	39	17	38	41	11	11	23	32.5	27	44	37.5	32	32	28	35	35	35	35	29	582	30.63

Perfect	Average	Percent
55	30.63	55.69%

Root Cause Failure Analysis — Rank: 5th

Item	R1	R2	R3	R4	R5	R6	R7	R8	R9	R10	R11	R12	R13	R14	R15	R16	R17	R18	R19	Total	Average
1	5	4	4	1	4	3	4	4	4	5	4	4	4	4	4	4	4	4	4	74	3.89
2	5	1	3	1	1	4	1	4	1	4	4	4	3	4	4	4	4	4	4	60	3.16
3	5	1	3	4	4	4	1	4	1	3	4	1	4	4	4	4	4	4	4	63	3.32
4	5	1	4	4	4	3	1	4	1	1	3	4	4	4	4	4	4	4	4	63	3.32
5	5	1	4	4	1	2	1	3	1	1	4	4	4	3	3	3	3	3	3	53	2.79
6	3	1	4	3	4	4	5	3	1	1	4	4	2	2	2	2	2	2	2	51	2.68
7	5	1	4	4	4	3	1	4	1	1	4	4	4	4	4	4	4	4	4	64	3.37
8	5	3	3	4	4	2	3	4	3	1	4	4	4	4	4	4	4	4	4	68	3.58
9	1	1	4	4	4	2	1	3	1	1	4	1	4	4	4	4	4	4	4	55	2.89
10	5	4	4	4	4	3	4	4	4	1	4	1	4	4	4	4	4	4	4	70	3.68
Total	44	18	37	33	34	30	22	37	18	4	39	31	37	37	37	37	37	37	37	621	32.68

Perfect	Average	Percent
50	32.68	65.37%

Reliability Improvements — Rank: 6th

Item	R1	R2	R3	R4	R5	R6	R7	R8	R9	R10	R11	R12	R13	R14	R15	R16	R17	R18	R19	Total	Average
1	5	1	4	1	1	1	1	4	1	4	4	4	1	4	3	4	4	4	4	55	2.89
2	1	1	3	4	1	1	1	3	1	4	3	3	4	3	3	3	3	3	3	48	2.53
3	1	1	3	1	1	1	1	3	1	1	1	1	4	3	3	3	3	3	3	38	2.00
4	1	1	3	4	4	1	1	3	1	4	3	1	4	4	4	4	4	4	4	55	2.89
5	5	3	4	4	4	1	3	3	1	5	4	4	3	4	4	4	4	4	4	68	3.58
6	5	5	4	4	3	1	5	4	5	4	4	4	4	4	3	4	4	4	4	75	3.95
7	5	5	5	4	2	1	5	3	5	4	4	4	4	3	3	3	3	3	3	69	3.63
8	5	4	4	4	3	1	3	3	1	5	1	4	3	3	3	3	3	3	3	59	3.11
9	5	5	4	4	1	1	5	4	5	4	4	4	4	4	4	4	4	4	4	74	3.89
10	5	5	4	4	1	1	5	4	5	5	4	4	4	4	4	4	4	4	4	75	3.95
Total	38	31	38	34	21	10	30	34	26	40	32	33	35	36	34	36	36	36	36	616	32.42

Perfect	Average	Percent
50	32.42	64.84%

Condition-Based Maintenance — Rank: 2nd

Item	R1	R2	R3	R4	R5	R6	R7	R8	R9	R10	R11	R12	R13	R14	R15	R16	R17	R18	R19	Total	Average
1	5	5	4	4	3	4	5	4	5	1	4	4	4	4	4	4	4	4	4	68	3.58
2	5	5	3	4	2	3	5	4	5	3	4	4	3	3	3	3	3	3	3	68	3.58
3	5	1	3	4	2	2	1	3	1	4	2	3	4	4	4	4	4	4	4	59	3.11
4	5	5	3	4	2	3	5	4	5	4	3	4	4	4	4	4	4	4	4	75	3.95
5	5	1	3	4	1	3	1	4	1	4	3	3	4	4	3	4	4	4	4	60	3.16
6	5	1	4	4	1	1	1	4	1	1	4	3	4	4	4	4	4	4	4	58	3.05
7	5	5	3	4	1	4	5	4	5	4	4	4	4	4	4	4	4	4	4	76	4.00
8	5	5	3	4	1	3	5	4	5	4	4	3	3	4	4	4	4	4	4	73	3.84
9	5	1	3	4	1	2	1	3	1	4	3	4	4	4	4	4	4	4	4	60	3.16
10	5	1	4	4	1	2	1	4	1	4	4	3	4	4	4	4	4	4	4	62	3.26
Total	50	30	33	40	15	27	30	38	30	33	35	35	38	39	38	39	39	39	39	659	34.68

Perfect	Average	Percent
50	34.68	69.37%

CMMS — Rank: 3rd

Item	R1	R2	R3	R4	R5	R6	R7	R8	R9	R10	R11	R12	R13	R14	R15	R16	R17	R18	R19	Total	Average
1	5	5	5	4	4	4	5	4	5	4	5	4	4	4	4	4	4	4	4	82	4.32
2	5	5	4	1	4	4	5	4	5	4	5	4	4	4	4	4	4	4	4	78	4.11
3	5	5	4	4	4	4	5	4	5	4	5	4	4	4	4	4	4	4	4	81	4.26
4	5	5	4	1	3	3	5	4	5	1	4	4	4	4	4	4	4	4	4	72	3.79
5	1	1	1	1	2	1	1	2	1	1	3	1	4	4	2	4	4	4	4	42	2.21
6	5	5	4	4	4	4	5	4	5	1	5	4	4	4	3	4	4	4	4	77	4.05
7	5	5	3	1	1	3	5	3	5	1	4	1	4	3	3	3	3	3	3	59	3.11
8	1	1	4	1	1	1	1	3	1	1	4	1	1	1	3	1	1	1	1	29	1.53
9	5	5	2	4	2	3	5	4	5	1	4	3	4	3	4	3	3	3	3	66	3.47
10	5	5	2	1	4	3	5	4	5	1	1	2	4	4	4	4	4	4	4	66	3.47
Total	42	42	33	22	29	30	42	36	42	19	40	28	37	35	35	35	35	35	35	652	34.32

Perfect	Average	Percent
50	34.32	68.63%

	R1	R2	R3	R4	R5	R6	R7	R8	R9	R10	R11	R12	R13	R14	R15	R16	R17	R18	R19	Total
Summation	487	330	455	430	303	286	343	434	343	397	424	389	427	416	395	425	427	427	420	7,705
Average	3.78	2.55	3.53	3.33	2.34	2.22	2.66	3.36	2.66	3.08	3.28	3.01	3.31	3.22	3.06	3.29	3.31	3.31	3.25	59.73

Average Performer: 405.53

Figure 9.22: Summary of Results on Assessment Sample Part 2

COMPANY ABC OVERALL MAINTENANCE ASSESSMENT

No.	Dicipline	Questions	Perfect	19 People	PGPC Score	Percent	Rank
1	Training Skills and Education	12	60	1,140	695.7	61.02%	9th
2	Maintenance Indices and KPI	10	50	950	680.7	71.65%	1st
3	Autonomous Maintenance	12	60	1,140	706.7	61.99%	7th
4	Basic Equipment Condition	11	55	1,045	644.0	61.63%	8th
5	Preventive Maintenance	11	55	1,045	709.0	67.85%	4th
6	MRO Spare Parts and Storeroom	12	60	1,140	638.0	55.96%	10th
7	Life Cycle Management	10	50	950	501.0	52.74%	12th
8	Lubrication Management	11	55	1,045	582.0	55.69%	11th
9	Root Cause Failure Analysis	10	50	950	621.0	65.37%	5th
10	Reliability and Continuous Improvement	10	50	950	616.0	64.84%	6th
11	Condition-Based Maintenance	10	50	950	659.0	69.37%	2nd
12	CMMS and Automation	10	50	950	652.0	68.63%	3rd
	Company ABC Overall Score	129	645	12,255	7,705	69.37%	
	Company ABC Overall Rating	405.53			Average Performer		

Figure 9.23: Summary of Results on Assessment Sample Part 3

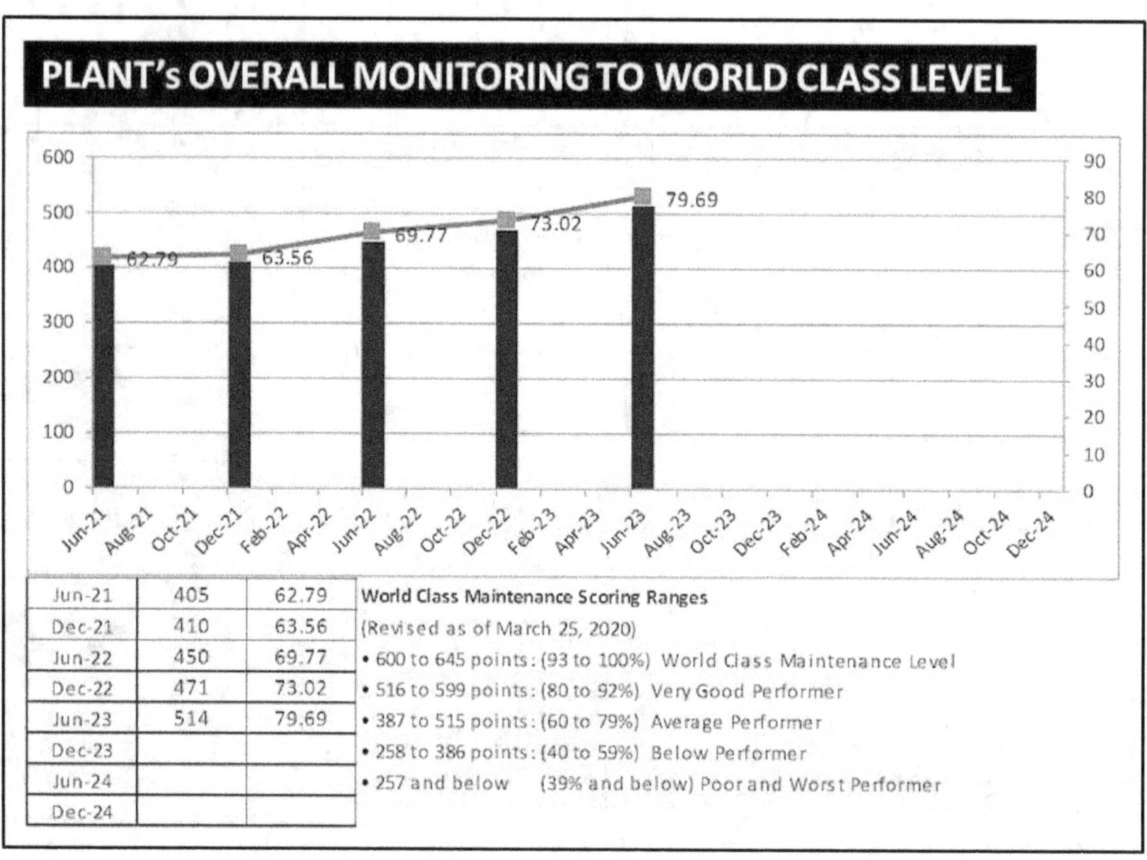

Jun-21	405	62.79	World Class Maintenance Scoring Ranges
Dec-21	410	63.56	(Revised as of March 25, 2020)
Jun-22	450	69.77	• 600 to 645 points : (93 to 100%) World Class Maintenance Level
Dec-22	471	73.02	• 516 to 599 points : (80 to 92%) Very Good Performer
Jun-23	514	79.69	• 387 to 515 points : (60 to 79%) Average Performer
Dec-23			• 258 to 386 points : (40 to 59%) Below Performer
Jun-24			• 257 and below (39% and below) Poor and Worst Performer
Dec-24			

Figure 9.24: Regularly Monitoring the Scoreboard on the Journey to World Class

It will usually take a couple of hours to complete the maintenance survey. Once the survey has been completed, someone should summarize the results. If 20 people took the survey, the overall results will be compared to the scores and rating on World Class as in figure 9.24. This maintenance assessment can be done every 6 months to 1 year and the results should be the higher the better and the more the industry is confident of achieving a World Class Maintenance Level as in figure 9.24.

9.6: Measure Performance Along the Way

Each of these 12 disciplines can be measured independently as in figure 9.25. The first and last step in implementing a World Class Maintenance Management Structure is to define and measure performance indices that tell us that we have achieved our goals and whether or not we are headed forth in the right direction. This will answer the question of how would we know we have arrived at our destination. Measurements should be defined at the very beginning of implementation and must be reviewed regularly. The success or failure of all our efforts depends entirely on these measurements. Measurement has always played an important role in our day-to-day lives. Still, the most important reason for measuring these maintenance indices and metrics is to improve the organization's performance. By knowing our performance, we ask ourselves, what exactly should we do differently to improve and achieve our goals?

RECOMMENDED MEASUREMENT FOR THESE DISCIPLINES	
THE 12 DISCIPLINES	**RECOMMENDED MEASUREMENTS**
1. Training & Skills	• MTTR, NUMBER OF LEVEL 4 MAINTENANCE
2. Measure Performance	• OEE, MTBF, MTTR, BDO, MTBA, ETC.,
3. Autonomous Maintenance	• NUMBER OF CERTIFIED OPERATORS PER STEP
4. Basic Equipment Condition	• ABNORMALITIES DETECTED VERSUS CORRECTED
5. Preventive Maintenance	• REDUCTION IN BREAKDOWN, PM EFFECTIVENESS
6. Lubrication Management	• REDUCTION IN LUBRICATION COST & FAILURES
7. Life Cycle Management	• SAVINGS GENERATED
8. Spare Parts Management	• INVENTORY ACCURACY, SPARES COST
9. Root Cause Failure Analysis	• BREAKDOWN RECURRENCE, MACHINE DOWNTIME
10. Reliability Improvements	• MAINTENANCE COST, DEPENDS ON THE IMPROVEMENT
11. Condition-Based Mtce	• MAINTENANCE COST, MACHINE DOWNTIME
12. CMMS	• RETURN ON INVESTMENTS

Figure 9.25: Recommended KPIs for Each Discipline

As previously discussed, there is no single measurement, indicator, or KPI for maintenance that will tell us the whole story. We need at least a minimum of five measurements, and this should be defined well at the very beginning of the journey on World Class Maintenance Management. While it is vitally important to measure the results, we should also look into the process of how the results were achieved. Focusing on the process creates better communication, more participation, more involvement, high morale, and motivation among the maintenance people. When managers focus on both processes and results, the team knows the management cares about how the results were obtained. When people focus on the process, the results will follow into place. Therefore, if we want to improve the results, we also need to improve the process.

Achieving a World Class Maintenance Management level can now be measured and can be part of your plant's overall KPI and indicator. The advantage of having this survey done in our industry is knowing where our weaknesses lie so that we can focus and improve on them. Improving each of these disciplines will require a different strategy. Once maintenance knows their weaknesses based on this survey, they can deploy the correct strategy to improve them. After implementing the strategy with success, perform the assessment once again after 6 months or 1 year to see if there is an improvement. Make this part of your indicator so industries will know how far or near they are in achieving a level of World Class Maintenance.

Chapter 10

Strategies to Improve These Maintenance Indices

> *It is Okay not to hit the goal, but it is not Okay to challenge and improve our previous performance. What is important is that we show the true figures rather than manipulating them so that it looks nice to the management.*

10.1: Do Not Overwhelm Yourself with Too Many Improvements

Each of these Maintenance Indices and KPIs we have covered so far will require different strategies to improve them. The thing is to prioritize whatever improvement strategy will be implemented and not overwhelm ourselves with too many improvements in the plant. This will just end up in having too many groups and too many meetings that will eat up your time especially when you will be assigned as the champion of that particular initiative.

When an improvement drive is introduced in the plant, people tend to observe, and the initial feeling is that people do not want to be a part of it as the thinking goes that it will just add up to their workload. In short, expect resistance especially during the beginning. When other priorities emerge, this improvement initiative takes a halt, and when nothing improves in the next few months, then it dried up and fades away just like some music that is no longer played today but definitely not the music of the Rolling Stones (I love this Band). In other industries I worked with, especially if it is dynamic, people move from one function to another completely abandoning the project.

Have you ever experienced coming to work in the plant starting with a morning meeting with operations, then after which you need to attend another meeting and another and another. At the end of the day, all you have done was to attend all these meetings? One day the manager called you and discussed problems about housekeeping in the plant and on that day forward you were called the 5's Champion. Cool! What you are going to do is create an ad hoc committee composed of different people from different departments of the plant and at first, you and the committee members will be excited about this new initiative. You and the team agreed to meet weekly to find out the progress of the 5's initiative. The following morning there was a near-miss accident and the safety officer called a meeting on each department and asked for a champion on each department to ensure the safety of the people in the workplace. The Quality Department initiated a Quality Improvement Circle composed of operators, maintenance, and process people to ensure the quality of the product and these improvement circles will again be meeting once a week. The President of the plant wants the OEE to be tracked and monitored regularly on every single piece of equipment and assigned you once again to spearhead an OEE committee and again there will be a weekly meeting because of this. The maintenance manager attended a course on RCM and was convinced that he needed RCM to be implemented in the plant so he organized the team and the team will be meeting twice a week. Hey wait, you still have the human resources and they got this cool initiative which is called the employee of the month award and you were lucky or perhaps unlucky to be selected as one of the committee members to derive the procedures and criteria in which the bottom line is that you need to attend this once a week meeting. Not to mention the operations meeting, toolbox meeting, supervisors meeting, which only end up in having a lot of action items to complete. Are we employed just to attend these meetings? There was one manager I knew who went to Japan and attended training with Masaaki Imai on Gemba Kaizen. After the training, he was so convinced of its philosophy and after returning to the plant, he assigned a champion to head the Gemba Kaizen project, but wait, isn't it there was already a 5's champion? Hey, we still got the Karate people from Six Sigma Black belts and I can go on and on and I hope you got my point now. The bottom line is that you have accomplished nothing for the day because your time was consumed by attending too many meetings.

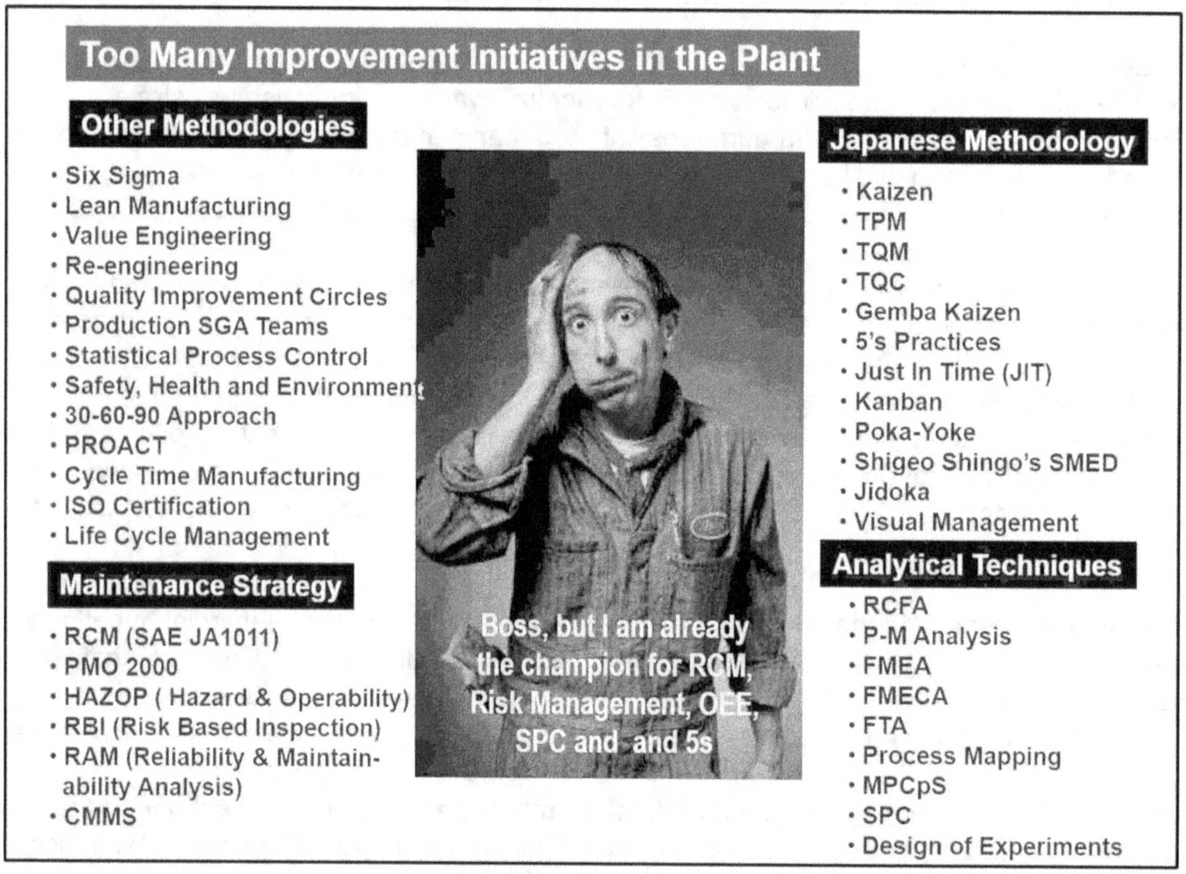

Figure 10.1: Too Many Improvement Initiatives in the Plant

While their initiative is good, the problem is that the same members will be part of the team or committee or whatever your industry called them. The main problem is that these plant initiatives were never consolidated in the first place. What will happen is that a lot of corrective actions will be generated as a result of these initiatives wherein what happens, in the end, is that you finally forgot your purpose and why the industry hired you as your time was consumed on attending all these meetings. The more meetings you attend, then the more problems and corrective actions that need to be done. What is important in this case will be to list down all the initiatives taken by your industry and try to consolidate everything just like a barbeque or satay. I love to eat that dish especially when I am in Malaysia. Have the main umbrella and that all improvement initiatives will be under one umbrella. The next will be to assign which department or group is best to handle that initiative. Refer to figure 10.2.

Figure 10.2: Consolidate All Improvements in your Plant

The point I would like to make is that do not confine your industry with so many improvement initiatives as there may be cases where some of them may try to conflict or even contradict each other. I remember when I was still working in the semiconductor industry, they organized a group called the BKM group or Best Known Method, in which the role of this group was that if a modification was performed on one of the equipment, it should be fan-out or replicated in all similar equipment in the plant since there are different departments with the same types of equipment. I truly contradict this because these people do not understand what the equipment's operating context means. RCM contradicts this. This means that if the problem is not present in other equipment even though they are identical, then the modification should not be replicated so as not to induce new problems on the equipment. This means that if the equipment is stable, even if they are of the same equipment type. The best thing to do is do not disturb stable equipment. Period. This group will ensure that the modification will be carried out to all similar equipment in the plants even if other similar equipment poses no problem whatsoever. Disturbing equipment which is stable may induce a new set of problems in the long run. When the equipment is stable, the best thing to do is to leave it alone and let it run.

Improvement drives should be a Top-Down Approach. One thing clear in TPM (Total Productive Maintenance) is that this improvement methodology is always a Top-Down Approach, This means that this should be initiated by the highest (not the tallest) person in the plant which is your President or CEO. This is very made clear in the 12 TPM Developmental Steps. One good point about TPM is that it tries to consolidate all other improvement initiatives in the plant as in figure 10.2. In the TPM book on World Congress, it states that their study indicates that among 3000 industries that implemented TPM, only 10% succeed. Although there may be different reasons why 90% failed. Research indicates that the main reason for the failure has something to do about Management Support. I tell industries that if you hire me on implementing this methodology, what you need is not just support but a commitment for a start, but the surest way to make this successful is for Top Management to own the program. What we need is management ownership and not just merely support. Management can support but never commit or even own the initiative and this is why most TPM implementation fails.

10.2: Performing PM Effectively and Efficiently to Improve Its KPIs

Improving these indices on Preventive Maintenance can easily be done by manipulating the figures so that they will look good to the boss, however, that is not the point of this book. There are two important recommendations I would propose to improve our PM KPIs.

Reviewing the PM Intervals: Different PM tasks will require different intervals and frequencies. Although the majority of industries have maintenance planners who will derive the plan and activities to be done on the equipment, the question is what is the basis of concluding that this is the correct interval for the tasks to be done? Are we doing the tasks too soon or too late, especially for activities in which we will touch the equipment such as overhauls and replacements? I have written one complete chapter on this in my 4th book on Cutting-Edge Maintenance Management Strategies, the thing is overhauls and replacements should only be done on age-related failures. Although originally, the P-F Curve (Potential to Functional Failure) is used for Predictive Maintenance tasks, parts to be overhauled and replaced under Preventive Maintenance likewise possess a potential failure that people involved should be closely watching, and that is a wear-out mode. This must be the pattern we should look for on parts and assemblies that either need to be replaced or overhauled. This means that the stress has finally exceeded the strength of the material. Signs of wear-out will be seen evidently on the part as a very tiny metal fracture will be evident. Planners must not only plan the tasks but should also be familiar with the correct interval on when to replace parts and perform overhauls. The PM plan is not always perfect, and most intervals I have seen are just merely based on guestimates and JIC or Just in Case it might fail. That is why maintenance planners must always be in constant communication with the people executing the PM activities. This is also one of the main reasons why Preventive Maintenance is costly because we replace parts that are still in working condition. This means that if a bearing is scheduled to be replaced as indicated in the PM tasks, and the bearing is still in good condition, then this should not be replaced, and the tasks need to be revised by the planner.

Determining the correct frequency of replacements and overhauls for Preventive Maintenance activities should be based upon the useful life and not on the average life of the part, which inhibits an age-related pattern. Parts that do not have a useful life should not be replaced on a time-based dominated frequency because their pattern is different. Only parts that will age because of stress-induced upon the part should be the subject for parts for PM replacement and overhaul. This simply means that the stress finally exceeds the strength of the material of the part. Aging is the process where certain parts of the equipment deteriorate because of stress-induced upon a part over a given period. Any part that conforms to this process will be the subject of Preventive Maintenance intervention.

[9]In the absence of data and historical records, using our experience to guess the task interval, or just following OEM recommendations is the only option. One proven method can be adapted to determine the correct task interval for overhauling and replacement, which is called the age exploration method. When we first try to overhaul or inspect the condition of the equipment on parts where aging and wear-out are thought to be possible and reveal that no wear-out signs exist, we can automatically increase the interval by 10% during the next schedule. Repeat this process until we can finally see the normal wear-out pattern on the part being inspected.

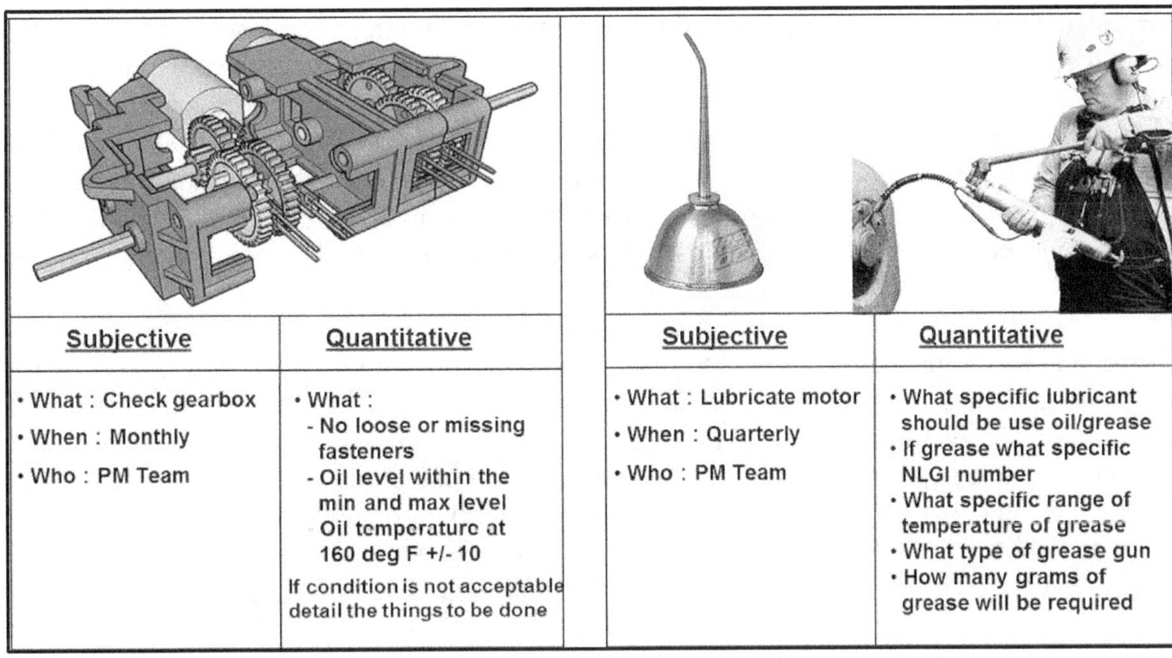

Subjective	Quantitative	Subjective	Quantitative
• What : Check gearbox • When : Monthly • Who : PM Team	• What : - No loose or missing fasteners - Oil level within the min and max level - Oil temperature at 160 deg F +/- 10 If condition is not acceptable detail the things to be done	• What : Lubricate motor • When : Quarterly • Who : PM Team	• What specific lubricant should be use oil/grease • If grease what specific NLGI number • What specific range of temperature of grease • What type of grease gun • How many grams of grease will be required

Figure 10.3: PM Tasks Should be Precise and Quantitative

This means that if the equipment is due to be overhauled after one year and reveals that upon thorough inspection, there are still no traces of wearing or aging on suspected parts, then

---

[9] Smith, Anthony, *Reliability-Centered Maintenance, Gateway to World Class,*(Mc Graw Hill Inc, 1993), Page 126

we can automatically increase the interval by 10% during the next overhauling schedule. This means the next PM overhaul will happen on the 13th month. If there are still no signs of wearing out, we increase it again by 10% until a wear out is evident on the parts.

Application of Precision Maintenance on PM: The second problem is that Preventive Maintenance will always include risks. There is always an assumption that everything will be put back in place well together after a PM initiative. Likewise is the possibility of Human Errors which leaves to infant Mortality of newly installed components which eventually leads to additional failures after endorsing back the equipment to production. It is important to conduct a thorough review regarding the activities that are being performed on the equipment during a Preventive Maintenance shutdown and that understanding that most failures are not related to the operating age of equipment. Majority if not all cases of infant mortality failure have one thing in common and that is somewhere, there is a human error involved. Most human errors that occurred on maintenance may either be a slip or a lapse. A slip is when you perform the wrong sequence which means that if we want to perform Step 1, Step 2, Step 3, Step 4, and Step 5, what we did was Step 1, Step 2, Step 3, Step 5, and Step 4. On the other hand, lapse is when we missed out something in a planned sequence of events, this means if we need to perform Step 1, Step 2, Step 3, Step 4, and Step 5, what we did was Step 1, Step 2, Step 3, and Step 5. In this case, we missed out on Step 4.

[10]According to James Reasons and Alan Hobbs, the author of Managing Maintenance Errors, people who wrote the manuals and procedures rarely if ever carried out the activity under real-life conditions. Regardless of the industry, human errors on PM will happen mostly during replacement and overhauls especially the latter. For overhauls, this will include two activities, the removal, and disassembly and the reassembly process. Reassembly is more vulnerable to human error than disassembly. Planners must communicate with the PM executioners, how easy or difficult certain tasks will be done and the most important is how should the tasks be done in detail that it can be followed easily whoever is performing the tasks, whether it will be done by the most or least experience of the craft. If possible, make the tasks more quantitative.

Precision Maintenance describes the maintenance culture in a facility. This means that the maintenance should yield the exact same results no matter who is performing the tasks. The concept of Precision Maintenance is simple, whoever is executing the Preventive Maintenance tasks whether this person is the most or the least experienced of the craft should yield the same results and outcome. This will definitely reduce infant mortality failures experienced by operators right after the PM group completed their PM tasks and endorse the equipment back to operations, where the operators are having a difficult time running the equipment.

10.3: Improving the Breakdown Rate and MTBF

Although there are several strategies to reduce the breakdown rate and increase MTBF, one of the proven and long-term strategies designed for this type of equipment loss is adopting the

---

[10] James Reason and Allan Hobbs, Managing Maintenance Error – A Practical Guide, Ashgate Publishing Company, Page 1

TPM's 4 Phases of Planned Maintenance.   Although there are variations depending on the consultant on how these will be done, the main objective of this TPM Pillar is two-folds.   The first 2 Phases will attempt to improve the equipment by reducing unplanned breakdowns on the equipment, while the next 2 Phases will aim to improve the system on how maintenance is carried on the equipment so that breakdowns and failures will not revert back.   What we want in the later Phases is to create a sustainable process and this is perfectly where we can integrate the concept of Reliability-Centered Maintenance.   What TPM is doing is that to be more effective, let us address the basic equipment condition which refers to the small problems first before implementing RCM on the equipment.   The simplest explanation I can provide in this case is thinking about your equipment when it was newly installed and commission on your plant. If you cannot recall just try to Google your equipment on the internet, after which, look at your equipment as it is right now. Is there any difference? Are there some missing parts, dents, scratches, gauges that are too stained to read, and so on?

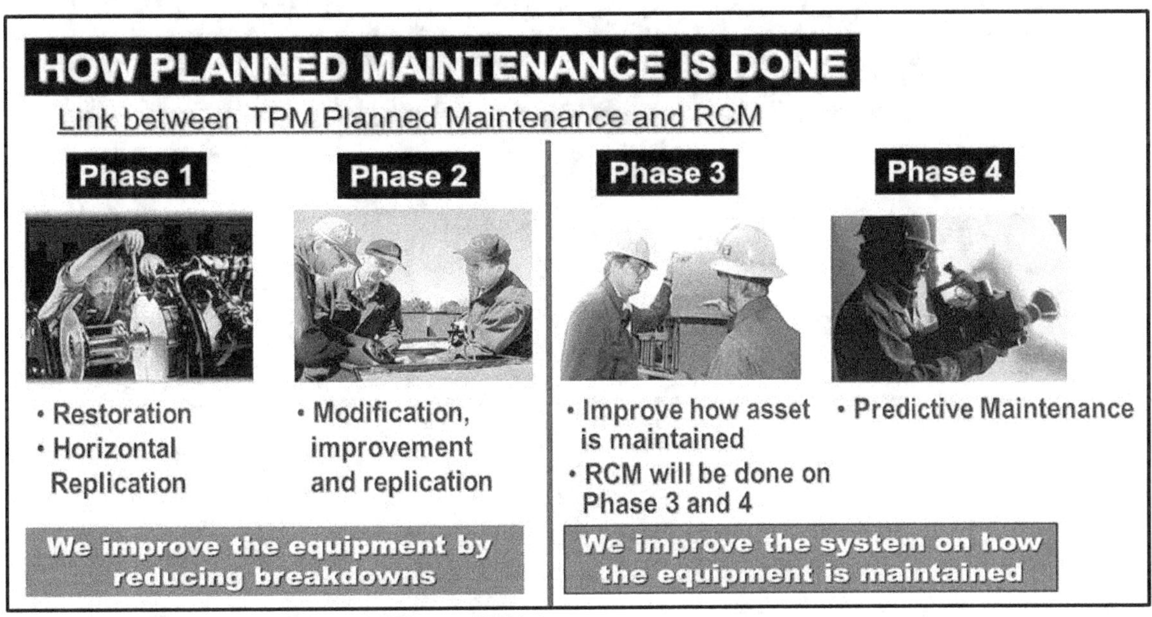

Figure 10.4: Concept of the 4 Phases of Planned Maintenance

Phase 0: Planned Maintenance Preparatory Stage: Planned Maintenance Phase 0 details the requirements needed to set up an effective Planned Maintenance structure in the plant. These are the basic requirements needed for a solid foundation. The preparatory stage usually takes around 1 to 3 months or even more depending on the size of the industry, number of equipment, assets, system, sub-systems, components, and manpower in the plant.   It is important not to ignore this stage since this will serve as the very basic foundation of Planned Maintenance activities.   Phase 0 of Planned Maintenance will be the foundation of building a solid Planned Maintenance structure. The team will aim to complete this Preparatory Stage by determining a timeframe of its completion which is called the Master Plan.   One important activity in this phase is to conduct a machine ranking on every single asset and equipment in the plant. Only equipment classified as Rank A and Rank B will undergo Planned Maintenance. Rank A will be considered as the worse equipment.   The aim of Phase 0 is to plan and forecast ahead of the activities of Planned Maintenance by creating a timeframe of completion for all equipment to undergo the different phases of Planned Maintenance. The key activities for Planned Maintenance Preparatory Stage will include the following:

• Establishing the Planned Maintenance Vision and Mission
• Defining the roles and objectives of each Planned Maintenance member
• Defining what will eventually constitute a breakdown and not a breakdown
• Updating current equipment inventory and conducting a machine ranking
• Preparing a Master Plan of completion projected with a minimum of 5 years
• Pilot equipment selection
• Getting initial KPI such as MTBF and breakdown occurrence on the team's pilot equipment
• Establish initial training needs required for the Planned Maintenance pilot team

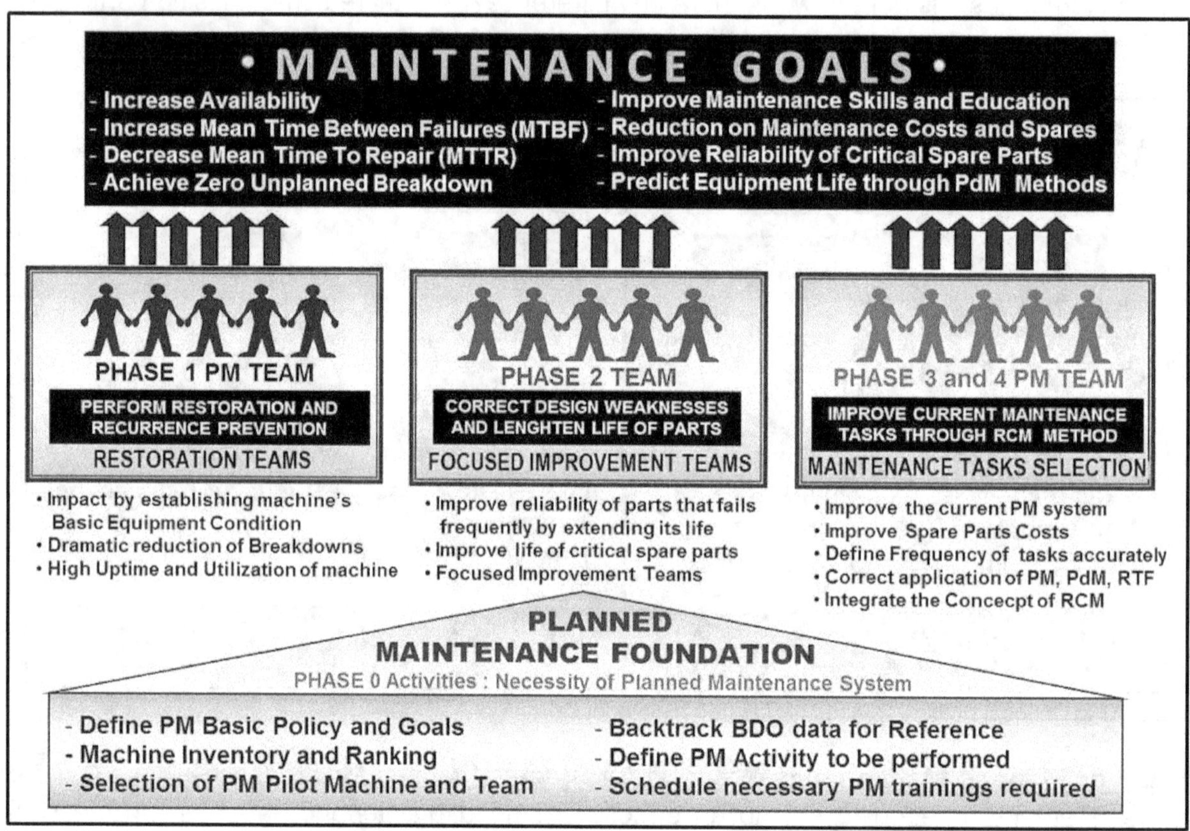

Figure 10.5: How the 4 Phases of Planned Maintenance is Deployed

Phase 1: Stabilized MTBF through Restoration: Both Autonomous Maintenance and Planned Maintenance will begin by accepting the concept that equipment is always a shared responsibility for both of them. These two TPM pillars will address abnormalities and deterioration side by side while maintenance restores the equipment back to its original basic equipment condition. As Autonomous Maintenance focuses more on the exterior part of the equipment, the Planned Maintenance teams will address the interior parts of the equipment. The goal of establishing the Basic Equipment Conditions hand in hand with Autonomous Maintenance on their equipment is to transition from a stage where accelerated deterioration is rampant to a stage of natural deterioration in which the parts that are age-related in nature will finally reach their natural lifespan. The objective of Phase 1 on restoration is to reduce the number of unplanned breakdowns experienced on the equipment. What TPM believes is that breakdowns are simply an accumulation of small problems that had been neglected over time. Once the basics had been addressed, then it is expected to have lesser breakdowns on the equipment. Our previous Japanese Consultant, Mr. Ogake San once told us that if you perform Phase 1 religiously, expect your breakdowns to be reduced in half. He was totally wrong since when we implement this phase the breakdowns were reduced more than 85%, other machines' breakdowns even reached a reduction of 90% just by addressing the basics.

Phase 2: Addressing Design Weaknesses through Redesign and Modification: Once accelerated deterioration has been reduced in Phase 1, there are parts in the equipment that will wear out and suffer only from natural deterioration which means that the part has reached its natural lifespan. There are spare parts on the equipment that will deteriorate naturally concerning time. The team will expose and analyze several parts of the equipment with an inherently short lifespan and study to correct design weaknesses by improving either its part's dimension, shape, the strength of materials, construction, dimensions, or measurements. The goal of Phase 2 is to lengthen the lifespan of these parts through modification or redesign. A Maintenance Prevention design is usually used for this activity and later on looped back to the Initial Flow Control Activity Pillar (IFCA)or Early Equipment Management (EEM) pillar of TPM so that when the industry decides to purchase new equipment in the future, these improvements are discussed with the designers and OEM to be included in the new equipment purchase. A cycle must be established where MP Design improvements must be looped back to IFCA or Initial Flow Control Activities of TPM. Correcting design weaknesses can prevent major breakdowns from recurring unexpectedly. The Planned Maintenance teams will be trained on several analytical and problem-solving tools such as P-M Analysis for a more detailed approach in dealing with chronic breakdowns. The team on Phase 2 will initially be trained on problem-solving and analytical techniques such as Root Cause Failure Analysis, FMEA/FMECA, Fault Tree Analysis, 8 - D, and other analytical and problem-solving tools as all of these breakdowns are caused by human errors, both operators and maintenance must upgrade their skills to reduce human errors. Application of Poka-Yoke solution may solve human errors but not necessarily improve the level of understanding of the mistake or error caused by the person involved. Once these parts with design weaknesses have been addressed in Phase 2, their lifespan will be prolonged, hence, maintenance needs to have a thorough review on how the asset is going to be maintained. This is accomplished with the aid of a maintenance algorithm which will be covered in the next phase of Planned Maintenance. RCM refers to this as a decision diagram. A thorough study of the different maintenance tasks must be done to establish the most suitable maintenance tasks together with the interval for every single failure mode which is likely to occur and affect the equipment. Another important activity in Phase 2 is that parts that are under modification should be well communicated with the storeroom and purchasing people so that the replenishment of these parts especially if these are fast or slow-moving can be managed and controlled.

Phase 3: Periodically Restore Deterioration through RCM: The idea of Phase 3 is to create a sustainable process to preserve the equipment from failures and breakdowns. Since the basics had already been established at this stage, this phase will be done by adopting the concept of Reliability-Centered Maintenance (RCM). Take note that not all tasks will need to undergo Preventive Maintenance. This is accomplished by having an understanding of each of the failure modes that can affect the equipment. Each part will have its own unique failure characteristic pattern. The key to this activity is to understand the six failure patterns so that maintenance can derive the correct maintenance tasks for each failure mode through the aid of an algorithm or decision diagram. The decision diagram will guide the team to make a decision on the most feasible maintenance tasks to be carried out to address every single possible failure modes that can likely affect the equipment. In this activity, we are no longer interested in reducing the number of breakdowns but rather we are now focusing on how to improve the way we maintain the equipment. This is done to create sustenance on the activities done on Phase 1 and 2 where the team will decide which tasks should undergo Preventive, Predictive Maintenance, Run to Fail, Failure Finding Tasks, and Modification. The decision as to which tasks to undertake will depend upon the consequences of the failure itself. This Phase can be done in parallel with Phase 2 of Planned Maintenance or when the Planned Maintenance team completed Phase 1 restoration activities and have established the basic equipment condition. A separate team which is usually composed of the plant's Rolling Stones or the most experienced maintenance people in the plant will spearhead this Phase 3 on Planned Maintenance. I am referring here to people with white hair or those nearing retirement so that we can capture what they know on the machine before they leave the plant permanently for good.

We can also integrate Reliability-Centered Maintenance (RCM) on this phase to create a sustainable activity on the restoration (Phase 1), and improvement activities (Phase 2) done on the equipment as we do not want the equipment to revert back to the same condition before the restoration activities. We want both operator and maintenance to work for hand in hand to address both abnormalities and restore deteriorated parts. This means imagining what the equipment looked like when it was new and how it looked today. The goal of Phase 3 is to create a sustainable process on the equipment and what better way this can be done with the application of RCM. Implementing the classical approach on RCM means adopting RCM based on the systematic methodology on how the airline industry has first adopted it. Reliability-Centered Maintenance finds its roots in work done by the international commercial aviation industry. Driven by the need to improve reliability while containing the cost of doing maintenance, this industry develop a comprehensive process for deciding what maintenance work is needed to keep aircraft airborne. This process evolved steadily in its early beginning during the 1960s. By definition, Reliability-Centered Maintenance is a process used to determine the maintenance requirements of any physical asset in its operating context. A lighter definition is Reliability-Centered Maintenance is a process used to determine what must be done to ensure that any physical asset continues to do whatever its users want it to do in its present operating context. The 7 classical RCM questions include:

1. What are the functions and associated performance standards of the asset in its present operating context?
2. In what ways does it fail to fulfill its functions?
3. What causes each functional failure?
4. What happens when each functional failure occurs?
5. In what ways does each failure matter?
6. What can be done to predict or prevent each failure?
7. What should be done if a suitable proactive task cannot be found?

Phase 4: Predict Equipment Lifetime: In Phase 4, we now introduce the concept of Predictive Maintenance or Condition-Based monitoring techniques. This is similar to using the human senses in a much higher perspective with the aid of non-destructive precision instruments. These instruments are just a higher form of our human senses. Many failure modes will provide signs or symptoms that they are on the verge or process of failing, and those failure modes can be predicted through the use of these specialized diagnostic instruments. This is done by checking the actual condition of the equipment rather than basing maintenance on a scheduled mode. The key in using this technique is to determine the P-F interval. P stands for potential failure and F for functional failure. If a failure mode will show signs or symptoms that they are on the verge of failing, then these parts will be a good candidate for using Predictive Maintenance. For example, a bearing may produce noise, an increase in temperature, or an increase in vibration. These are symptoms or potential failures indicating that a failure is likely to occur or is on the verge of occurring. Therefore, maintenance can schedule a greasing activity, or replacement on the equipment before the actual failure can eventually take place. One of the advantages of using Predictive Maintenance is that the life of these parts are maximized, and we will only replace parts that are on the verge of failing. If nothing is wrong then the decision is to allow the equipment to continue running. The approach in doing Planned Maintenance is simple; we address the basics first before we can attempt to go for Predictive Maintenance. It is useless to buy top-of-the-line Vibration Monitoring if your equipment lacks bolts and fasteners. Also, an important point to consider in Phase 4 is that maintenance will be investing in two things, first will be the acquisition of these Predictive Maintenance instruments, and second the certification of the users.

10.4: Improving Overall Equipment Effectiveness

As discussed in Chapter 3, OEE is composed of three components which include Availability, Performance Efficiency, and Quality Rate. Each of these components represents its own unique equipment losses as reflected in figure 10.6. Each of these losses will require a different strategy and can be measured independently using the indicators in figure 10.6. Improving OEE is all about understanding the losses that the equipment is suffering and providing the right choice of strategies to improve them. Each of these losses can also be measured independently and may require a unique team to address them individually.

IMPROVING OVERALL EQUIPMENT EFFECTIVENESS

Components	Losses	TPM Team	Indices	Strategy
Availability	Breakdown Loss	Planned Maintenance	MTBF, BDO	- 4 Phases of Planned Maintenance, RCFA
	Set-up and Changeover	Focused Improvement	MTTS	- Single Minute Exchange of Dies (SMED)
	Cutting Tool Change	Planned Maintenance	Downtime	- Modification, Redesign
	Start-up Losses	Planned Maintenance	Downtime, MTBF	- RCM, Precision Maintenance
	Shutdown Losses	Planned Maintenance	PM Effectiveness	- RCM, Precision Maintenance
Performance Efficiency	Minor Stoppages	Autonomous Maintenance	MTBA, MTTA	- 7 Steps of Autonomous Maintenance
		Focused Improvement		- Modification, Redesign
	Design Speed Loss	Focused Improvement	Cycle Time	- Modification, Redesign
Quality Rate	Defect and Rework Losses	Quality Maintenance	Yield	- Conventional Problem Solving Tools - P-M Analysis for Chronic Defects

Figure 10.6: Improving OEE Will Require Different Strategies

Reducing Breakdown Losses: As explained in Section 10.4 of this book, deploying a Planned Maintenance team and implementing the 4 Phases of Planned Maintenance will definitely reduce the breakdown losses of the equipment. For Chronic Failures, performing a Root Cause Failure Analysis investigation will reduce these chronic losses. Root Cause Failure Analysis is about trying to understand why something went wrong with our equipment and assets. It will try to identify the basic source or the origin of the problem so that industries can finally learn from the failure and the things that go wrong themselves. A good indicator to track breakdown loss will be MTBF, Number of Breakdown Occurrences, Availability, and also Machine Downtime caused by these breakdowns.

Shortening Set-up and Adjustment Losses: This type of equipment loss will likewise experience downtime on the equipment. Set-up, adjustment, or others called this changeover will be the time to convert from one product to another using the same equipment. Application of SMED techniques (Single Minute Exchange of Dies) developed by Shigeo Shingo will help reduce the set-up time as explained in Chapter 2, Section 2.2 of this book. According to Shigeo Shingo, the target for set-up time in manufacturing will be 10 minutes and below. Set-up and adjustment loss should be part of the continuous improvement process as improving this type of loss is not a one-time deal.

Lengthening Cutting-Tool Change: Other equipment will experience a machine downtime caused by changing the cutting tool, dies, punches that come in direct contact with the product. These cutting tools are changed either due to premature failure or it has deteriorated or worn out to a point that the product is producing quality problems. Those that prematurely wear or fail can be included in the breakdown loss while cutting worn-out tools will be a subject for a Preventive Maintenance replacement. Improving the performance of the cutting tool simply means prolonging the lifespan. Phase 2 of Planned Maintenance is all about lengthening the lifespan of parts by addressing their design weaknesses. Most activities involved in this case will include redesign and modification. Modification, in this case, will include either a change in shape, size, dimension, measurement, or a change in the strength of the material. Lengthening the lifespan of cutting –tools must be evaluated thoroughly especially if this involves changing the materials since changing the materials of the cutting tool to a harder material can pose new problems as it will make the cutting tool more prone to a brittle fracture.

Shortening the Time for PM Shutdown: Among all the 8 losses on the equipment, shutdown loss is a Planned Downtime and will include stopping the equipment for a periodic or scheduled Preventive Maintenance activity. Improving the shutdown loss means shortening the time to perform these maintenance tasks, removing intrusive or redundant activities on the equipment, and making our Preventive Maintenance more efficient and effective. A strategy for improving the shutdown loss will include the implementation of Reliability-Centered Maintenance which can be tied up on Phase 3 and 4 of Planned Maintenance. Improving this type of loss can also be done by studying the interval of performing the task which means that we are doing the tasks not too soon or too late. An indicator to measure the effectiveness of our maintenance tasks is PM-Effectiveness.

Shortening the Time for Start-up losses: These are losses experienced when starting the equipment for production to get to a steady running operating condition, after a planned or unplanned shutdown. Hence, this may include a warming-up period or cases of infant mortality failures that represent patterns D, or F of the 6 failure patterns. Strategies to reduce infant mortality failures can include implementation of Reliability-Centered Maintenance and the integration of Precision Maintenance on Preventive Maintenance activities. A major cause of infant mortality failures is due to human errors. Implementation of Precision Maintenance means that whoever is performing the task whether the person is the most or the least experience should yield the same results and outcome. This will also reduce human errors which is the main reason for infant mortality failures. Indicators to monitor start-up losses can include downtime, MTBF, PM-Effectiveness, or determine the frequency of start-up losses.

Reducing Minor Stoppages: Addressing the basic equipment condition can reduce the number of minor stoppages occurring on the equipment. This will include activities on keeping the equipment clean, completing bolts and fasteners with the correct torque, applying the correct lubricant, and addressing any form of leaks on the equipment. Dirt can cause the workpieces to jam which can cause errors. Implementing Autonomous Maintenance can reduce minor stoppages. However for minor stoppages that persistently keep on recurring, either a Focused Improvement or Planned Maintenance team can be deployed to address the problem. Improvement activities to reduce minor stoppages will include redesign or modification. A good indicator to track Minor Stoppages will be measuring the MTBA of the equipment every month.

Improving the Design Speed Loss: Improving this type of loss will require a change in the design of the equipment to increase its speed which will not compromise the quality of the products produced. Improving the design speed loss will improve the Performance Efficiency and cycle time to produce a product. An example of an improvement in the design speed includes changing the speed ratio of the driver and driven gear so that the equipment can run at a faster speed. Care must be taken in improving the design speed of the equipment so that no further problems will occur especially on the quality of the products.

Reducing Defects and Reworks: Quality defects and reworks are often chronic by nature. The word chronic means recurring and the causes are always multiple and in many cases a combination of several complex causes. Eliminating a single cause will not solve chronic defects. A recommended approach, in this case, is to use the conventional analytical problem-solving tool such as Ishikawa, or Fishbone Diagram, Pareto Analysis, and other similar tools to bring the defect to a minimum. Once the defect is already at a minimum, then this is where we need to apply P-M Analysis to finally zero out these defects completely.

10.5: Reducing Maintenance Costs

Although all industries want to reduce their maintenance cost, not all have been successful so far. There are two ways to reduce the maintenance cost, first, we have the right way, and second, we have the wrong way. The sad part is that most industries are doing the latter. The wrong way includes cost-cutting. Cost-cutting refers to measures implemented by a company to reduce its expenses and improve profitability at all means. Cost-cutting measures are typically implemented during times of financial distress for a company or during economic downturns. These cost-cutting schemes are usually short-term initiatives and can cause more harm than good for the company. In the majority of cases, these cost-cutting schemes will be initiated by the Decision Makers and Finance people. They can include:

• Laying off employees
• Delayed Appraisal for employees
• Removing certain employee's benefits
• No Christmas Bonus for the year due to Covid 19
• Purchasing going for the lowest bidder
• Cutting training budget for employees especially maintenance
• Abolishing Continues Improvement functions
• Terminating non-value activities
• No Overtime Policy

Case 1: A mining industry has a heavy consumption of lubricants. It was spending around 2.4 million PHP on its lubricants every month. The cost of repairs on their pumps, motors, mobile equipment spares is also high. The purchasing manager decided to source other vendors who can provide the lowest cost of lubricant. Three vendors submitted their proposal, Vendor A cost is at PHP 18,000.00 per drum, Vendor B, costs 24,000.00 per drum, and Vendor C's proposal was at 26,500.00 per drum. Without any consultation from the technical people from the plant, the Purchasing went on to award the oil to the lowest bidder, Vendor A. After a year of using the cheaper brand of hydraulic oil, their cost of lubricant increased by 25%, and the failure on their hydraulics seemed to increase. Upon investigation, the Flashpoint of the lowest vendor seemed to be the lowest at 180 ° C. While for Vendor B, the Flashpoint was at 210 ° C and Vendor C at 205 ° C. What does this mean? The oil is said to have "flashed" when a flame appears and instantaneously propagates itself over the entire surface. Flash Point means the temperature at which oil gives off vapors that can be ignited with a flame over the oil. The lower the flashpoint the greater the tendency to suffer vaporization loss at high temperature and to burn (min. 400 F). In layman's terms, the lower the temperature of oil, the easier the oil can burn and evaporate. When maintenance keeps adding oil, there might be a chance that the oil has a low Flash Point.

Lubricants Used	Average Monthly Purchase (liters)	Average Monthly Consumption (liters)	Unit Costs in PHP per liter	Purchased Costs Phil. Peso (PHP)	Consumption Costs Phil. Peso (PHP)
Hydraulic Oil (Tellus T-68)	24,000.00	15,838.00	57.14	1,371,360.00	873,270.62
Transmission Oil (Shell Rimula X-30)	24,000.00	20,872.00	52,19	1,252,560.00	1,089,309.68
Engine Oil (Shell 15W-40)	12,000.00	2,316.86	52.84	634,080.00	122,422.88
Drilling Oil (Torcula 150)	12,000.00	6,915.86	49.14	589,680.00	339,845.22
Total	**72,000.00**	**45,942.72**	**159.12**	**3,847,680.00**	**2,424,848.40**

Figure 10.7: Monthly Costs of Lubricants for A Mining Industry

Most of these cost-cutting schemes degrade the morale of employees and their enthusiasm at their work, creating more stress on the people. The correct way to reduce cost is not to focus on it but rather to apply the correct strategy on maintenance. This will automatically reduce the cost of doing maintenance on our equipment and assets. First, what is important is to know what is contributing to your maintenance expenses.

Determined to reduce the cost of lubricants and repairs of pumps and motors, the maintenance manager allocated a budget on training their maintenance people on Lubrication Strategy and Oil Contamination Control. The mining industry deployed a plant-wide Oil Contamination Control, and subject their lubricants for Oil Analysis. Maintenance people are made aware of how oil is contaminated and they were taught how to extract contaminants in the oil and maintenance applied a cleanliness standard based on ISO4406 cleanliness codes. By-pass filtration was installed, oil-filtration carts were purchased and included in the Preventive Maintenance activities. The maintenance people also improve their greasing practices based on the correct ways. After a few months of implementation, their cost of lubricant decreased together with their machine downtime, and the maintenance manager treated his crew for a well-deserved dinner.

PART	DESCRIPTION	PART NUMBER	QTY	UNIT COST	TOTAL COST
MTOR-042	Motor, Hydraulic	391012-7	1	15,701.30	15,701.30
MTOR-043	Motor, Hydrostatic	300-38811759	3	188,687.40	566,062.20
MTOR-045	Motor, Wagner	MV22-4012 563881	2	157,750.55	315,501.10
MTOR-045	Motor, Wagner	MV22-4012 563881	1	139,428.87	139,428.87
MTOR-061	Motor	367586	1	21,217.61	21,217.61
MTOR-066	Motor, Hydrostatic	64123057	1	186,277.00	186,277.00
MTOR-095	Motor, Hydraulic Reel	65531010	1	24,915.73	24,915.73
PUMP-097	Pump	66555691	1	156,287.78	156,287.78
PUMP-098	Pump, Charge	403343	1	31216.42	31216.42
PUMP-104	Pump, Hydraulic	900-3083169 680316	3	42,507.10	127,521.30
PUMP-104	Pump, Hydraulic	900-3083169 680316	4	31,672.70	126,690.80
PUMP-104	Pump, Hydraulic	900-3083169 680316	1	31,216.42	31,216.42
PUMP-104	Pump, Hydraulic	900-3083169 680316	1	32,468.78	32468.78
PUMP-105	Pump, Hydrostatic	900-3092582	6	198,554.67	1,191,328.02
PUMP-105	Pump, Hydrostatic	900-3092582	1	189,733.75	189,733.75
PUMP-105	Pump, Hydrostatic	900-3092582	1	217,678.57	217,678.57
PUMP-110	Pump, Wagner	MV21-2072 567418	2	117,700.27	235400.54
PUMP-110	Pump, Wagner	MV21-2072 567418	2	117,396.85	234,793.70
PUMP-110	Pump, Wagner	MV21-2072 567418	1	149,264.51	149,264.51
PUMP-112	Pump, Assembly	347915	1	64,776.40	64,776.40
PUMP-115	Pump, Hydrostatic	24VDC 6655755	1	347,042.70	347,042.70
PUMP-120	Pump, Assembly	371077	1	40,229.96	40,229.96
Overall Average Monthly Costs of Pump Motor Failures			37	2,501,725.34	4,444,753.46

Figure 10.8: Monthly Costs of Repair for Pumps and Motor

Case 2: The Bhopal India Case: When Everything Else Fails: Perhaps one of the classic cases where cost-cutting can go wrong is the case of Bhopal India. What happened in Bhopal,

India, is that at midnight of December 2, 1984, 40 tons of highly poisonous methyl isocyanate gas leaked out of the Union Carbide pesticide factory Bhopal. A cloud of methyl isocyanate gas escaped from the Union Carbide plant in Bhopal. A declogging operation was performed in the pipes, which cause the water to flow inside the MIC tank. A procedure should have taken place where the people or contractor involved should have placed a slip-line or a stopper between two pipes so that the flow of water can be averted going through the MIC tank. However, this procedure was not done, causing the water inside to flow until it reached the MIC tank. A mixture of water and methyl isocyanate gas created a very violent reaction in the tank and turned the MIC liquid into gas which escaped the plant where the aftermath initially killed more than 6,400 people, while many from the effects died several years after the disaster. 30,000-40,000 were seriously injured in the world's worst chemical disaster. Over 500,000 men, women, and children have been exposed to the poison clouds, and at least six thousand people died within the first week of the disaster. The current death toll was estimated to reach over 16,000 and is still rising. There were a lot of people that died of cancer depending on where the wind blows. Those women who were pregnant at that time were miscarried, while those who survived, their babies were stillborn. The poison affected some people's lungs and nervous system, while others became totally blind. This tragedy in Bhopal India is said to be the worst Chemical Disaster of all time.

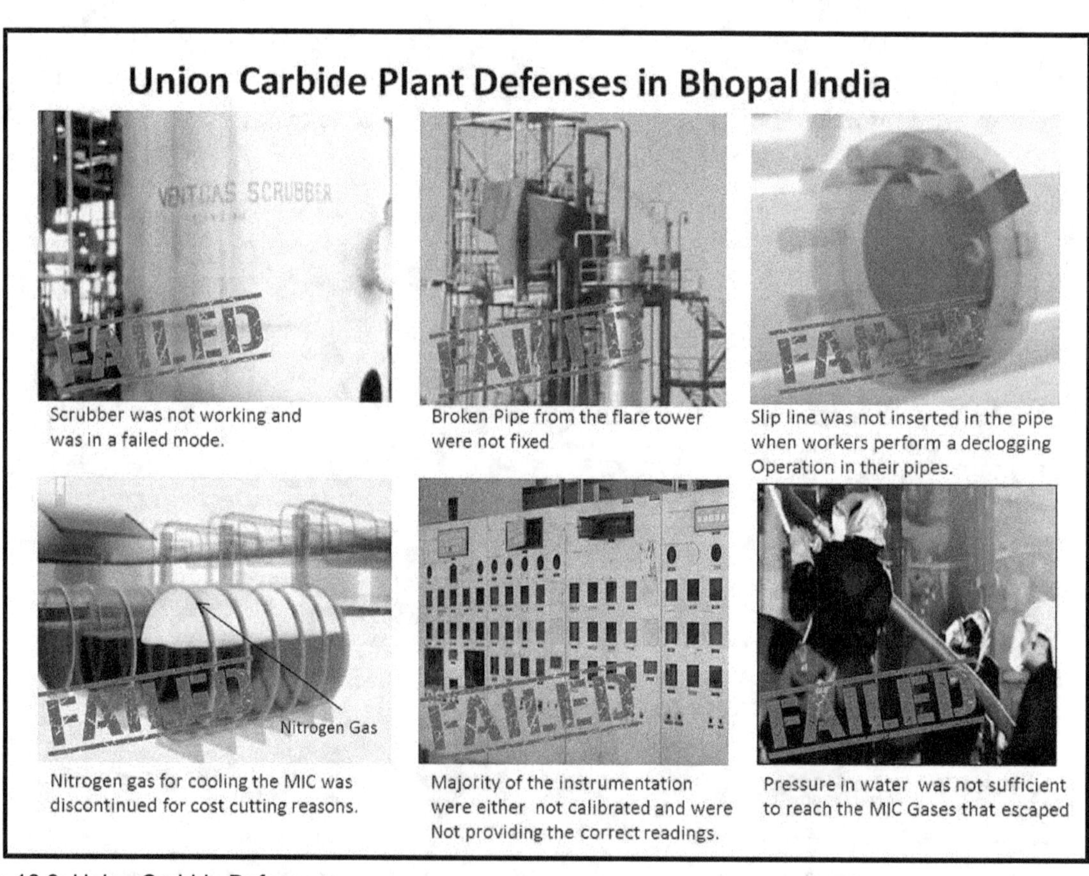

Figure 10.9: Union Carbide Defenses

• The maintenance manager was laid off from his work months before the tragedy.
• Instrumentations were not calibrated due to a lack of resources and cost-cutting.

- The scrubber was not working during the time of the tragedy for several weeks. Although, the scrubber can only contain a fraction of the leaked MIC gas that escaped.
- Gauges and instrumentation were uncalibrated and provided false reading. There was no one to check since the maintenance manager was laid off.
- The corroded pipe was not replaced on the flare tower. Should the flare tower be operational, it might have averted the tragedy.
- Emergency water hose spray insufficient pressure to reach the vapor exiting the exhaust stack.
- Nitrogen gas was removed from the MIC tank. It serves as cooling for the MIC. If the Nitrogen gas was not removed, perhaps we can either prolong the leak, or the leak would never happen as it cools the MIC. It was removed due to a cost-cutting scheme by management.
- A declogging operation took place before the tragedy. A slip-line should be placed before their declogging operation. However, it was not done.
- A proposal was initiated from one of the people from EHS to conduct awareness on the people living nearby the plant on what to do in case of a situation like this tragedy which again was denied by management. The person who initiated this proposal resigned from the plant.

Union Carbide plant was located in Bhopal where many people were residing near the plant. Most of these people lived in slums. There was no orientation done to the people living near the plant just in case some emergency cases, happened. One of the employees from the EHS was proposing to provide an orientation to the community but was rejected by his superiors. The person resigned from Bhopal. If people just knew what to do, perhaps even if the MIC gas escapes then there will be fewer casualties. Since MIC gas is denser or heavier than air, then people can inhale it. If you are in your home, you just need to close down the windows and door. Place a wet towel on the bottom of the window or door, the MIC gas that escaped will be absorbed by the towel and will not be inhaled by the people inside the house. This was actually what one of the survivors did when she called the person who knew what to do in case of the MIC gas escaped and survived the tragedy.

If we conduct a survey, ask each and every reliability and maintenance manager in every industry which of these two items is more important, reducing costs or improving reliability? Almost everyone will agree that improving reliability is more important than reducing costs simply because if reliability on the equipment starts to improve then cost will definitely go down; however, it is not the other way around. In fact, in many cases, reducing costs on maintenance will create problems with reliability. But if we conduct an actual survey in the real world on what most organizations and industries are focusing on, I think it is more on cost-cutting or having some form of cost reduction program which is a dangerous game to play. This is the name of the game most industries play today. Maintenance often realizes a cut or slash in their maintenance budget by 10, 20%, and the figures increase every year. Headcounts are reduced as a result of a study on man to machine ratio. And what happens to those affected or classified as excess, sad to note that most of these good people will be laid-off. There are cases when old maintenance people are forced to retire, and industries replaced them with a new breed of the younger generation of people with less experience on the asset being maintained. Spares budgets have to be cut, and some of them need to be localized to cope with the plant's cost-cutting program. Purchasing people are busy looking on the yellow pages or sourcing the internet for vendors that can provide the lowest possible cost compromising the quality of the spare part.

Maintenance people are always in a never-ending battle against production people in getting the equipment for scheduled PM work. The budget for maintenance training is often cut-short, and maintenance can barely attend to the training. In short, maintenance is left in a never-ending war against being reactive with sudden breakdowns which are all around. Maintenance is definitely going to lose this war. The pressure is building on the maintenance side, and as pressure increases, maintenance people are in a state of confusion, and when maintenance is confused, it only boils down to one thing, "CHAOS," and that is the life of a maintenance person. It doesn't have to be this way since there is always a better way. If your industry is dead serious about reducing costs on maintenance, why don't you consider the following, especially if you are from the management side?

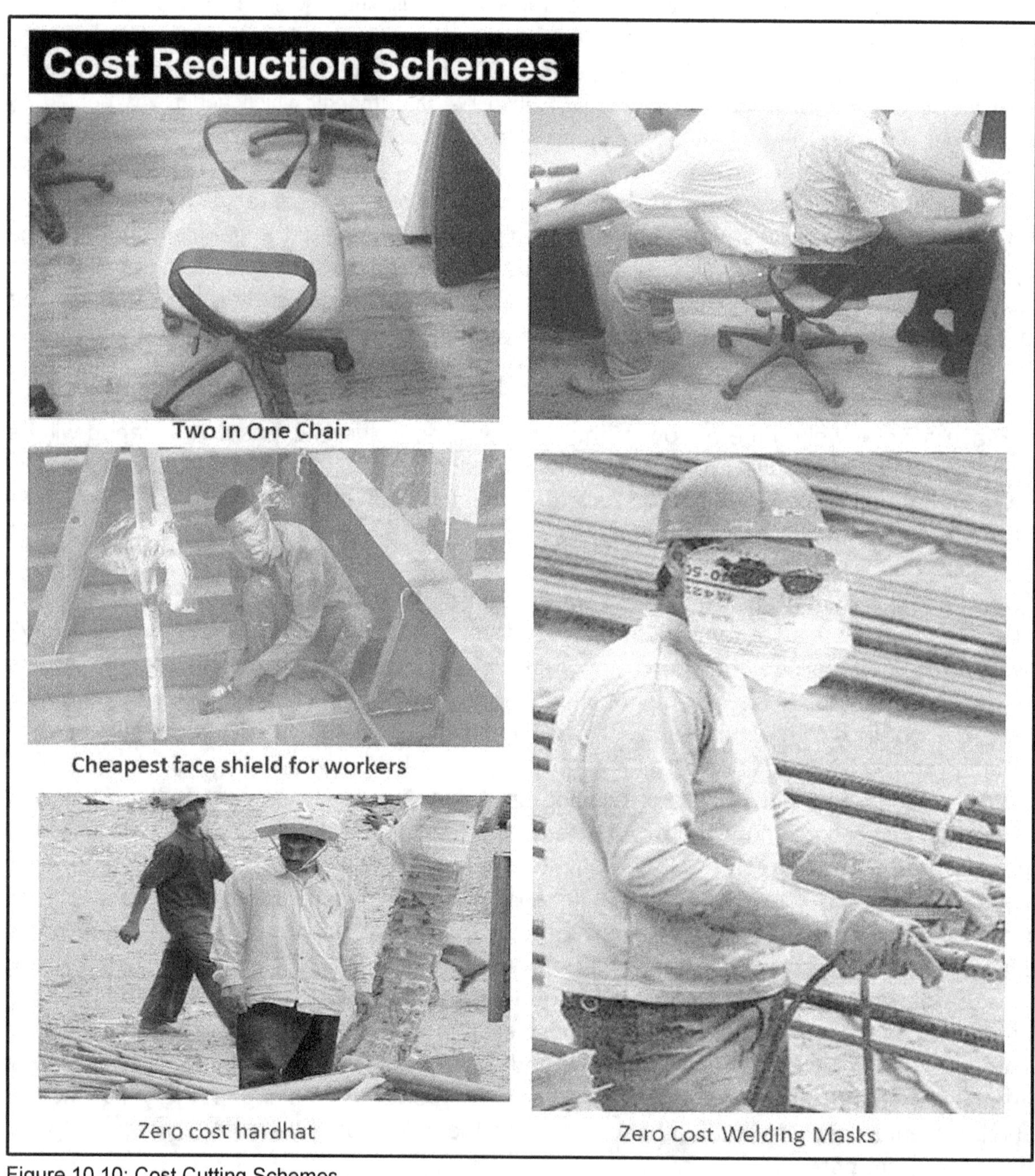

Figure 10.10: Cost Cutting Schemes

10.5.1: Where Maintenance Should Focus to Reduce Costs

There are many ways to reduce the maintenance cost the right way and the best way to reduce it is not to focus on it. What do I exactly mean by this? There will always be an investment in doing maintenance and we cannot escape this fact, but the good thing is that if we apply the correct (not the best) practices, then it will yield returns just like business in which additional cost can either be avoided or saved. What I am trying to say is that the focus of maintenance must not be on reducing cost but on both improving and sustaining the reliability of the equipment. This means that if the equipment's reliability improves, automatically whether we like it or not, the cost of doing maintenance will decrease, but if we focus on reducing cost at all means, then this is just temporary and may produce more harm than good for the equipment.

Address the Basics Condition in your Equipment: If there is one thing I believe, big problems, catastrophic failures are just an accumulation of small problems frequently neglected. The process of addressing the basic equipment condition simply means keeping the equipment clean, applying the correct lubrication, tightening bolts, and equipment with no form of leaks whatsoever. These are very simple tasks to perform, yet the impact will be felt on having lesser failures occurring on the equipment. It is useless to conduct vibration monitoring on the equipment if the equipment lacks bolts, nuts, and fasteners. Remember that it will only take one loose or missing bolt to create a chain of destruction and havoc in our equipment.

Take a Look and Study Your MRO Spare Parts Management: Instead of laying off people in your maintenance organization, why not let 2 to 3 people improve your storeroom and organize a Spare Parts Management System in your plant, or why not man your MRO storeroom for 24 hours if your plant is running for 24 hours. Most industries have a stockroom to stock parts, but not all have Spare Parts Management in their plant. Let me just give you some hints. If you have a form of computerization for your spare parts, does the physical quantity always match the system's quantity in the system? If they don't match, then you got a big problem. Just imagine going to your stock room to get some parts you needed for a repair only to realize that the part has no more physical quantity left even if the system says that there are at least a dozen more of them. How would you feel? What do you do with the spare parts whose equipment had already been decommissioned or in the process of decommissioning? What do you do with those parts that are classified as obsolete? Or how about the same parts with different part numbers because they came from different vendors and are used by different departments? How do you control and consolidate them? These are just some of the questions that can be raised in your stock room. I am pretty much sure that if we can utilize these people to improve your spare parts, the savings they can generate will go far out weight the time they spend in improving their stock room. Remember that the goal of the Spare Parts Management team is to provide the right part at the right time when maintenance needs it most. The longer the spare part is acquired from the storeroom, the longer the downtime on the equipment can be realized.

Consider the Study of Life Cycle Costs of your Spares and Equipment: For industries, there is always a temptation to purchase equipment or spares based on the lowest or cheapest source. This might not be such a good idea since purchasing based on the initial costs only tell us one side of the story. The true costs can be seen based on their performance. In my experience, this is the problem with most procurement and purchasing departments since they make the decision to purchase parts based on the lowest bidder or lowest possible costs without involving the engineering or technical people in the plant. The savings these departments claim are insignificant since both operations and maintenance can encounter many failures due to this. They always look at the initial cost of the part and not the cost of problems the part may give the user in the long run. This is the problem with cheaper parts and equipment that fails every day and will likely yield a much larger amount of cost in a period compared to slightly higher equipment. The true cost of the equipment can be felt from the time the equipment was commissioned to the plant until the time it is decommissioned or disposed of. Life Cycle Management not only pertains to the life cycle of the equipment but can also be used on spares.

Let me give you an example of two oil filters; one is rated as a nominal filter with a Beta Rating of 2. This filter has an efficiency of 50%, while the other filter is considered absolute and has a Beta Rating of 1000. The efficiency of this oil filter is 99.9%. The efficiency of the filter refers to the number of solid particles or contaminants that the filter element can remove. Let us say that we have a 5-micron size filter with an efficiency of 50%; this means that if there are around 100,000 contaminants in the size of 5 microns and above that will pass through the upstream of the oil filter of 5 microns, the filter can only capture 50% of the contaminants or around 50,000 out of 100,000. The remaining half will travel downstream of the filter together with the oil and back again in the system which can be trapped in tight clearance, and cause the start of abrasion.

Figure 10.11: Which Oil Filter Will Save You on Cost?

On the other hand, if you have an absolute filter of 5 microns and a rating of 99.9%, this means that the filter is guaranteed to trap all five-micron size contaminants and larger. This means that 99,500 out of 100,000 will be trapped by the oil filter. Obviously, the filter that can trap more contaminants will provide you more reliability and smooth operation than a nominal rating of the filter. Now you must be wondering about the cost of this high-efficient filter. If I told you that the cost of both filters is 450. Definitely, you would like to purchase the absolute filter, isn't it? But there is one big difference in the costs, although both filters cost 450, the absolute filter cost 450 USD while the nominal filtration filter costs 450 Philippine Pesos (1 USD = 46 Pesos). This means that the costs of the nominal filter will actually be 10.65 USD compared to the filter with a beta rating of 1000, which is 450 USD. But if you can compare the other costs, which include downtime costs, replacement costs, spare parts costs, overtime costs, and other costs, the nominal filter costs much more compared to the absolute filter in the long run. Again if we ask the maintenance people on which filter they prefer, I am pretty much sure that they will choose the oil filter with a higher efficiency rating, but perhaps the sentiments of the procurement or purchasing people will be different since they will select the nominal oil filter because its initial cost is much cheaper. Although initially, the purchasing will be saving on the initial cost, in the end, they will leave maintenance with the agony and despair of experiencing a lot of problems and failures on their equipment since this nominal filter is only designed to remove 50 % of the contaminants which will leave the other 50 % to be trapped on tight clearances on mechanical parts that will surely end up in failures and breakdowns. There will be many cases where purchasing the cheapest item or part may be expensive, and purchasing expensive parts and items will be cheaper in the long run and will definitely save maintenance on cost. What is important is that the decision should be based on the life cycle.

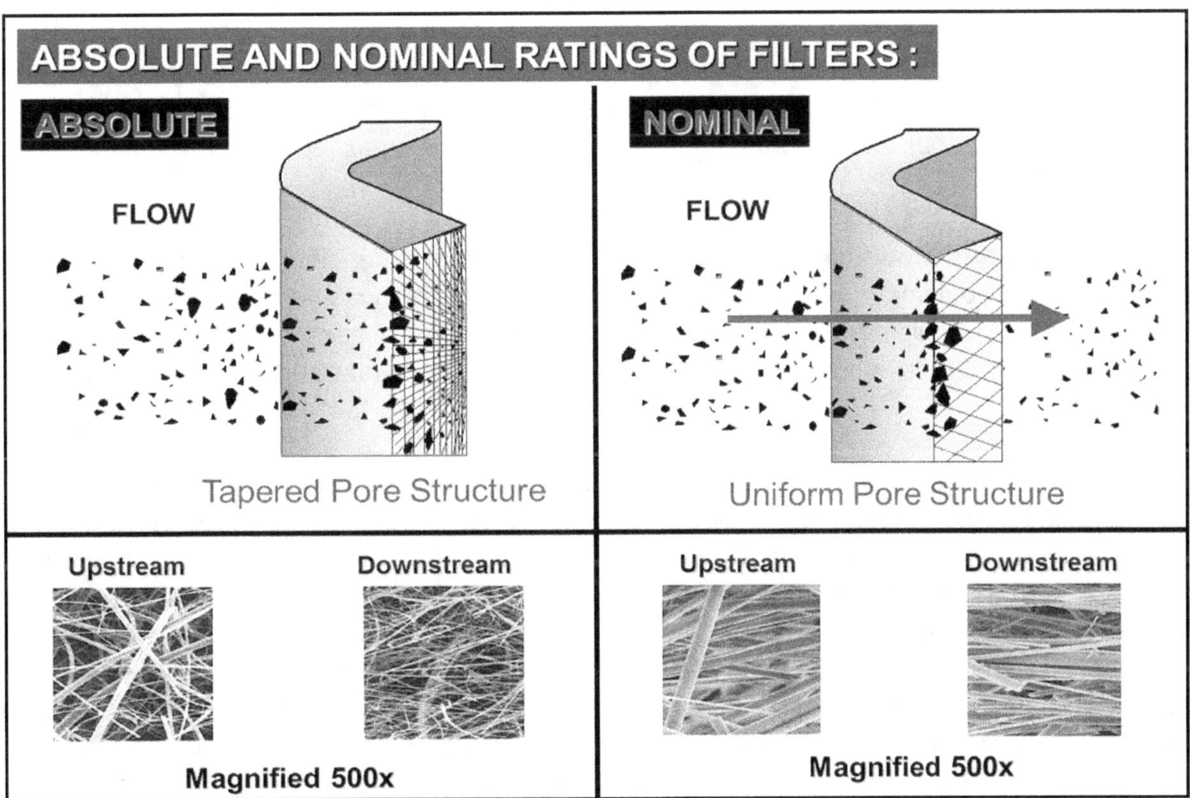

Figure 10.12: Difference Between Absolute and Nominal Filtration

Establish a Lubrication Strategy in your Plant: Most of the time, lubricants are just taken for granted. Oil is just oil; if the level is low, the maintenance will add oil or grease to our equipment. If there are just too many types of equipment around, then maintenance decides which equipment gets the oil. That is all we do because, from the maintenance point of view, that is all they know. Putting too much lubricant or less lubricant will not be good for your equipment, so the question raised here is how much quantity of lubricant is sufficient enough to be placed on each and every piece of equipment. If we are greasing a bearing on the motor, for instance, does maintenance know what type of grease to place in the bearing, how much quantity is required, how do they know that the grease is reaching the raceway of the bearing, what type of grease gun to use and all that sort of stuff? These are some of the important questions that should define a precise answer because, in most PM activities, the level of information is insufficient. Most PM checklists state what equipment is to be re-greased, how frequent and who should be responsible but as to what type of grease, the correct amount in grams, the NLGI number, type of grease gun to be used, special accessories such as high-pressure digital meters, and the hearing aid is not explicitly known.

Another important aspect to consider in your Lubrication Strategy is Oil Contamination Control and Awareness, how are your lubricants being stored. Dispensing methods, methods of transferring oil to equipment, disposal, oil analysis test performed, addressing oil leaks, and so on. The smaller the contaminant, the more destructive and abrasive it can turn out to be. These tiny contaminants, often invincible to the naked eye, are the main cause of problems that affect our equipment. They can be controlled if we know the ways and means to address them.

Address All Secondary Functions or Function Reduction Breakdown: A function reduction breakdown is any failures, breakdown, or malfunction of secondary functions in which the primary function is still being fulfilled. This means that despite these failures, the machine or equipment is still capable of providing the production needed. Examples of this include protective devices that are not working, defective gauges, by-passed sensors, busted bulbs, and others.

Improvements and Modifications: One of the roles of maintenance is not only to sustain but to observe their equipment, machines, and assets and identify those parts with inherent design weaknesses, No equipment is perfect by nature unless God himself designs the equipment. There will always be flaws in the design. We need to identify them and improve these design weaknesses.

Invest in your People: Training is where we acquire knowledge and knowledge is the very basic and first step in developing and building the skills of our maintenance people so that they can perform their function right the first time around. Skills must not be based on trial and error but should be based on the right knowledge. As a start, when our boss or manager send their people to training. Managers should talk to their people and set expectations on why they are being sent to the training. Let the people know that they are sent because we want the learnings to be used and spread among our people. Remember that people can only be considered the greatest investment if they are equipped with the right knowledge on how to do their work correctly. One of my inspirations in writing all my books on reliability and maintenance is based on my frustrations. One of my biggest frustrations as an author and maintenance teacher lies in the fact, that many maintenance people are deprived of knowledge they need to do their work correctly since their management decline these good people to attend to training, and those who attend training do nothing except post their certificates on LinkedIn telling his connections to view their certificate, or perhaps to upgrade their CVs and resumes so that it will look cool to the Human Resources when they are seeking greener pastures. But there will be always the few, that despite all these odds will find the knowledge through any means, adapt it so that their industry can benefit. I salute these people.

So the next time that your industry is going to lay off some maintenance people in your organization for cost-cutting reasons, why not convince the decision-makers in your organization to place some of your maintenance people permanently on areas that will improve your maintenance most. This is a win-win; the savings generated will go far out weight their salaries. Some of them are explained above. Maintenance is not only a department; it is just more than an activity or task. These are your very own people that look upon the welfare of your assets and equipment. Maintenance is a business, and in any business, there is always an investment. Maintenance is more than a repair function but sad to note that most industries perceive maintenance to be exactly that. Worst are industries that don't value their maintenance people because they are inclined to think that they can be replaced anytime. I tend to agree to some extent but kindly take note that the skills, knowledge, experience, camaraderie will take a much longer time to replace. All I would like to share in this book is that try to value your maintenance people, provide them with the right tools, the right knowledge, training, skills, and most especially the respect they deserve, and I am pretty much sure that these people will deliver with all humility the best they can to preserve their equipment and assets. Remember that it will be the people that will improve the equipment and not the equipment or technology that will improve the people.

10.6: Improving the MRO Storeroom Should be Done Both Inside and Outside

This chapter explains how to improve the MRO Storeroom and Spare parts from inside of the storeroom. However, improving the MRO spare parts and reducing the inventory should be done both inside and outside the storeroom. This means that parallel activities must be deployed by the maintenance organization. If the number of breakdowns and failures is reduced, then we do not actually need to stock all these parts inside the storeroom immediately, sure they will fail but on a longer time frame. If these strategies can be implemented correctly with the right people, and right methodologies, then these people can help maintenance and the storekeeper manage their spare parts more intelligently. If these parts can be predicted or prevented, then most of these parts can be ordered just in time when the user actually needs them. Being proactive on maintenance can lessen the inventory that is kept inside the storeroom.

One of the underlying issues confronting the storeroom is, to whom they should report. If the storeroom people in the plant reports to different functions of the organization such as Purchasing, Warehouse, Finance, Supply Chain, or whatnot, their goals, objectives, and direction may entirely be different from the goals of the maintenance, engineering, or reliability function of the industry. The line of communication must always be open at all times between maintenance and the people from the storeroom. If the storeroom people do not report directly, then they can report indirectly to the maintenance function. That is all we could do, but it is not a smooth road to that, as logistics, supply chain or warehouse may have different things in mind from the maintenance function. It is difficult to play basketball with two coaches telling you opposite things. The users of the MRO spare parts are not the warehouse or logistics, but the users are maintenance. If one or two pieces of equipment will be scheduled to be retired or decommissioned, a major shutdown will be done next month, or major modification will be made, will the current original stock, still be used in the equipment. Storeroom people should be aware of these kinds of things. Maybe others will disagree, but my stand is simple, for the best outcome, MRO Storeroom people should be reporting to maintenance (sorry for sounding like a broken record as I have reiterated this point so many times) and not to other functions because maintenance people are the users of MRO. They know the parts better than anyone else in the organization. They know what works well and what will not with their equipment. As industries become less reactive, the need for spares will become less. The role of the storekeepers is not only to receive, or issue parts but also to analyze and understand how to maintain the spares that are stocked inside the storeroom.

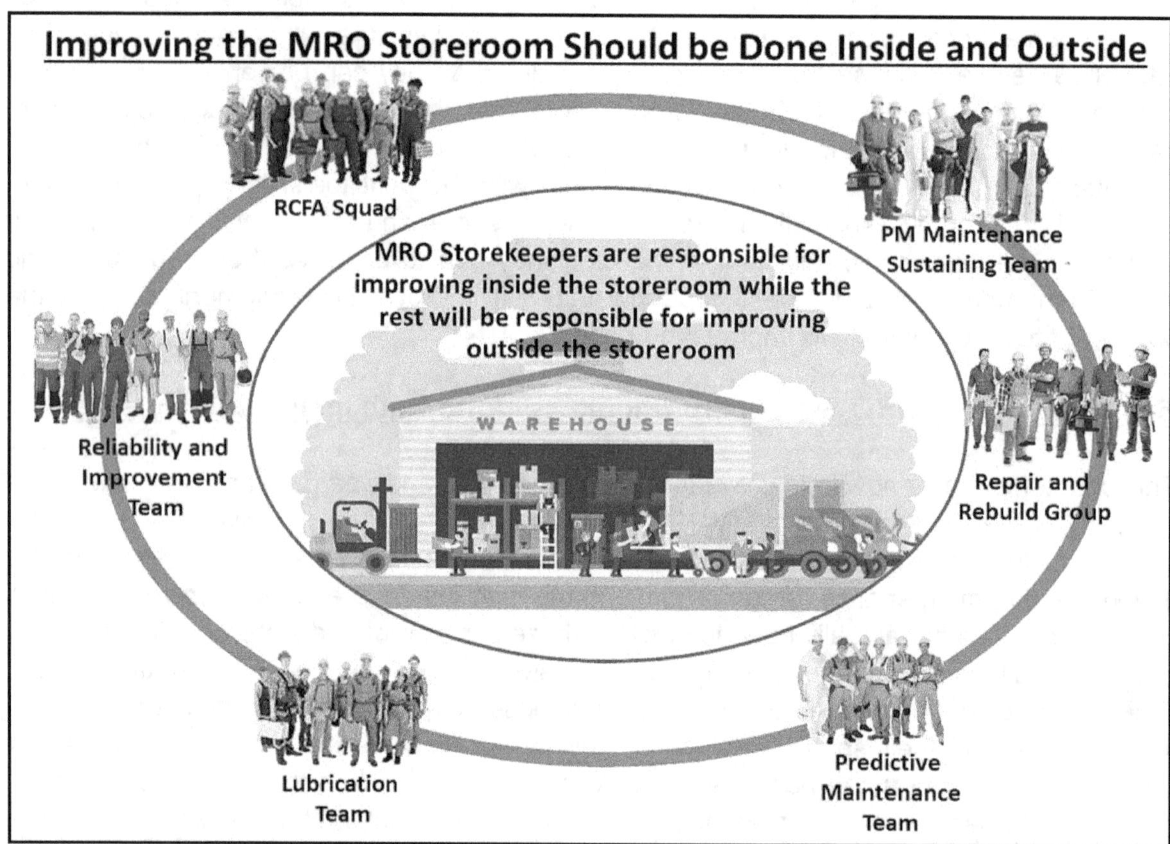

Figure 10.13: Improving the MRO Storeroom should be done Inside and Outside

Maintenance, reliability, and engineering people have a role in the storeroom as all their efforts are meant to improve what is outside the storeroom. If the team responsible for modification has lengthened the lifespan of the part that fails frequently, then this means that replenishing or reordering the item will be lengthened so as the rest of these improvement teams. MRO Spare Parts and Storeroom will always be a part of the overall maintenance strategy of the plant. Making our equipment more reliable simply means lesser failure and downtime on the equipment. Moreover, the word reliability means trustworthy, which means that we can trust the equipment to run smoothly. This means that parts are reaching their lifespan. Moving from a reactive to a proactive environment is not an easy task for industries as most of their efforts are just focused on the short-term plan. All maintenance strategies must always be focused on both the short-term and the long-term plan, which will not happen overnight. It will require having the correct knowledge, discipline, a change of mindset, and a change of heart, which is not easy to do as most industries since these industries have been accustomed to doing what they are doing for such a long period even decades. Industry's c-level people and decision-makers must provide their people initially with the knowledge and education they need so that they can do their jobs right the first time around.

The majority of the problems in industries can be solved, as mentioned in this book if industries are willing to make changes in how they do things in the plant. Industries that achieve a level of World Class Maintenance were not born that way. They were also reactive in the past, but the leaders have a change of heart and propelled their workforce in a new direction so that they can stand off from the rest and compete globally in this fierce world of competition. Achieving World-Class Maintenance level is not having the best software in town or having the latest smart sensors, Predictive Maintenance, and all these technologies but having the right people doing the right things in the plant, and storeroom, and I think that's all I have to say about that.

Chapter 11

FAQ, and TIPS, on Maintenance Indicators

> *Measuring our maintenance performance indices and KPI's clarifies the need to focus on long term goals and strategies by comparing our actual performance against the goals we set forth. We can only be in control of the situation if we measure what is meaningful and important to us because measurements allow us to make the right decisions in our organizations.*

11.1: Frequently Asked Questions on Maintenance Indices

Here are lists of frequently asked questions and answers on Meaningful Measures of Equipment Performance. Some of these questions have been raised in my previous training classes.

Figure11.1: MMEP FAQs, and Tips

1. **If an industry is aiming for the TPM excellence awards, and the OEE of their equipment dropped since it was not used due to this COVID 19 lockdown or for any other**

reasons such as low demand, will it affect our chances of getting the certification? The answer is no. JIPM/JMA Assessors will understand the situation and just indicate in your graph the reason for the decrease in the OEE and the JIPM auditors will understand this.

2. **Our equipment is old but is running perfectly. Our problem is that we do not know the ideal or design UPH of our equipment. What will we use as the ideal or design UPH (units per hour) in our case, how should we compute the Performance Efficiency on OEE?** Answer: You may get an IE and conduct a time and motion study or you can use the highest output ever produced by an operator on that machine. This would serve as the ideal UPH of the machine.

3. **Our equipment is not dedicated and is used to produce several products. The ideal or design UPH varies from one product to another. What will be the UPH we will use in this case?** Answer: The cleanest way is to determine the time each of the products runs in the equipment and specify the ideal UPH for each product for the respective time it ran on the equipment.

4. **The equipment was converted to another product and took the maintenance 2 hours for the changeover time. However, the maintenance forgot to remove the Allen wrench, and when the operator started the equipment something was damaged that took the maintenance another 2 hours to repair the equipment? Will the total downtime be charged to breakdown or on set-up or changeover?** Answer: This will depend on the indices being measured. If we are measuring availability, then we need to include both the downtime caused by set-up and breakdown. If we are calculating MTTS or Mean Time To Set-up, only machine downtime caused by set-up will be included. The same goes for computing MTBF, only machine downtime caused by breakdowns will be included.

5. **Some manufacturing equipment encounters what you call short stoppages or assists. TPM term this as chokotei or minor stoppages, where the equipment stops, the operator resets, and the equipment runs again. Usually, it takes seconds or minutes to address this kind of problem. Will minor stoppages or short stoppages be included in the MTBF calculation or not?** The answer is no. Assist are different from breakdowns; some industries include them as breakdowns when waiting time exceeds six minutes and above. This is entirely wrong and should not be included in the MTBF calculation since what should be included in the MTBF calculation will only be downtime caused by breakdowns and failures. Suppose you insist on including them in the MTBF calculation; in that case, they might as well change its name from MTBF (Mean Time Between Failure) to **MTBFAA** or Mean Time Between Failure and Assists. Again, assists or short stoppages should not be included in the MTBF calculation.

6. **The Predictive Maintenance or CBM group spotted a spike or increase in vibration on one of the equipment, which is a clear sign of a potential failure. Finally, the group decided to schedule the equipment for an overhaul and replacement, which took 4 hours of downtime. Will the downtime be included in the MTBF calculation or not?** Yes, if we follow the formula that MTBF is equal to operating time divided by the number of failures, operating time is equal to the loading time minus machine downtime caused by breakdowns and failures. And loading time is equal to the available time minus planned downtime. Hence, the equipment's maintenance repair will be included in the planned downtime on the MTBF calculation. However, the denominator of MTBF is the frequency of failures or breakdown, this

should not be added in the frequency of failures since this is a Planned Downtime, only Unplanned Downtime will be included in the denominator of the MTBF. The rule of thumb to follow is that if the maintenance tasks were done first before the failure, then this will be considered as a Planned Downtime, while if the repair happen first and the maintenance reacts and repair, this will be considered as unplanned and will be included in the denominator of MTBF which is the frequency of failures.

7. **A safety officer audited one of your equipment to have a severe leak and was provided a non-conformance ticket for safety reasons. The machine was not allowed to be used for safety purposes unless the leak was corrected. The machine was finally repaired for leaks, which took the maintenance 3 hours to fix. Will this be included in the MTBF calculation or not?** The answer is Yes, in this case, it should be included in the MTBF calculation, although there was actually no breakdown that occurred in the equipment. This will be included as a planned downtime of the MTBF calculation. This situation is actually the same as in Case number 4. Let me give you a concrete example so you can follow it.

<u>Given:</u>
• Time to fix the leak is 3 hours, which is considered as a planned downtime
• Available time in one day is 24 hours.
• Actual breakdown in this situation is zero

• MTBF = Operating Time / Breakdown Occurrences
• MTBF = (Loading Time - Machine Breakdown) / Breakdown Occurrences
• But Loading Time = Available Time - Planned Downtime = 24 - 3 = 21 hours.
• MTBF = (21 - 0) hours / 1 = 21 hrs.

In this situation, we compute MTBF for that given day, and in the case where there is no actual breakdown, two options will be required to calculate the MTBF value. First, lengthen the time to get the MTBF until an actual breakdown is encountered, or second, assume a denominator of 1 to obtain the numerator; otherwise, an answer of infinity will be achieved.

8. **In some manufacturing plants, there are machines used to process more than one product. If the equipment is used to run a different product, some parts will need to be replaced for the next product to run. This is called the conversion time or set-up time. Suppose some of the toolings were not fitted correctly, which cause breakage on the equipment, which occurred right after the conversion. Will this be included in the MTBF calculation or not?** In this case, an actual breakdown had occurred on the equipment caused by poor set-up and conversion. Therefore, this will be included in the actual MTBF calculation. Note that only when the equipment was down caused by the breakdown will be included, but the downtime caused by the conversion and set-up time will not be included in the MTBF calculation.

9. **A question perked up during one of my training on OEE where a student asked that since in their plant Changeover and Conversion is usually a planned downtime, hence the question is Set-up, conversion or changeover a Planned or Unplanned Downtime.** Even if the conversion was planned to be performed at some time, this is still considered as Unplanned or Machine-Related Downtime. According to TPM books, this is considered one of

the major equipment losses, therefore whether you industry accept it or not, it is considered as unplanned or machine-related downtime.

10. **If a piece of equipment achieves 85% OEE or higher, is the equipment already considered World Class?** Although in some books they will state that achieving an 85% OEE will consider the equipment World Class. My response to this is not exactly. Why, as explained in Chapter 3, OEE is only a measure of the primary function of the equipment, but every single piece of equipment will have its own primary and secondary functions. There are cases where the failure of a secondary or function reduction breakdown will be far more serious and dangerous than the failure of a primary function. Second, OEE is not only about achieving 85% but ensuring that all three components are satisfied. If we have an Availability of 99%, a Performance Rate of 98%, and a Quality Rate of 90%, this will have an OEE of 87.13% which is already beyond 85%, but this equipment is suffering from a large number of defects and scrap which is simply unacceptable.

11. **In one manufacturing firm, during A-shift Machine 1 was able to produce 2000 products but 20 products were rejected and reworked. These 20 rejections were again reprocessed on the machine, the question is do we include them in the calculation on Quality Rate?** The answer is Yes, this will be included in the Quality Rate even if the products were once again reprocessed. The quality rate will include those products that were scrapped, rejected, reworked, and those that had been returned by the customer.

12. **Assuming that there are replacement activities on Preventive Maintenance that have not been done since the Predictive Maintenance measurement indicates that it is still working, should this be reflected in the PM Compliance?** My answer to this question is yes. This will depend on the type of industry. Assuming that the scheduled shutdown happens in a power plant in which the next shutdown will happen in the next couple of years, then, technically we need to replace the part even if it had been Predicted unless otherwise, the system has a complete backup or redundancies and I do not know of any Predictive Maintenance instrument that can predict the failure 2 years in advance. However, for scheduled maintenance that will not be done this long, people in charge of performing Preventive Maintenance can provide feedback to the planner to include the replacement under the watch of Predictive Maintenance

13. **If we buy-in on the formula on Reliability in figure 4.12, it is dependent on MTBF, if breakdowns and failures are completely eliminated, can we declare that the equipment is reliable?** If we based our definition that Reliability is the probability that an item will operate without failure throughout a specified interval and that the item will perform its intended function under specified operational and environmental conditions under a given and specified time, the answer is yes, however, declaring the equipment reliable should not solely be based on breakdown alone as other losses may also be present. If equipment has a capacity of 1000 units per hour, and the operator can only produce 920 units per hour without any downtime, then we have a design speed problem. Although the formula will not indicate this since it is only based on failures and breakdowns. What is important is we also consider the other equipment losses. The second point, we also need to consider the secondary functions, or function reduction breakdowns in the equipment. Meaning both primary and secondary functions should be working to consider our equipment reliable.

14. **If a maintenance task includes changing the oil, and the oil was not changed because**

the Oil Analysis report such as RULER (Remaining Useful Life of Oil) indicated that the oil is still fit to be used. Do we consider this as a maintenance backlog? My point of view on this is that even if the maintenance executing the tasks did not follow the maintenance plan for whatever reasons, this will be considered as a backlog. However, the best thing to do is to write a Memorandum, Engineering change, Revision of Standards, or whatever you might call this document in your industry and have this signed by the Maintenance or Engineering Head. Once, the document is signed, a copy should be given to the planner and Quality Control especially if these people are involved in auditing the tasks. In this case, we can consider the previous maintenance backlog to be null and void, and the planner should revise his future plans based on the modification. Unless this is not done, then not doing the change oil will be considered a maintenance backlog.

15. **Should MTTR be included in the maintenance appraisal?.** Although I know some industries that are doing this, I would not recommend this practice. Ideally, the faster the maintenance can repair the equipment, the better for the MTTR indicator. However, we also need to understand that repair varies depending on the failure. Repairs may include tightening a bolt to completely dismantle the equipment. Hence, if Charlie is just tightening some of the loose bolts and Bob's repair requires dismantling the whole equipment, Charlie would have a better MTTR which is quite unfair. Or how about the case where Charlie repairs so fast that he has a good MTTR, but the equipment always fail when Charlie works on it, on the other side, Bob takes his time to repair the equipment and ensures that he will not be called back again to repair the same failure on the same equipment. In this case, Charlie is efficient (Fast) but not effective.

16. **Can OEE be applied to power plants or similar plants in which their assets are system-based?** Originally, OEE is designed for manufacturing plants that produce a product. Power plants do not produce products but services. OEE can still be used by power plants but only using the component on Availability and Performance Efficiency. Quality Rate will only be used if the power plant can quantify its service. For example, I have a client which is a geothermal plant in which they have a laboratory that can measure the quality of the steam they provide to their customer, hence, in this case, Availability, Performance Rate, and Quality Rate are being measured. If this will not be possible, then OEE for power plants will only be a product of Availability and Performance Rate.

17. **With the many variations on the formula on Availability, which should we use?** Although in the book of Seiichi Nakajima, the founder of TPM used the formula Availability is equal to the Loading Time minus Machine Downtime divided by the Loading Time. This is actually utilization and not availability since the denominator for availability should be the total plant's available time. Hence, the availability should be equal to the available time minus all downtime which includes Planned and Unplanned divided by the available time. Although this would provide a lower value compared to Nakajima's formula, it also opens more challenges for improvement. Also, Nakajima's formula is not Availability but more of utilization.

18. **Our industry made it a policy that all minor stoppages which took more than 10 minutes to address will be included as a breakdown. Is this right or wrong?** This is entirely wrong since a minor stoppage is different from a breakdown. Although for both minor stoppage and breakdown, the machine will stop. The difference is that for breakdown

something failed, and the machine needs to be repaired. For minor stoppages, there is no failure or breakdown, the machine just stopped, and typically, whatever error experienced will not be repaired but corrected, once it is corrected the machine will be restarted. If we mix both breakdown and minor stoppages, then we will be having variations in calculating MTBF, MTTR, and other indices.

19. **Are there Maintenance Indicators that relate to those secondary failures, or those function reduction breakdowns?** TPM has a measurement for this which is about the total abnormalities detected versus those corrected and expressed in percentage. These abnormalities can be anything such as broken pressure gauges, missing bolts, leaks, and others. Or perhaps we can be more precise as to determine all those secondary functions that are not working in operations for the totality of all equipment and assets.

20. **If the equipment is not experiencing any failures and breakdowns, can we declare that the equipment is reliable?** Not exactly, although the formula is directly linked to the failure rate which is the reciprocal of MTBF, it does not mean that the equipment is reliable if it is breakdown free since failures and breakdowns are just one of the equipment losses as discussed in Chapter 2 of this book. Second, to consider the equipment reliable, both primary and secondary functions of the equipment should be functioning and not in a failed state.

21. **Is it true that Reliability cannot be improved but can only be sustained as other experts in the field claim?** My answer is both yes and no. If we consider maintenance to be a task, then this statement is true and correct since maintenance can just sustain the reliability of the equipment. But if we consider maintenance as a human being, reliability can be improved as maintenance people are capable of generating improvements, modifications, and redesign on their equipment and assets.

22. **Should maintenance measure utilization?** I would not recommend this measurement for the maintenance function because this is an operations measurement, and maintenance has no control over this measurement. What maintenance can control is the availability. If the utilization is low since there is no raw material or for some other reasons, this is way beyond the scope of the maintenance function, and maintenance should not be blamed for the low utilization.

23. **If an oil and gas plant or similar industries have a lot of redundancies, what will be the MTBF of those idle and redundant components?** This will depend on how the plant measures its MTBF. This means that if the plant is measuring their MTBF on a system based and a failure was experienced on one of the motors in which the back-up automatically started, then if we measure the MTBF for that particular day, it will be perfect at 24 hours. However, if we measure MTBF on a component or asset level that composed the system, then all idle, redundant, and backup equipment's MTBF is zero since they were not used.

24. **Who are the best people to make an impact and monitor OEE in the plant?** The best people to tract OEE are the people who can improve this index and also it depends on the losses suffered by the equipment. For example, if failures and breakdowns are the only losses that contribute to OEE, then the best people to track OEE are the maintenance function. If Quality Rate is the main contributor to having a low OEE, then the Quality people should be tracking OEE. The answer to this question is it depends on the equipment losses. Usually, a

cross selection of people called a Focused Improvement team is deployed to address these losses, hence they should be tracking OEE.

24. **If the equipment is not loaded since there was no raw material to process, is the availability considered 100%?** As in our example previously about the car in Chapter 3 of this book, if it takes 1 hour to travel from home to the plant and another 1 hour from the plant to travel to your home, we have utilized the car only for 2 hours during that day. The rest of the time the car was parked and available, hence availability is at 100%. Although this may sound simple but not in industries, since if the equipment is left idle or is not operating since there is no raw material to process, then whether we like it or not this will be considered as a Planned Downtime even if there is actually no downtime.

25. **Is maintenance backlog a good indicator?** The opinion of the author of this book is that I would not recommend this indicator since this indicator has a lot of flaws. First, the maintenance backlog is based on the due date and not on the criticality of the failure. If there are unfinished tasks on maintenance that have accumulated after some time, the decision on which to prioritize should be based on the criticality of the task and its consequences of failure, and not on the due date of the tasks.

26. **What is the best way to reduce maintenance costs for industries?** The best way to reduce maintenance costs in industries is not to focus on it at all. The focus of maintenance should be to preserve, sustain, and improve the reliability of our equipment and assets. Once we are successful in doing this, then expect costs to reduce automatically. This is the correct way it should be done. Remember that there will be times that focusing on cutting-cost will affect reliability, a lesson we should reflect upon. Having a low maintenance cost is always a consequence of good maintenance practice.

11.2: Tips on Maintenance Indices

- There is no question about having highly skilled people in our maintenance team. Technically speaking, the goal of MTTR is to reduce the repair time, but the real goal of maintenance is to analyze failures by addressing the root cause of the problem. Remember, maintenance is not measured by how fast they repair but by how they were able to analyze the failure.

- MTBF should not be used as a baseline for determining the frequency of replacement of parts and overhauls in our Preventive Maintenance since MTBF is only about the average. Determining the correct frequency of replacement and overhauls should be based on the useful life of the part or item, which will only be applicable for age-related failures.

- Measuring performance should be done at the beginning of any reliability improvement initiative. Key Performance Indicators are necessary to track the performance of maintenance to determine if they are moving on in the right direction.

- Have the right reasons to measure equipment performance. Set goals on what you want to achieve on these indices and aim your people to challenge them. Determine what reliability and maintenance strategies should be implemented and executed to ensure that goals will be achieved. Once the goals are achieved, set new goals, and continuously improve them.

- We can only improve if we measure what is important to us. What cannot be measured cannot be managed. Remember, if we do not measure performance, then we are just another person with an opinion, and in the real world of maintenance, opinions don't last a lifetime.

- Decide what indicators should be measured by both operators and maintenance. No one KPI or indicator will tell you everything. Have a minimum of at least 5 KPIs.

- Whether these indices will be generated by software or tracked down manually, it is important to determine the frequency or period of generating these indices. This means that these indices will be tracked either on a daily, weekly, or monthly basis. What is important is the trend of each of these indices.

- For maintenance, determine which will be your leading and lagging indicators and measure them consistently. Management should not only look at the results but at how the results were achieved.

- Never manipulate these indices to look good to management as we will never know the problem itself. What is important is that if the goals were not met, maintenance should evaluate the situation determine the causes of why these goals were not met, and do something about it.

Figure 11.2: Visibility of KPIs to Everyone

- KPIs and indices should not only be known by the manager and those who have laptops, or those that have access to CMMS but should be cascaded to everyone including the shop floor people. Electronic boards or activity boards should be visible to everyone. Location of these electronic, bulletin, or activity boards should be strategic where every employee can see like the company's cafeteria or canteen.

- These indices should be deployed to all levels of the maintenance function, from the highest down to the hands-on people on maintenance. Everyone in the organization should be aware of the company's goals and targets.

- It is recommended that teams that contributed to the goals should be recognized. It is highly recommended to recognize the team and not the individual.

- Although the operator's indices are intangible and will be difficult to assess, they can contribute to the lagging indicator. Allow these operators to showcase their improvements to their top management. This is to let them show that they are part of the improvement process of the plant.

- In calculating MTBF, only machine downtime caused by failures and breakdowns should be considered. For MTBA, only downtime caused by minor stoppages, assists, or errors should be included in the machine downtime. For MTTS, only downtime caused by the set-up, conversion, or changeover should be included. The thing is do not mix these machine downtime when calculating a specific Mean Time Indicator

- The different maintenance functions, engineering, operations, quality, EHS, and other functions should sit down and define what will constitute a breakdown or not so that we can have a consistent definition of breakdown, If this will not be done, each of us can have our own definition of failures and breakdowns which can cause variations in calculating indices such as the number of breakdowns, MTBF, MTTF, MTBA, and others.

- Have a clear distinction on what will be included and those that should be excluded as breakdown which must be made known to all users, maintenance, and other functions of the organization. The keyword here is to be consistent.

- Minor stoppages should be separated from breakdowns as these two equipment losses are entirely different. A minor stop will always be a minor stop despite the waiting time from the technician or maintenance.

- In calculating OEE, if the design speed is unknown, either perform a Time and Motion Study or base the equipment's UPH (units per hour) on the highest output obtained by the operator.

- Different functions from maintenance as well as other functions should sit down and discuss what will we include and exclude as Planned and Unplanned downtime.

- When trending MTBF on a graph on a weekly or monthly tracking, do not represent MTBF as a cumulative, but rather there should be a cut-off. An example of this is if we calculate MTBF every week, assuming that MTBF for January is 50 hours and February is 45 hours. The calculation for January's MTBF will begin from January 1 to 31, while for February it will be from 1 to 28. Do not add January's MTBF of 50 hours in February's MTBF at 45 hours making it 95 hours which is cumulative. It will give a higher value for February where the fact lies that it is actually lower.

Chapter **12**

The Conclusion on These Meaningful Measures

> *The best way to reduce maintenance cost is not to focus on it. What maintenance should focus is sustaining and improving the reliability of their equipment and assets. Remember if reliability improves, then automatically cost will go down, and it cannot be the other way around. Focusing or cost-cutting schemes are merely temporary and can have a detrimental effect on our equipment and assets permanently. This is one thing C-level people and decision makers should understand.*

12.1: Treat Maintenance as a Business

Although every industry's dream is to reduce their maintenance cost to a minimum, industries at times may be reducing their maintenance cost with the wrong means just like the management cost directive on having a cost-cutting program on everything. Hence, budgets on maintenance training are slashed, purchasing going for the lowest bidder, or not investing in Predictive Maintenance instruments and user's certification. This type of cost-cutting scheme by management usually ends up in more problems on maintenance.

Just like any business, it will require an investment, but the goal of any investment is not to produce a profit and revenue as these can be achieved by any means and even in the wrong manner. Profit must just be the secondary objective. Our primary objective is to satisfy our customers and clients. Make them happy so that they will stay and remain loyal to your company. Just imagine investing in a restaurant business, and at the end of the day, you got some leftovers, so what you do is just mix them up on the next day's dish thinking that no one will find out about it. Although, the owner might think that the leftovers were taken care of, what would you think will happen if the customers who will dine and taste your dish will say about that. Would you think they will come back again? Not only will you lose that customer, but expect them to spread their experience not only by word of mouth but also with the influence of social media. Here are some quotes on why profit should not be the main goal of any business.

- Profit is a reward for the appreciation given by a customer, it should not be the ultimate goal of the company, because if profit is the goal, then any means to get the profit can be made. From Konosuke Matsushita
- If you take care of the Quality, the profits will take care of themselves. From Masumasa Imaizumi, Musashi Institute of Technology
- A satisfied customer is your best advertisement. From Promot Batra, Management of Thoughts
- If you do build a great experience, customers tell each other about that. Word of mouth is very powerful. From Jeff Bezos, Founder, and CEO at Amazon
- The goal as a company is to have customer service that is not just the best but legendary. From Sam Walton – Founder of Wal-Mart
- Profit in business comes from repeat customers, customers that boast about your product and service, and that bring friends with them. From Dr. Edward Deming
- Business must be run at a profit or else it will die, but when anyone tries to run a business solely for profit, then also the business must die for it no longer has a reason for existence. From Henry Ford

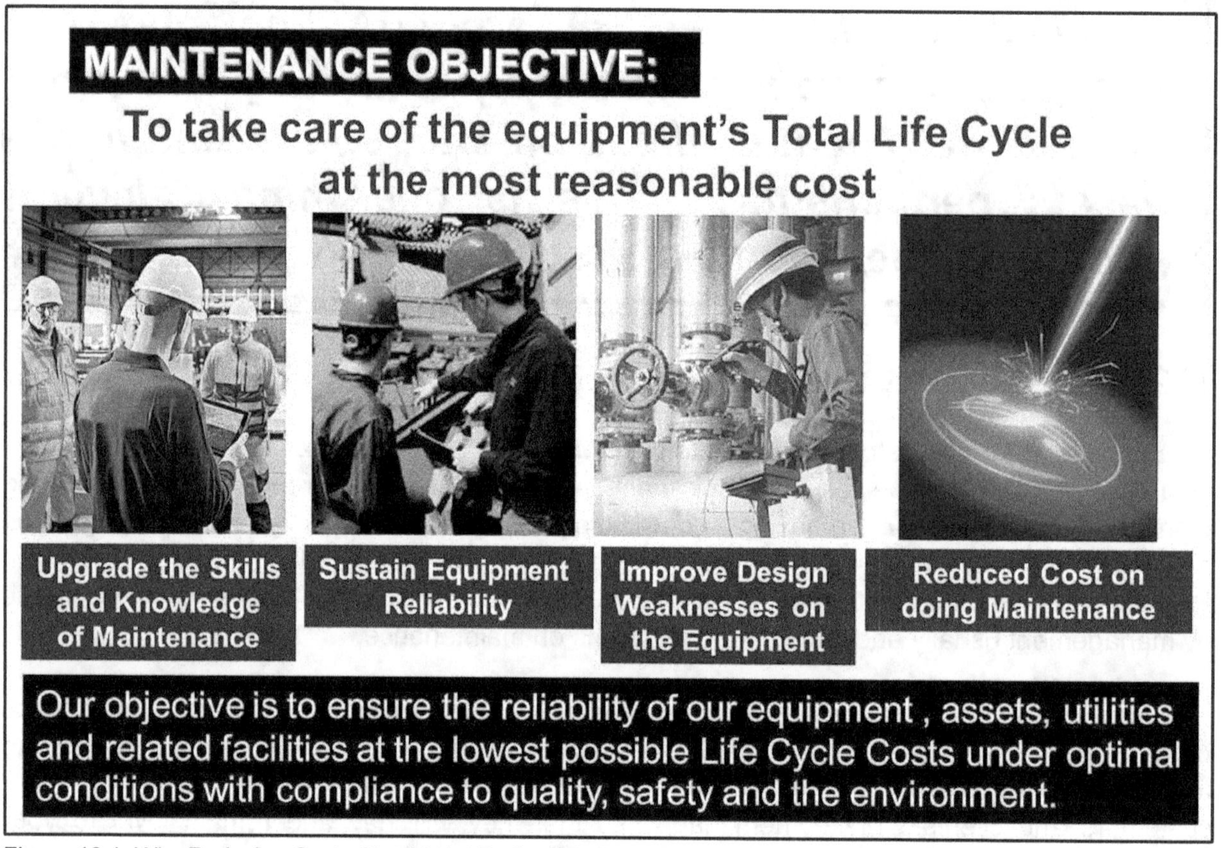

Figure 12.1: Why Reducing Costs Should Not Be the Priority on Maintenance

Just like any business, the goal of maintenance is not to reduce cost but to sustain and preserve our equipment and assets. Every maintenance should focus on preserving and improving our equipment and not reducing costs. Why? Because if we can improve the reliability of our equipment and assets, then the cost will definitely go down; it cannot be the other way around. Remember that there will be times that focusing on cost-cutting schemes will hurt reliability, a lesson we should reflect upon. Having a low maintenance cost is always a consequence of good maintenance practice.

Our primary goal in maintenance is to sustain and preserve our equipment, but everything will start on adopting the correct practices on maintenance. Knowledge is the key to everything. Knowledge will be the fundamental seed that is needed so that maintenance can develop the skill to perform their jobs correctly. This is what every industry needs to sustain and preserve its equipment and assets. Once we know how to preserve our equipment, then we move to the next level and improve the design flaws and weaknesses of our equipment. Doing this will automatically reduce our cost of maintenance. This is the correct way of reducing our maintenance costs. This process cannot be reversed, yet sad to say that in other industries, they remain reactive and slash every budget on training that their maintenance people need.

I have met many experienced maintenance people in my lifetime. During my maintenance training class most of them remain silent, but looking at the wrinkles on their faces, I know that they know the ins and out of their equipment. During coffee breaks, they approached me and tell me how wish they have met me 10 or 15 years ago. When I ask why is that? They tell me that they are already scheduled to retire next month. Their industry never bothered to invest in them during their working career, and that breaks my heart. I also met a lot of people who attended my class whom I learned later that they paid for their own training expenses and not their industry since their industry has no budget for any form of training.

Industry must value their people. To improve the equipment we need to remember one simple thing, people will be the ones that will make this happen and not technology. We cannot reverse this process One of the most important message I learned from doing TPM is that if TPM is 80 % people and 20 % equipment. Focus on the people and the people will focus on their equipment.

12.2: People are the Key In Improving These Maintenance Indices

There is a saying that people are the company's greatest asset. I tend to disagree with this statement. To me, the right people are the company's greatest asset, and the wrong people are simply called liabilities. And we can only have the right people if they are equipped with the right knowledge to do their jobs correctly right the first time around. Our maintenance, engineering, and reliability people should be both effective and efficient in performing their jobs which should have a direct link with the company's goals and targets.

Although the word effective and efficient are used interchangeably, however, there is a slight difference between the two. The word effective is that the people we hire should be capable of producing the desired result we wanted. On the other end, the word efficient is also about hiring people that can provide results, but with lesser time, money, and resources. The main difference between these two is that both can produce the result we desired, but being efficient is the capability of doing things in the simplest manner with fewer amounts of resources and energy as possible. But being effective and efficient is not the only trait that we want with our people. There is one more trait that is needed and that is being loyal and can be trusted. Meaning even if we have both effective and efficient people in our organization, however, if they have some other plans such as aiming for another industry with greener pastures, then for sure, you will lose this person in due time.

I recalled in 1994 when I was applying to a big semiconductor industry in the Philippines, I was interviewed by the operations manager. He only got 1 question for me, which I believe made all the difference on why the company hired me. The operations manager asked me, you know Rolly, this building has 4 floors, what if we go to the roof, look down on the ground, and suddenly I told you to jump, will you jump? The question caught me for a while and I looked at him in the eye and said, if you tell me to jump, then I will jump. The Operation Manager smiled at me and asked me if I will be willing to start on a Monday? I still recall vividly that interview was on a Friday. I said yes sir. As I was about to leave his office, he told me Rolly, what you don't know is that I placed a rope at your back and even if you jump, I got you. I just nod and left the room and started my first day on a Monday. I spent 8 years on that semiconductor leaving my legacy on TPM Planned Maintenance. I left that company because they abolished TPM, my story is written in my first book on World Class Maintenance Management – The 12 Disciplines. There are 7 industries in the Philippines that have passed the most prestigious JIPM TPM Awards, and I have served 4 of these seven industries. My second book, Maintenance – Roadmap to Reliability tells my story about that. That is why in my first book, the cover has a logo that states that to achieve a World Class Maintenance level, you need to have a plan, you also need to have a team, and the most important part is that your team should have a big heart in completing the journey on World Class just like any distant runner completing the iron-man race.

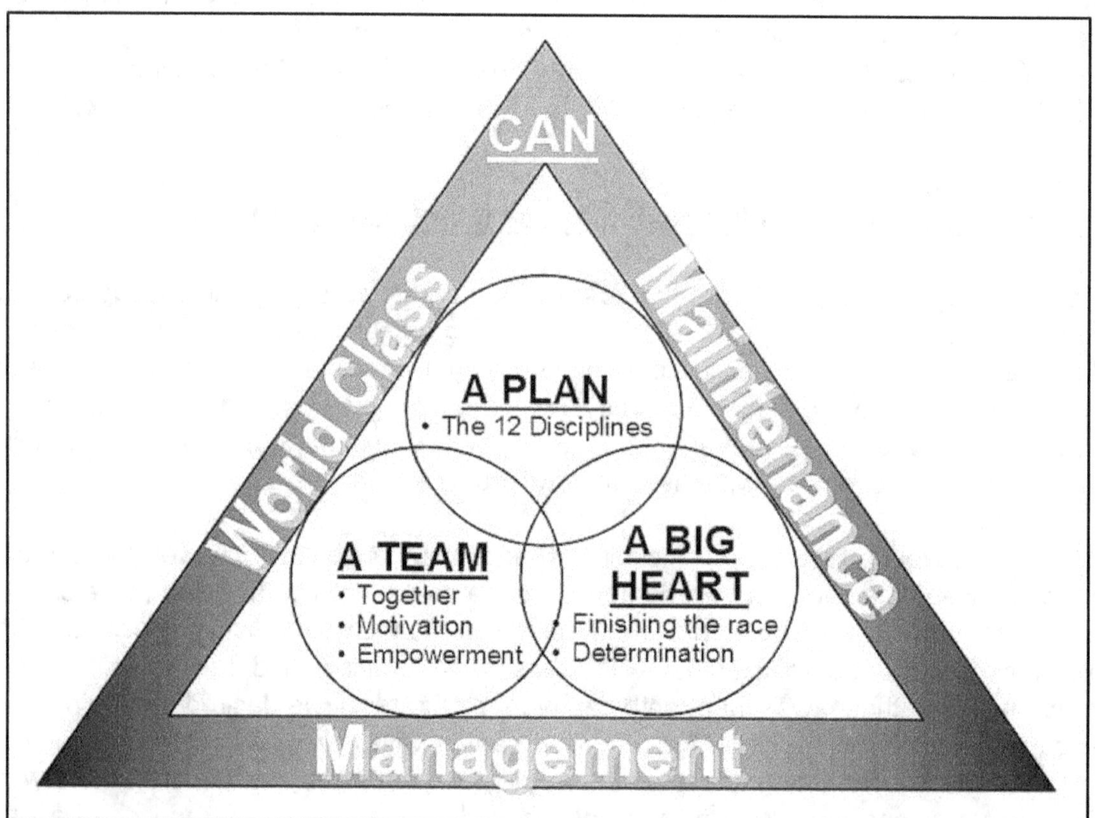

Figure 12.2: A Plan, A-Team, and A Big Heart

Hence, if I asked the reader, which would you prefer, people in your maintenance that are both effective and efficient or a trusted and loyal maintenance that will stay and complete the race together with you no matter what the odds are? I have seen and experienced many people from Predictive Maintenance who achieved level 2 or 3, then suddenly disappear leaving the plant for good because they were offered a better package by other industries. When Michael Jordan stayed with the Chicago Bulls, they won 6 NBA Championships, but when things went sour with the Bulls Management, Michael Jordan retired for the second time in 1999. He made a comeback in January 2001 and joined the Washington Wizards but never won an NBA Championship game. Jordan has no interest in winning a championship with a team other than Chicago, and he even admitted that to his Flight School campers. Michael Jordan used to host his annual Flight School camp for kids in Santa Barbara, California. In one of the Q and A events with the kids, a kid asked him if he also have the desire to win a championship with another team, and Michael Jordan replied, What team are you thinking about (crowd laughs)? No, I have never really thought about it, Jordan said. You know, I've always played in Chicago. I'd like to remain in Chicago. I'd like to keep winning championships in Chicago. No, I'm not coming to LA. No, I'm not going to Golden State. I'm sorry. That is why what good will it does when people will be leaving you midway into the process. I would prefer to work with people who will stay, complete the race, and leave a legacy. What I believe with all my heart and soul is that invest in your people as your people will be the ones that will improve their equipment, assets, and process, this cannot be reversed.

12.3: The Challenges of Today's Maintenance in This Global Pandemic

Performing maintenance today is quite challenging especially in these Pandemic times and I believe this problem will still be us for many years to come as new variants emerge from the original **Covid 19** virus before everything gets back to normal once again. This means that industries will have lesser manpower resources compared to my time working in industries when things were normal. We may not be having all our maintenance resources or if industries are working in skeletal forces or working from home.

Although in my first book, I consider acquiring the CMMS should be the last part, most industries today have their maintenance software in place, but most CMMS or EAM is only as good as what we populate in the system. And many industries have not maximized the use of their software. This would be a good time in maximizing the use of our maintenance software by populating the system with information we need so that important and relevant data needed can be retrieved with accuracy, speed and explore the boundaries on what our current CMMS is capable of doing. By coordinating and communicating with IT or those involved in the CMMS, we can update our CMMS regarding the following revisions, updates, apps, or data that needs to be populated on the system. These may include the following:

• Revisions in Procedures
• Data needed for KPIs
• Detailed Maintenance tasks for each equipment type
• PM Plans for the maintenance planner such as work details, Bill of Materials, Duration of tasks, etc.,
• Equipment Schedules for PM tasks
• Adjustments in the PM Frequencies and Intervals
• Predictive Maintenance Tasks Monitoring
• E-learning for maintenance if this exists in the plant

• Online Learning through 3rd party consultants or industry's SMEs (Subject Matter Experts)
• MRO Analysis such as FNSO (Fast, Slow, Non-Moving and Obsolete parts)
• EOQ Calculation for Fast and Slow Moving Items
• Approval process for Purchase Requisition
• Training Needs Analysis per employee and number of training hours needed
• Automating meetings through Online Platforms like Zoom and Ms. Teams
• MRO Spares Automation such as the use of Barcoding or RFID
• Software capability of generating Barcodes
• Detailed maintenance tasks for every equipment type and interval
• Detailed procedures for conversion per product type
• Videos on how to repair or trouble-shoot a particular failure
• Videos on how to perform PM for different equipment types
• Videos on how to perform greasing correctly
• Work Orders and Backlogs
• Others that we think is important that needs to be on the system
• Upgrades and Apps
• Others

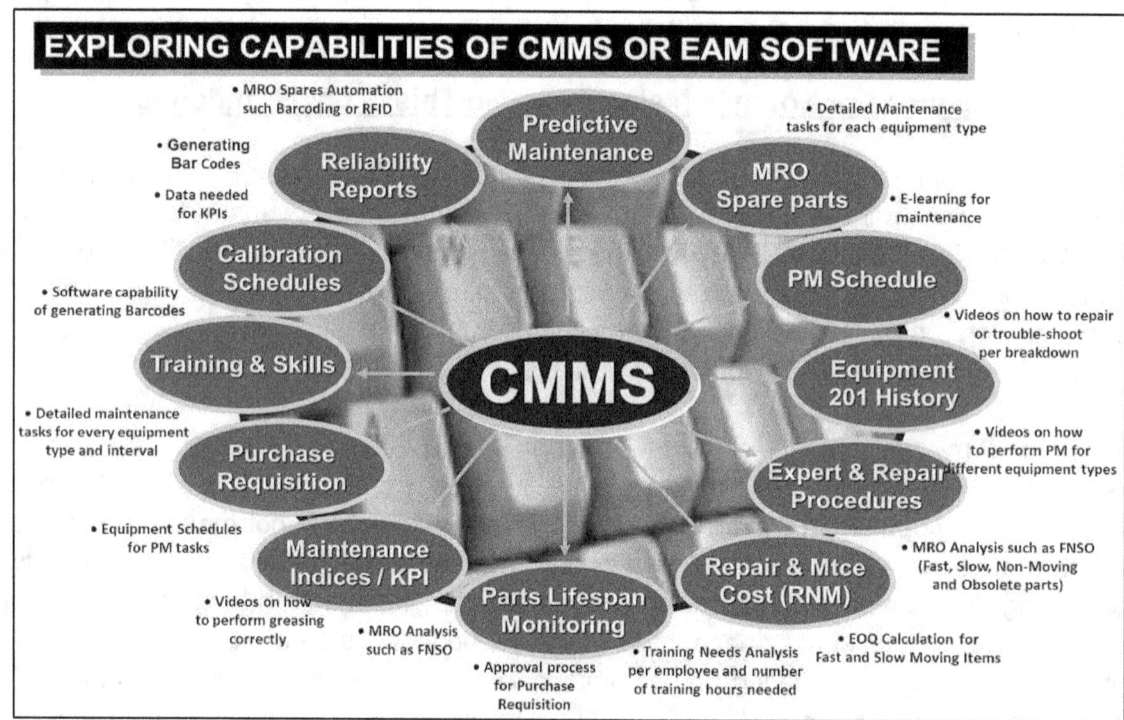

Figure 12.3: Exploring the Capabilities of Our CMMS / EAM Software

Most CMMS used by industries today are underutilized. There are many useful things that we can automate and what we need to know is to explore if our software is capable of doing it. Maintenance can brainstorm using the online platform on what they think is relevant to automate so they can simplify and streamline the overall process. Although many industries today have some form of software in the form of CMMS or EAM, the majority if not all of this software is underutilized or perhaps some maintenance or even the IT may not be aware of its entire capability, and features of what their CMMS can actually do. Perhaps it is a good time to explore the boundaries of what our software can and cannot do.

Industries must also realize the needs of their maintenance organization. As I have always said in my engagement and have written in my previous books that training is an investment that industries must provide their people. This serves as the backbone of any change initiative in the plant. This is where everything will start as we build the skills of our maintenance workforce.

12.4: The Difficulty in Standardizing These Measurements

As discussed in Chapter 3, section 3.2 of this book, the formula on Availability and Utilization have many versions. There are also variations on how we calculate the mean indicators depending on how we separate the Planned and Unplanned Downtime as well as what do we consider to be a breakdown. And the fact lies that this variation exists among the different types of industries. Even other consultants, myself included have their own versions, and terms. I am not saying that what I am writing in this book is 100% correct, but this is what I have been taught previously and how we implement and experienced using these indices. So the question is, why is there no Universal Accepted Standard for Industries? This will be a very difficult question to answer because, first, there are many types of industries. It is easy for a power plant, oil, and gas, or plants in which their equipment is interconnected or system-based to measure MTBF and when to declare a breakdown. On the other hand, for manufacturing industries, this is not the case. Almost all manufacturing industries have different types of losses that can be experienced on the equipment. For example, the SEMI E10 - Specification for Definition and Measurement of Equipment Reliability, Availability, and Maintainability (RAM) and Utilization which is usually used by semiconductor industries indicate that Set-up or conversion time will be included as a Planned or Scheduled Downtime, however, if we go to the TPM books, this is included as Unplanned or Machine Downtime. In this case, I would prefer using the original concept from TPM that set-up is a machine downtime and not a non-machine or scheduled downtime because this was what I have been taught and this was what we have been using.

The only way to standardize these measurements and formulas is if there will be a sit-down just like the La-Cosa Nostra (Mafia Mobsters) whenever the boss has some disputes. Well of course, after that, they kill each other. We just borrow the concept of the sit-down and not the killing. This means that there will be representatives of different experts if possible from all the 7 continents from around the globe. An example is the ISO 55000 Asset Management, the development of this standard was the output of 31 countries which includes its participating members from Argentina; Australia; Belgium; Brazil; Canada; Chile; China; Colombia; Czech Republic; Denmark; Finland; France; Germany; India; Ireland; Italy; Japan; Korea (Republic of); Mexico; Netherlands; Norway; Peru; Portugal; Russia; South Africa; Spain; Sweden; Switzerland; United Arab Emirates; United Kingdom; and the United States of America. There were also 11 observing members from Armenia; Austria; Hong Kong, China; Iraq; Israel; Malaysia; Morocco; New Zealand; Slovakia; and Thailand. But still, there are 195 countries all around the globe. If we include the 31 plus 11 observing countries this is just 21.5% of the whole country around the globe. Perhaps if this figure can be doubled or even 50%, then additional key and relevant information can be added to the standard. Although ISO have their standards, and a review board, consideration must also be given that there are many types of industries. If an expert or consultant from an Oil and Gas plant submitted a standard, which later on will be approved by ISO I'm sure that other experts from other countries will also have their respective invaluable inputs and contribution. Or perhaps a particular standard may only apply to a certain industry and not to all industries as there are so many types of equipment, machines, and assets. For example, TPM is originally designed for manufacturing and process plants, however, today, I know of several power plants that also implement TPM on their own. I am just not too confident if they can apply Autonomous Maintenance Initial Cleaning on their sub-stations, or for a distribution pole.

Having a standard for these maintenance measurements will take a long deliberation process just like passing a bill to congress and the senate, but this will be a sure one, and that is why I believe we can come up with a Universal Standard for these measures and KPIs if and only if there can be a consensus among these experts from around the globe.

12.5: The Integrity of Our Maintenance Indices

This book might not be 100% perfect, but this is based on what I know, what we have implemented, what I teach, and what I have experienced. This book is not just about knowing the different Meaningful Measures of Equipment Performance we can use on maintenance as indicators but going down to the very nitty-gritty details on what to include and exclude in these indicators. Even if our industry has an automated system to track these indices with just one press of a finger, the question is how did we come up with these indices such as;

• What formula are we using on Availability, especially when calculating OEE?
• If a shipment was returned to the plant because of defects, is the Quality Rate adjusted on the date the product was manufactured?
• If the machine is not loaded and sitting Idle, is the No-Load included as a Planned Downtime?
• Is maintenance in agreement with this that no load is a Planned Downtime?
• Is set-up and changeover a Planned or Unplanned downtime?
• Have we separated on what to include and exclude as Planned and Unplanned downtime?
• Have we agreed on what we shall consider a breakdown and not a breakdown? Have they been separated from the system?
• Is everyone clear on the definition of breakdown and what do we include as breakdown or not especially in calculating MTBF, and MTTR where the denominator is the frequency of failures and breakdown.
• For manufacturing industries who suffered minor stoppages in their equipment, are these losses clearly separated from the breakdown especially when the downtime is prolonged due to no available technician.
• Is it really worth measuring PM backlogs, as this indicator is based on the due date and not on the criticality of the tasks. Are we certain that the interval of performing these tasks is correct or we might be doing it too soon?
• Is it clear to the operator who encodes the codes on the downtime to separate downtime caused by breakdowns and minor stoppages if both of these losses are experienced on the equipment.

Having an accurate measurement of these maintenance indices is not just about using the correct formula or equation. It is about having a thorough understanding of what we include, and what we exclude in these indices we measure. If different people have different opinions or definitions then, one thing is for sure, we will be having different variations on the overall value of the indices we measure. One good example of this is the OEE measurement, while everyone knows that this is the product of three components which are Availability, Performance Rate, and Quality Rate, the question is, what formula do we use for Availability? Are the maintenance people in agreement that if the equipment is not loaded, then this is a planned downtime? When we compute for the MTBF, are we crystal clear on what will be included or excluded as a breakdown? And the list goes on. By being consistent on what each indices include or exclude, we can focus on how we can improve them. Even if our equipment and machines are fully automated where every downtime is recorded directly to the system, this will just be plain useless since we do not know the cause of the downtime as all we have is the time the machine is down. Even if the machine has a code for every failure, the problem is if we mix these minor stops from breakdowns, then it will not provide us an accurate MTBF since MTBF is all about the machine downtime caused by breakdowns only.

Finally, measure what you think is important with your maintenance organization. Different industries will have different measures and indices. The true value of KPIs comes from knowing not how the number is calculated but on what actually comprises or composes that particular indicator and this is made clear to all the people involved. As the late Peter Drucker quoted, the problem with management is they are measuring the wrong things. The first metric you used must drive the right behavior positively. Measuring these Meaningful Measures of Performance is important for any organization as downtime translates to a loss of productivity, lost revenue, and loss of customer confidence. Downtime in one business segment can easily have a direct impact on your customer's business. Industries can change for the better, but change is never an easy thing as most feel safe in their comfort zone. It takes guts to change. Many industries remain reactive and they will continue to remain that way unless they see a change in themselves. There is one thing I believe, and there is still hope for industries. If you can see the cover of this book, or on all my books, there is a light on the upper right side of the book cover, it symbolizes hope for industries that they can do better. While many industries will still remain trapped in the reactive world, there are a few that will start to change for the better.

Top management, C-level people, and decision-makers must understand what maintenance and reliability are and their role in their industry. Improving the way we do maintenance in the plant can be done by the maintenance function, but improving the overall reliability is not only the sole responsibility of the maintenance, engineering, and reliability people. This will include everyone in the organization starting with the Top Management people, and the only way to do this is if we can lessen the silos and barriers from each function of the organization. As my good friend R. Keith Mobley wrote in his book An Introduction to Predictive Maintenance, 17 % of problems on the equipment are caused by the maintenance function, which means that the bigger percentage of 83 % that contributes to equipment problems, defects, and failures on our equipment are beyond the maintenance function. These different functions must understand that they too affect the reliability of our equipment and assets. I hope that industries realize this message before it's too late for them, and I think that is all I have to say about that.

Appendix A: Summary of Maintenance Indicators

Summary of Maintenance Indicators

No.	OEE Indicator	Formula	Unit	Trend
1	Availability	Availability = (Available Time) - All Downtime / Available Time in %	Percentage	Higher the Better
2	Utilization	Utilization = Loading Time - Unplanned Downtime / Loading Time in %	Percentage	Higher the Better
3	Performance Efficiency	P.E. = Total Output / Operating Time x UPH	Percentage	Higher the Better
4	Quality Rate	Q. R. = Number of Items Produced - Rejected / Number of Items Produced in %	Percentage	Higher the Better
5	Overall Equipment Effectiveness (OEE)	OEE - Availability x Performance Rate x Quality Rate	Percentage	Higher the Better
6	Total Effective Equipment Performance (TEEP)	TEEP = OEE x Utilization	Percentage	Higher the Better

No.	Mean Time Indicator	Formula	Unit	Trend
7	Mean Time Between Failure (MTBF)	MTBF = Operating Time / Number of Failures or MTTR + MTTF	Hours	Higher the Better
8	Mean Time To Fail (MTTF)	MTTF = MTBF - MTTR	Hours	Higher the Better
9	Mean Units Before Assist (MUBA)	MUBA = Total Units Produced / Number of Assists	Hours	Higher the Better
10	Mean Time to Repair (MTTR)	MTTR = Total Repair Time / Number of Failures	Hours	Lower the Better
11	Mean Time to Set-Up (MTTS)	MTTS = Sum of Set-Up Time / Number of Set-up	Hours	Lower the Better
12	Mean Time to Assists (MTBA)	MTBA = Operating Time / Number of Assists	Minutes	Lower the Better

No.	Other Maintenance Indicators	Formula	Unit	Trend
13	Failure Rate	Failure Rate = 1 / MTBF	Failures / Hour	Lower the Better
14	Reliability	$R(t) = e^{-\left(\frac{1}{MTBF}\right) x \, Time}$	Percentage	Higher the Better
15	Machine Downtime Due to Breakdown	Sum of All Downtime Caused by Breakdowns	Hours	Lower the Better
16	Machine Downtime Due to Set-Up	Sum of All Downtime Caused by Set-up	Hours	Lower the Better
17	Machine Downtime Due to Assists	Sum of All Downtime Cause by Minor Stoppages	Seconds / Minutes	Lower the Better
18	Breakdown Occurrences	Total Occurences on Breakdown	Frequency	Lower the Better
19	Set-up Occurrences	Total Occurrence on Set-up	Frequency	Lower the Better
20	Minor Stoppages Occurrences	Total Occurences on Minor Stoppages	Frequency	Lower the Better

No.	Preventive Maintenance Indicators	Formula	Unit	Trend
21	PM Compliance	PM Compliance = Number of PM Tasks Completed / Total PM Tasks Listed	Percentage	Higher the Better
22	PM Effectiveness	Compare Number of PM Tasks Completed vs. Corrective Maintenance	Hours	PM High / CM Low
23	Ratio of PM to Breakdown Maintenance	Compare umber of PM Tasks Completed vs. Number of Breakdowns	Frequency	PM Tasks High / BM Low
24	Maintenance Backlog	Backlog = Total Sum of All Maintenance Work / Sum of All Man hours	Hours	Lower the Better
25	Maintenance Cost	Sum of All Incurred Cost and Expenses on Maintenance	USD ($)	Lower the Better
26	Maintenance Unit Cost	Total Maintenance Cost / Total Products Produced in %	Percentage	Lower the Better
27	Percentage of Maintenance Cost to RAV	% Maintenance Cost to RAV = (Annual Maintenance Cost ($) x 100) / RAV ($)	Percentage	Lower the Better
28	Predictive Maintenance Percentage	% PdM = Number of Potential Failures Detected / Corrected in %	Percentage	Higher the Better

No.	MRO Spare and Storeroom Indicators	Formula	Unit	Trend
29	Inventory Cost	Inventory Cost = Sum of All Inventory Cost Inside the Storeroom	USD ($)	Lower the Better
30	Number of Stock-outs	Stockout = Number of Times a Inventory of a part becomes zero	Frequency	Lower the Better
31	Total Cost of Obsolete Parts	Total Cost of All Obsolete Parts in the Storeroom	USD ($)	Lower the Better
32	Total Cost of Non-Moving Parts	Total Cost of All Non-Moving Parts in the Storeroom	USD ($)	Lower the Better
33	Total Cost of Fast Moving Parts	Total Cost of All Fast Moving and Consumables in the Storeroom	USD ($)	Lower the Better
34	Lead Time (Exponential of Average)	Lead Time = Difference between Number of days ordered and delivered	Days	Lower the Better
35	Inventory Turnover	Turnover = Total Value of Storeroom / Total Cost Withdrawn	Months	Higher the Better
36	Carrying Cost	Carrying Cost = (Purchase Order x Unit Cost x Carrying %) / 2	USD ($)	Lower the Better
37	Emergency Purchases	Total Cost of Items Orders on an Emergency Basis	USD ($)	Lower the Better

No.	Autonomous Maintenance Indicators	Formula	Unit	Trend
38	Percent Suggestions	Percent Suggestion = Total Suggestions Implemented / Total Suggestions	Percentage	Higher the Better
39	Number of Question Lists Generated	Question Lists = Total Questions Asked / Total Questions Answered in %	Percentage	Higher the Better
40	Percent Abnormalities	Percent Abnormalities = Total Abnormalities Corrected / Total Detected	Percentage	Higher the Better
41	Pecent OPL	Percent OPL = Total OPL Taught to Operators / Total OPL Generated	Percentage	Higher the Better
42	Percent Kaizen	Percent Kaizen = Total Kaizen Implemented / Total Kaizen Activities	Percentage	Higher the Better
43	Percent Contamination	Percent Cont. = Total Contamination Corrected / Total Contamination Sources	Percentage	Higher the Better
44	Percent Difficult to Clean	Percent Difficult = Total Difficult to Clean Corrected / Total Difficult Listed	Percentage	Higher the Better
45	Percent Missing Bolts	% Missing Bolts = Total Missing/Lose Bolts Corrected/Total Detected	Percentage	Higher the Better
46	Percent Leak Detected	% Leaks = Total Leaks Corrected / Total Leaks Detected	Percentage	Higher the Better

No.	World Class Maintenance Indicator	Formula	Unit	Trend
47	World Class Maintenance Indicator	Refer to Chapter 9, Section 9.5 of this book	Percentage	Higher the Better

Figure App. A1: Summary of Maintenance Indicators

Appendix B: Answers on MMEP IQ Quiz

Take Quiz on Meaningful Measures Part 1

1. a	11. b	21. d
2. Bonus	12. d	22. c
3. a	13. a	23. d
4. d	14. b	24. a
5. b	15. a	25. d
6. b	16. b	
7. a	17. c	
8. b	18. b or d	
9. c	19. c	
10. c	20. d	

Take Quiz on Meaningful Measures Part 2 (True of False)

1. b	11. a	21. b
2. b	12. b	22. b
3. a	13. a	23. a
4. b	14. b	24. a
5. a	15. a	25. b
6. a	16. b	
7. b	17. a	
8. a	18. b	
9. b	19. b	
10. b	20. b	

Chapter 2.11: Take Quiz on Equipment Losses

1. b	6. a	11. f
2. a	7. a	12. c
3. d	8. e	13. d
4. f	9. b	14. c
5. c	10. d	15. d or e

Chapter 3.11: Take Quiz on OEE

1. b	6. b	11. b
2. a	7. a	12. a
3. b	8. a	13. a
4. a	9. b	14. a
5. a	10. b	15. b

Chapter 4.2: Take Quiz on Will You Consider This a Breakdown or Not

1. a	16. a	31. b
2. b	17. b	32. b
3. b	18. a	33. a
4. a	19. a	34. b
5. b	20. b	35. a
6. b	21. a	
7. a	22. b	
8. a	23. b	
9. a	24. b	
10. a	25. b	
11. b	26. a	
12. b	27. a	
13. b	28. b	
14. b	29. a	
15. b	30. a	

Appendix C: RSA Maintenance Courses

RSA Reliability and Maintenance Consultancy Firm have been around for 17 years, and through these years, we have developed more courses suited for our reliability and maintenance people in industries. Here is a complete list of maintenance courses and services that we offer in-house, public or online training in your plant which your industry can avail of.

RSA Courses on Total Productive Maintenance

1. Total Productive Maintenance (3 days)
2. Planned Maintenance 4 Phases to Zero Unplanned Breakdown (3 days)
3. Understanding Autonomous Maintenance, Operators 7 Steps to Empowerment (3 days)
4. Understanding Focused Improvement-Kobetsu Kaizen (1 day)
5. Relationship between OEE and Equipment Losses (1 day)
6. Advance Maintenance Strategies on Planned and Autonomous Maintenance (2 days)

RSA Courses on Reliability and Maintenance Strategies

7. Lubrication Strategy-Understanding Tribology, the Importance of Oil Contamination Control (2 days)
8. Reliability-Centered Maintenance for Industries (3 days)
9. Condition-Based Maintenance, Total Approach to Failure Prediction and Analysis (2 days)
10. Root Cause Failure Analysis-Understanding Equipment Failure (3 days)
11. Optimizing Equipment Reliability-Streamline RCM Approach (3 days)
12. Optimizing Preventive Maintenance Strategy (2 days)
13. World Class Maintenance Management - The 12 Disciplines (3 days)
14. Understanding MRO Spare Parts and Storeroom Management (3 days)
15. Failure Mode and Effects Analysis (FMEA/FMECA) (1 day)
16. Practical Best Maintenance Practices (3 days)
17. Advance Maintenance Leadership in TPM, RCM, LUB, and RCFA (5 days)
18. Advance Maintenance Strategies on RCM and RCFA (2 days)
19. Advance Maintenance Strategies on Lubrication and CBM (2 days)
20. Cutting-Edge Maintenance Management Strategies (3 days)
21. Effectively and Efficiently Implementing Preventive Maintenance (2 days) New
22. Understanding the Concept of Life-Cycle Management (2 days) New

RSA Courses on Reliability and Maintenance Concepts

23. Meaningful Measures of Equipment Performance - Understanding MTBF, MTTF, MTBA, MTTR, MTTS, Failure Rate, OEE, and Weibull Overview (2 days)
24. Basic Maintenance Concept, Understanding Reactive, Preventive, Predictive and Proactive Maintenance (1 day)
25. Understanding Proactive Maintenance (1 day)
26. Maintenance Best Practices on LUB and CBM (2 days)
27. Advance Maintenance Strategies on RCFA and RCM (2 days)
28. Preventive and Predictive Maintenance Strategies (1 day)
29. Proactive and Precision Maintenance Strategies (1 day)

RSA Facilitation, Guidance, and Consultation Services Includes

- Facilitation and consultation on Total Productive Maintenance implementation
- Facilitation and consultation on Root Cause Failure Analysis
- Provide guidance on starting up a Predictive Maintenance strategy in the plant
- Facilitation and consultation on Reliability-Centered Maintenance
- Conduct initial assessment on maintenance Technical Training Needs Analysis
- Facilitation on Strategic Planning for Maintenance and Reliability professionals
- Provides assessment on World Class Maintenance-The 12 Disciplines
- Consultation on setting up an Oil Contamination Control in your plant
- Implementation of TPM Planned Maintenance Pillar
- Implementation of TPM Autonomous Maintenance Pillar
- Implementation of World Class Maintenance Management-The 12 Disciplines
- Maintenance Assessment to determine where your industry currently stands

RSA Reliability and Maintenance Consultancy Firm accept in-house training services for industries for local and international overseas countries. Special arrangements can be offered for overseas countries on any of our maintenance courses selected. These courses are what industries need to improve how we maintain and sustain our equipment and assets in the plant.

The following are lists of training I offered to plants and industries concerning reliability and maintenance courses. My mission is to uplift our maintenance human resources' technical competence in industries searching for ways to achieve maintenance excellence by capturing the industry's reliability and maintenance correct practices. These courses provide in-depth details and a wealth of information regarding maintenance correct practices from the most basic to the most advanced maintenance and reliability strategies. These powerful courses have been proven by industries that the best way to reduce their maintenance cost is to sustain their equipment's reliability. With consistent focus on output and productivity and secondary to maintenance and reliability, the latter results in frequent failures, costly unscheduled repairs and unexpected downtime, and an inevitable high cost of maintenance that calls for the adoption of a more rigorous and more effective maintenance strategy that is truly world-class.

Finally, it is also undisputed that every maintenance manager's challenge is maximizing the equipment reliability through a traditional and often self-designed and centered Preventive Maintenance system. This practice is why the approach to maintenance management seems to remain reactive rather than proactive. Truly, these courses are designed for all maintenance managers, engineers, and professionals whose mandate is to optimize their equipment capacity and reliability at the lowest possible cost. In the preceding light, we designed these courses that can be tailored fit and made available to be conducted in-house in your plant for your people's wider participation. Should you be interested in any of these courses, you may reach me at www.rsareliability.com or email me at rollyangeles@rsareliability.com.

Training Package Includes:

- Training Materials on Meaningful Measures of Equipment Performance in PowerPoint (Softcopy)
- Online Certificate of Attendance
- E-book (Readable Version) on World Class Maintenance Management – The 12 Disciplines Book (314 pages)
- Lifetime Access to RSA Reliability Newsletters

About the Resource Speaker:

Rolly, is a seasoned international reliability and maintenance consultant with 30 years of solid experience in the field. He had been invited in different countries and have conducted reliability and maintenance trainings in United Arab Emirates, India, Malaysia, Indonesia, Nigeria, Bangladesh, Botswana, Brunei, Thailand, China and South Africa. His portfolio of maintenance trainings include Maintenance Management courses on TPM, Lubrication, Tribology, Condition-Based Maintenance, RCM, RCFA, Planned Maintenance, World Class Maintenance Management, The 12 Disciplines, Oil Contamination Control, Maintenance Indices and KPI's, Maintenance Management Strategies and much more. Rolly previously worked w/ Amkor Technology Philippines, as a TPM Senior Engineer, an industry engaged in the manufacture of IC products and spearheaded their Planned Maintenance initiative compose of maintenance managers and engineers. He was also responsible for the dramatic reduction of unplanned breakdowns in their TPM Journey as well as RCM implementation on their Facilities AHU units and as well as their substation equipment. Rolly had written 9 books based on his original book on World Class Maintenance Management - The Twelve Disciplines. Each of the books Rolly wrote details a specific discipline on maintenance.

Brief Course Overview:

As maintenance professionals we know the importance of these KPI's to our organization goals and objectives. The difficulty is in translating the overall company strategies to a meaningful measure of performance. The old saying that "if you can't measure it, you can't manage it " is as true for performance as for anything else. Much of what is meant by performance often appears to be simply not measurable. It's OK looking at how things happened after the event & traditional measures can do this. What a manager needs is a much more dynamic, real time view of performance as it happens. People are central to this and qualities like motivation, confidence, leadership & perception are what needs to be understood. Getting quantifiable results and dependable measurements is crucial if resource and time are to be used to good effect. As the saying goes we are what we measure and if what we measure is important then our goals will definitely be rewarded.

Meaningful Measures of Performance
IK Academy Online thru MS Teams (June 29, 2021)

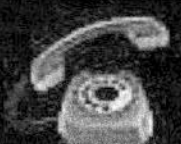

0919-451-1708
0920-482-8225
rollyangeles@rsareliability.com
marievic_angeles@rsareliability.com
http://www.rsareliability.com

Figure App. D1: MMEP Training Module

Meaningful Measures of Equipment Performance

Understanding MTBF, MTTF, MTBA MTTR, MTTS, OEE and Failure Rate

Objective of the Course :

Understand the need and importance of measuring performance of our equipment and assets.

Learn to sell performance measurement as a tool for improvement.

Learn to use meaningful measures of performance to improve equipment reliability.

Determine the right measures of performance for your industry.

Understand the different types of equipment losses and its individual relationship on Overall Equipment Effectiveness

Who Should Attend:

- Maintenance Managers
- Facilities and Utilities Managers
- Preventive Maintenance Group
- Predictive Maintenance Groups
- Reliability Engineers and Managers
- Operations and Production Managers
- Management and Decision Makers
- Continuous Improvement Groups
- People in charge of their assets
- CMMS and Spare Parts Group
- People in charge of lubrication
- Maintenance Planners
- CMMS Groups
- Decision Makers
- Maintenance Managers
- Reliability Managers
- Continuous Improvement Leaders
- Those in Charge of Reliability and Maintenance

Feedback from Delegates

We were able to understand more the very basic structure of OEE which I'm sure that will help us a lot in determining different factors affecting our equipment and its performance. *From Ryan Atinsod, EE, Sunpower Corp..*

The training is so much interesting. It will really benefit every maintenance personnel. Thank you so much to Mr. Rolly Angeles for giving his time and effort to share his knowledge related to the maintenance and sustaining the efficiency and effectiveness of machine/equipment. May God bless him even more. *From Michael Ballesteros equipment Technician, Sunpower Corporation*

I have learned the different means of getting the overall effectiveness of the equipment and how yo improve its availability thru the different methods of analysis regarding breakdowns and losses. *From Eutemar Poyaoan, Sunpower*

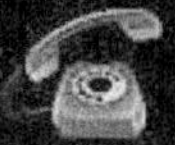
0919-451-1708
0920-482-8225
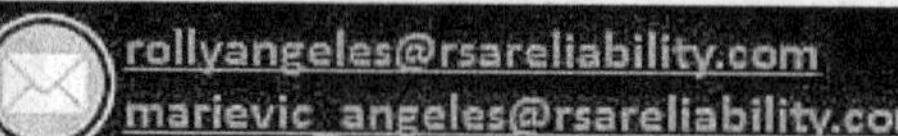
rollyangeles@rsareliability.com
marievic_angeles@rsareliability.com

http://www.rsareliability.com

Figure App. D2: MMEP Training Module

Figure App. D3: MMEP Training Module

Appendix E: Previous MMEP Training Classes Conducted

Figure App. E1: July 26, 2016, MMEP Training at Amkor Technology, Sucat, Philippines

Figure App. E2: May 30, 2017, MMEP Training at UPPC, Bulacan, Philippines

Figure App. E3: July 25, 2017, MMEP Training RSA Public Training, El Cielito Hotel, Philippines

Figure App. E4: November 30, 2020, MMEP Online Training with IK Academy, Malaysia

Appendix F: MMEP Feedback and Testimonies

• Very relevant to our needs especially when starting our operations for the start of FY 2008, helps us distinguish the difference between different measurement methods. *From Rowin Halayhay, Engineer 3*

• The facilitator shares his personal experiences with relation to the subject matter which makes the training more interesting. *From Ronald Campos, Engineer*

• Well verse and adept regarding the subject matter. Has a vast wide experience to support the discussion. *From Paule Alex Caesar Senior Equipment Engineer*

• Can easily understand the topics because of the good examples and experience of the facilitator. Good interaction between the trainer and the delegates. *From Joel Chan, Engineer*

• MTBF and MTTF were clearly differentiated and well understood. The quality loss was also defined clearly understood and differentiated from breakdown. *From Joseph Tomanan, Senior Engineer*

• The trainer knows the methodology of the subject matter and knows how to impart his knowledge to the trainees. The relevance of the subject to the employee is achieved in this training. *From Gilbert Manguerra, Engineer 3*

• Experienced speaker. Many true-to-life data were presented which are very good references and very professional presentation materials. *From Joselito Figuracion,*

• The facilitator discussed clearly the topic and he motivated us to apply the method of MTBF, MTTR, and other indices activities that will help our colleagues and company to improve further. *From Alex David, Engineer 1*

• Very good at imparting his knowledge about the subject matter and relates ideas in actual reality in manufacturing industries. His addition of humor attitudes is well appreciated as he gives examples and thoughts on the subject. Thank you. *From Ericson Caguioa*

• I have learned a lot of new concepts and proper ways of measuring the performance of the machine. With this training and seminar, my question before regarding our current proactive if it is right or wrong was answered. *From Arnel Hulipas*

• Topics are relevant to the maintenance function. Can predict/assume failure in the machine. Can accurately measure indices in machine failure. The resource person is very good at the topics. *From Mario Vergara, Maintenance Engineer, Amkor Technology*

• Good explanation of the difference between MTBF and MTTF. Enlighten us on how to use the MTTR for gauging the maintenance personnel. Good explanation on availability and utilization and who is responsible to measure it. Good explanation about MTBA and how it is properly used as an improvement tool. *From Rodante de Guzman, Amkor Technology*

• Very good, very knowledgeable, very alive. Can answer your question in every detail. You can listen to every word he said. Good teacher. Good job sir. *From Solomon Ultado, ME1, Amkor Technology*

- Topics are well explained and discussed. The facilitator is very competent and very knowledgeable in every maintenance aspect. The facilitator is also very passionate about maintenance aiming for a world Class Maintenance Management. *From Gilbert Valenzuela, Maintenance Engineer, Amkor Technology*

- Able to maintain my attention and interest throughout the topic. Learn many things regarding maintenance KPI. Very knowledgeable on the topics being discussed. *From Alfie Morales, Maintenance Engineer, Amkor Technology*

- This training helps me a lot to know and understand the basic function of maintenance and how to maintain a piece of equipment running in good condition over time. That it is not the sole responsibility of PM personnel to maintain the good condition of equipment but rather all production personnel have a part in it too. *From Marvin Bautista, ME Supervisor, Amkor Technology*

- Training is very relevant to maintenance organizations in attaining world-class indices. Key performance indices can be monitored properly through the given KPI. *From Ted Tanchuan, Engineer, Amkor Technology*

- Very informative and relevant to my job function as maintenance with practical samples useable and applicable in the line. Refreshing the basic guides in addressing our day-to-day maintenance frameworks. Interactive and practical samples shared with actual details. *From Jasper Buenaventura, Senior Engineer, Amkor Technology*

- Superb knowledge of the training module. Approachable on the questions asked and good in making the analogy. *From Jeremias Deduyo, Equipment Engineer, Amkor Technology*

- Able to understand clearly the maintenance indicators such as MTBF, MTTR, OEE, and others. *From Marvin Lustiva, Maintenance Engineer II, Amkor Technology*

- Machine maintenance and performance improvement is a continuous activity that may take time to achieve the goal. Learned that MTBF study should be performed by MEE team with the good focus of time and dedication. Root Cause Failure Analysis should have objective evidence. If a problem or trouble happens in the machine stop and escalates start investigation first before fixing to avoid destroying the evidence. *From Ricardo Ignacio, ME, Amkor Technology*

- Able to understand the usage of Mean Time Indicators that affects our equipment. This training helps me to apply this tool to our line. *From Alexander Guigab Jr., Maintenance Engineer, Amkor Technology*

- Topics apply to our current position. Understand and correct previous ideas about MTBA and MTBF. The resource person is an expert in the topics being discussed. *From Michael Pilueta, CEET, Amkor Technology*

- Superb visual presentation. Excellent ice breakers. Speaker has overall expertise in his lecture and the speaker has a great analogy to explain his subject. *From Richard Garcia, Engineer/Trainer, Amkor Technology*

- The subject matter covers all of the needed MEE indices that we need in our line to help us improve the line. This training helped me to distinguish the differences regarding the different

MEE indices and when to use them. The training also helped us what factors are needed in using these MEE indices. *From Brian Zerrudo, ME Supervisor, Amkor Technology*

• Facilitator explained well the concept and content of the seminar. All the topics discussed are very important that can be used to improve our KPI. Well and good interaction among attendees. *From Grace Navoa, ME Supervisor, Amkor Technology*

• Ability to maintain the attention and interest of the audience. Ability to explain and add more information. Can explain the slides more accurately. Can give the video for example to give more idea. *From Nelson Moreno, MEE, Amkor P3*

• Better understanding of how to define breakdowns and assists. Know the importance of tools to be used on analysis of fixing breakdown and frequent stoppages. Know the maintenance indicators and measurements. *From Cesar Debondon Jr., ME Test, Amkor Technology*

• Well explained and the trainer always connects to the audience. Trainer's good sense of humor. *From Conrado Laumoc, Engineer, Amkor Technology*

• Maintenance is not just about machine maintenance but it is also about equipment effectiveness. *From Ferdie Reocera, Engineer, Amkor Technology*

• Training and topics are very much related to our current position. Very enlightening. No dull moments. Good presentation materials. I became aware that not all activities can be answered by PM activities which became the practice of Amkor. *From Carina Borja, TSSOP, Amkor Technology*

• Well verse on maintenance best practices. Good presentation materials with audio and video presentations. Good interaction between trainer and trainees. Many trivia was gained. Keep up the good work. *Name withheld*

• Lively discussion and interaction with lots of informative topics and videos. Good speaker. *From Connie Lontok, MEE, Amkor P3*

• Very good training. Very relevant to our current job. Discussion of topics is laid well in detail. *From Shiela Mendoza, Amkor Technology P3*

• All modules regarding the topics are well discussed. All topics have a good visual aid to help the listeners easily learn the definition of each module. *From Ericson Bugay, ME 1, Amkor Technology*

• Able to discuss the topic clearly and differentiate each maintenance indicator of measurement. Very good presentation materials. Knows the topics very well. *From Teddy dela Cruz, SE Supervisor, Amkor Technology*

• Able to discuss the topic clearly. Able to have the attention of his audience. Good presentation materials. Able to answer questions clearly. *From Jesus Salustiano, Supervisor, Amkor Technology*

• The topic and training are clear. Information is useful in the line. Good and superb discussion. Very knowledgeable and expert on the training and he delivers perfectly. *From MSI Runo, Amkor Technology*

- Broad knowledge on the topic. Good sense of humor and very understandable. *From Raymond Guaca, Engineer, Amkor Technology*

- All topics that were discussed are applicable in the production line. Given examples are all related to the topic. Very good speaker and can speak clearly. *From Frederick Dizon, Technician, Amkor Technology*

- Deep and well verse on maintenance KPI. Appreciate your wisdom and insights on the actual state of ATP maintenance practice. *From Carlos Aspi, Section Manager, Amkor Technology*

- Trainer ensures trainees participate and interact. The trainer always got the attention of the trainees and participants. Trainer has deep knowledge of the training module. *From Edwin de eon. Senior Engineer, Amkor Technology*

- The facilitator explained the topic very well. He even cited examples in the simplest and very understandable way. Never had a dull discussion at all. *From Rochelle Villamor MEII, Amkor Technology*

- Good in delivering the topic. Good interaction with the trainee. Very good presentation. *From Arwin John Necio, Amkor Technology*

- Knows the topic very well. The training was very alive from start to end. There was no boring moment. Always have clear examples of every topic. *From Rolly Villanueva, Amkor Technology*

- Materials used in the program were clear and easy to understand. Superb theoretical information of materials. Maintain the attention and interest of all attendees throughout the training. *Name withheld*

- Experience of the facilitator makes him discuss the topics more in detail which is beneficial to the attendees. *From Angelo Santiso, Engineer III, Amkor Technology*

- Knows very well the topic discussed. Excellent technique teaching. Excellent training modules and materials. Excellent training experience. *From Sanny de Leon, Engineer, Amkor Technology*

- Facilitator is very good and knows the topic very well. I came from C Shift but I never got bored on this training. The topic is very interesting and applicable. *From Amiel Riel, Engineer II, Amkor Technology*

- Tools and techniques teaching this training are very good. Clear to understand all materials used in this program. *From Jon Bascao, MEE, Amkor Technology*

- Lively teaching. Very informative and the facilitator is very knowledgeable. *From Wallie Oliva, Amkor Technology*

- The seminar is new to me and I learned a lot. I would definitely use some of the good points discussed in the seminar. Salamat po Sir. *From Teofilo Manlongat, O&M, Meralco*

- Very good training and I am sure if my company will support having this Autonomous Maintenance, breakdowns will be decreased as well as the machine maintenance costs. *From Vidalleon Natividad, Plant Maintenance Manager, Motolite*

- Facilitator is very knowledgeable on the subject. Training materials are very helpful in understanding the subject discussed. *From Cesar Dipasupil, Engineering and Maintenance Manager*

- The facilitator was able to present the course with the best visual aid and related videos regarding the training materials. He was able to provide good examples related to Autonomous Maintenance. There was a pre and post-examination to rate the knowledge of the trainee. There was also a group activity as well which is also helpful for the evaluation of the trainee's knowledge gained. *From Edgar Anterola, Mechanical Supervisor, Kepco Ilijan*

- The training is very detailed considering the approach is detailed step by step yet the technique of Rolly is practical. *From Anthony Bathan, Assembly Manufacturing, On Semiconductor*

- Have understood the importance and benefits of having established performance monitoring metrics in a company as a best practice and as part of having a world-class maintenance strategy. Corrected the traditional thinking between the difference in RCA and RCFA in equipment failures. *From Jeffrey Nepumuceno, Head Reliability Engineering, EDC Generation and Engineering Group*

- Very interesting and lots of additional learnings about maintenance measurements which are clearly discussed. I have now understood MTBA, MTBF, MTTR, MTTF, and other maintenance measurements. *From Cris Amel Magnaye, Test Maintenance Engineer*

- The presentation materials were clear and concise. Easy to understand. The facilitator is well experience and clears in explaining the details. Give us an overview of true measures and use of MTBF, MTTR, MTBA, MTTS, and how to use each of them. *From Joel Resultay, Plant Engineering Manager, Concepcion Carrier*

- There is no dull moment during the training. All questions were entertained well. The training was designed based on the actual plant scenario. *From Francis Mark Ramos, Plant Engineering Planner, Concepcion Carrier*

- The training is very useful for an organization that wants to measure meaningful indicators to manage and better appreciate their maintenance performance for continuous improvement. This is an excellent course for those organizations who want to aspire to have what it takes to be world-class. The energy of the facilitator had not changed. Knowing Rolly previously, he is still at his best and this is not just about training but more of advocacy to maintenance practice. *From Gary Tongco, EDC*

- Really improved my understanding of the maintenance side such as what to use tools and measurements in improving the reliability of the equipment. *From Antonio Ramirez, Test Maintenance Engineer*

- Relevant in the field of maintenance. Topics and the introduction of maintenance tools are very useful and effective. It opens up a mindset of correct and effective maintenance strategy. *From Carlo Tiolo, Maintenance Engineer*

- All the topics are discussed well. The instructor gives out information about good and best maintenance activities. He has good slides and presentations. *From John Baulo Jose, Planner, UPPC*

- Informative, based on what is really happening at my sight. He kept the class lively and ensure that everyone understand before moving to the next topic. He has a good sense of humor. The learning gained and shared is helpful to my job. *From Loui Adriano, UPPC*

- Good speaker, No, he is great. Discussed with a heart. If you think that you are right then you are absolutely wrong. Learned that cutting costs can hurt reliability. *From John Paul dela Cruz, MED, UPPC*

- The training was lively. Easy to understand while talking. Discussed and explained the topic in a way that was easy to understand. *From Joseph Darrel Salamat, Engineer, UPPC*

- Learn more about the difference between our downtime and losses. Learn what approach we need to do if we encounter that kind of problem or trouble. The facilitator had a good training technique. *From Ma. Andrea Faustino, Maintenance Officer UPPC*

- Very informative. Able to explain the difference between MTBF, MTTF, MTTR, MTBA, and MTTS. No dull moment during this training. *From Karen Jose, Supervisor, UPPC*

- Great facilitator. All the topics were well explained. We fully understand the different maintenance indicators and measurements. *From Gary Vergara, Mechanical Maintenance Supervisor, UPPC*

- Very entertaining. Lots of actual experience regarding the subject that we can also relate to. Have a good sense of humor. The facilitator has a good knowledge of the subject matter. *From Jimmy Estrada, Reliability Engineer, UPPC*

- The topic is good enough and the facilitator is excellent. This is one of the best seminars that I attended. Full of jokes related to the topic. *From Lorenzano Mariano, Supervisor, UPPC*

- The examples used were based on actual data and our own experience of the trainer that we can easily relate to. *From Noriel Angel Tuazon, Supervisor, UPPC*

- Applicable to the industry I am working with is Medical Device Manufacturing. The trainer is knowledgeable and experienced. The training materials are complete. The training venue is good. *From Joseph Contreras, Maintenance Manager, Terumo Philippines*

- This training clearly discussed each topic. All topics are mastered by the trainer. *From Rommel Terrible, Supervisor, Terumo Philippines*

- Learned the different ways how to measure equipment breakdown, common mean indicators like MTBF, MTBA, MTBI, MTTC, MTTS, MTTF as well as new ways in organizing TPM. *From Alberto Macalino. Plant Maintenance and Engineering Manager, Rombe*

- Very relevant to our company's needs. A very good and effective speaker who is indeed a subject matter expert. Very clear and direct to the point explanation and true examples. *From Felix Adre, Terumo Philippines*

- Good point about this training is that this applies to all types of industries. The situation in attendees industries was considered when discussing each topic to be understood if applicable or not to attendees industry. *From Rodel Cantos, Maintenance*

- Training discussed almost everything I need to know which I can share with my teammates on MTBF, MTTF, MTBA, MTTS, and other topics on how to compute and get the component's reliability. *From Mark John Angelo Maclang, Process Engineer*

- Very clear and understandable. A very simple but in-depth explanation. *From Ayechelle Aquino. IAM Officer, Maynilad Water*

- This training helps us to measure the failure, breakdown, repair, etc., of equipment. Through this training, we will be able to know if our current or existing procedure is really effective to maintain our equipment as well as saving money. *From Majerraine Barrera, Diagnostic Manager, Manila Water*

- Identifying and knowing what you consider as a breakdown. Being able to analyze the overall effectiveness of our equipment in identifying the primary and secondary functions of the equipment. Learned the difference between availability and utilization. *From Ephraim Emmanuel Celedonio, Shift Head, Manila Water*

- There are a lot of points we can use to determine the life cycle and interval of PM replacement and the asset nearing failure. We have more than 1 KPI that we can measure and counter-check to validate the status of the equipment. *From Bonna May Estopido, Diagnostic Manager, Manila Water*

- We have learned how to calculate MTBF and MTTF of equipment and differentiate between MTBF and MTTF. It is important to have a Key Performance Indicator in maintaining our equipment to use in solving recurrence of breakdowns. *From Marnold Celis, Diagnostic Manager, Manila Water*

- The training program introduces a better approach to a well-rounded maintenance program for the company. It will enable people to involve to develop a more comprehensive system to an efficient maintenance agenda that is both reliable and economical. It could develop better policies in terms of processes involved in maintenance which includes technical capabilities, materials and equipment inventory, and better communication of people from maintenance and operations. The training is a wake-up call for us to realize the importance of a strengthened maintenance program that is home-based. It also encourages the development of a database that would monitor and track inventories and historical conditions and activities of facilities and equipment that is key to a better maintenance program. Attendees to the program and the other modules should include technical people from operations, maintenance specifically Plant Managers and operators. Cohesion to approaches is key. MAN-COM should include these programs in new approaches towards reliability and efficiency. *From Teddy Angeles, Asset Maintenance Manager, Manila Water*

- Forecasting the component failure for better maintenance costs. If we implement these maintenance approaches, our organization will have a lot of money saved in terms of repair and lower the costs of downtime and breakdown on operations. Organizing a committee or a group to lead RCM, to build a storage and warehouse to foster the gathering of defective parts we replace. *From Remmy Conrad Nonel, Shift Head, Manila Water*

- The need for the right instrument to record values for corrective maintenance. A partnership of maintenance people and operations people to correct breakdowns and improve the system.

Learn the importance of data needed in the computation of equipment reliability. Good maintenance eliminates breakdown. *From Excelsis Acacio, Reliability Analyst, Maintenance Services, Manila Waters*

- MSD can start a program for the maintenance activities for Manila Water. MSD can create historical data of equipment failures to monitor the efficiency of the equipment. *From Kiel Ella, Plant Manager, South Grid, Manila Water*

- Weibull distribution analysis, although we currently have no software, we learn to know and experience the forecasting analysis exists. For MTBF and MTTR, I am already 4 years in the maintenance industry but this is my first time to encounter there is a way to determine the probability of failure. Reading your book is good and understandable. *From Mark Paglinawan, Diagnostic Manager, Manila Water*

- Very useful and help us in our daily job activities. The trainer knows well about the subject. The trainer was able to elaborate on problems in our daily routine. Good skills in presentation and explanation. *Name Withheld*

- Learn how to calculate OEE, MTBF, MTTF, and implementation in our workplace. Understand the difference between operating time and loading time. *From Md Hajuperi Bin Md Ali, Tenaga Nacional Berhad, Malaysia*

- Sample explanation and easy to understand. The instructor shares his industrial experience. Argument on question and answer is very open, instructor delivers on funny words. The instructor is easy to reach. *From Muhammad Firdaus Bin Muhd Yusuf, Perodua Global Manufacturing Sdn Bhd. Malaysia*

- Clear and very good explanation on how to apply MTBF, MTTR, MTBA, MTTA, and MTTS in monitoring the performance of equipment, utilities, and facilities. Funny but very useful example. Excellent in experienced sharing in several domain industries. *From Zulfiki Ishak, Head of Engineering, Pharmaniaga Manufacturing Berhad, Malaysia*

- Good day! On behalf of the Dole team, I would like to extend our appreciation for your invaluable knowledge and time during our training. Your inputs and your willingness to enlighten us are very appreciated. For the documents, you want to give us and all the freebies. ☺ It would be a very big help for us especially since we are polishing our KPI using our newly implemented SAP PM module inline also in the ongoing WCM implementation company-wide. Actually, we are finalizing our SAP-based Maintenance KPI dashboard and we are into aligning the "legacy" computations as I told you during the training and to "what should be the correct one". Having all of these inputs from yours is a very big help for our team. And I am proud and thankful to say these: The best point about this training is that the subject matter is very helpful to our organization. They are all applicable and helpful. The Trainer is very generous and well-equipped and accommodates all questions and concerns. Our 2-day time is not wasted during this training and I really learn a lot. Fortunately, I have the opportunity to apply what I have learned from this training. That's why, I am very thankful to our trainer for imparting his knowledge and wisdom, as well as to our company and superior for allowing us to attend and acquire those learnings. Again, we are grateful for your wisdom. *From Miguel Gelig Jr. Dole Philippines Inc*

Serving Maintenance Mankind Worldwide

This book is dedicated to all "Maintenance" out there in industries. My mission is to reach out to industries searching for ways to improve their maintenance human resources. I believe that the key to improving reliability is to provide our maintenance people the knowledge and education they need to maintain their equipment and assets. At my small firm, we serve maintenance mankind worldwide.

Figure E: We're Serving Maintenance Mankind Worldwide

Bibliography

- Angeles, Rolly, **Cutting Edge Maintenance Management Strategies,** Central books Printing, Philippines 2020
- Angeles, Rolly, **Decoding Reliability Centered Maintenance for Manufacturing Industries,** Central books Printing, Philippines 2021
- Angeles, Rolly, **Lubrication Tactics for Industries Made Simple,** Central books Printing, Philippines 2020
- Angeles, Rolly, **Maintenance Roadmap to Reliability**, Central books Printing, Philippines, 2016.
- Angeles, Rolly, **Problems, and Solutions on MRO Spare Parts and Storeroom,** Central books, Printing, Philippines 2020
- Angeles, Rolly, **Reliability, A Shared Responsibility for Operators and Maintenance,** Central books Printing, Philippines, 2018.
- Angeles, Rolly S. **World Class Maintenance Management – The 12 Disciplines**, Central books Printing, Philippines, 2009
- Bazovky, Igor, **Reliability Theory and Practice,** Dover Publications Incorporated, 2004
- Campbell, John Dixon, **Uptime, Strategies for Excellence in Maintenance Management,** Productivity Press, 1995
- Moubray, John, **Reliability Centered Maintenance II, Second Edition**, Great Britain: Butterworth, Heinemann, 1997.
- Mobley, R. Keith, **An Introduction to Predictive Maintenance**, Butterworth-Heineman Elsevier, 2002, Heinemann, 2002.
- Nakajima, Seiichi, **Introduction to TPM, Total Productive Maintenance**, Portland Oregon: Productivity Press, 1984
- Reason, James and Hobbs Alan, **Managing Maintenance Error, A Practical Guide**, Ashgate Publishing Limited, 2003
- Robinson, Charles, and Ginder, Andrew, **Implementing TPM, The North American Experience**, Portland Oregon: Productivity Press, 1995.
- Shirose, Kimura and Kaneda. **P-M Analysis – An Advanced Step in TPM Implementation,** Portland Oregon: Productivity Press, 1995.
- Smith, Anthony, **Reliability-Centered Maintenance, Gateway to World Class**, Mc Graw Hill Inc, 1993
- Suzuki, Tokutaro, **TPM in Process Industries,** Productivity Press, Copyright by Japan Institute of Plant Maintenance, 1992
- Wireman, Terry, **Developing Performance Indicators for Managing Maintenance**, Industrial Press Inc. 1998

<u>Glossary on Maintenance</u>

This glossary is a collection and compilation of all my books on reliability and maintenance which includes the following;
• Volume 1: World Class Maintenance Management – The 12 Disciplines
• Volume 2: Maintenance – Roadmap to Reliability
• Volume 3: Reliability – A Shared Responsibility for Both Operators and Maintenance
• Volume 4: Cutting – Edge Maintenance Management Strategies
• Volume 5: Problems and Solutions on MRO Spare Parts and Storeroom
• Volume 6: Lubrication Tactics for Industries Made Simple
• Volume 7: Decoding Reliability-Centered Maintenance Process for Manufacturing Industries
• Volume 8: RSA Reliability and Maintenance Newsletter Vault Collection, Subscriber's Edition
• Volume 9: Investigating Equipment Failures through Root Cause Failure Analysis
• Volume 10: Maintenance Indices – Meaningful Measures of Equipment Performance

8-Disciplines is an analytical problem-solving tool designed to determine the probable or most likely cause of the problem. This method can be applied to defects and equipment-related failures. This method establishes a permanent corrective action based on data and provides the probable cause of the problem.

Abnormality can be defined as any deficiency, disorder, slight irregularity, defect, bug, flaw, or any unwanted condition that could lead to other equipment problems. This is addressed during Autonomous Maintenance Step 1 activities.

Abrasive Wear is a type of wear that can be categorized by a single keyword, cutting. Abrasive wear occurs when hard particles are suspended in a fluid or projections from one surface roll or slide under pressure against another surface, thereby cutting the other surface. Abrasive wear can be either two-body or three-body abrasions.

Absolute Filtration refers to the smallest size of particle that will be removed during filtration. This means that if the filter's size is 5 microns and the rating is absolute, almost all the contaminants in the range of five microns and above should be removed by the oil filter. The filtration efficiency for absolute rating usually is high at 99% and above, which is the capability of the filter to remove contaminants.

Absolute Viscosity measures the resistance to flow when an external force is applied like a spindle driven by a motor. For example, when we buy water-based paint and open it, it takes some force to stir the paint. However, if we add water and mix it with the paint, it is much easier to stir. This means that the viscosity decreases.

Additive is any blended chemical added to a lubricating oil or grease to improve its performance and properties to protect the base. It is also mixed with the oil to counter any negative effects of the oil from heat and contamination. Additives usually represent around 1 to 30 % of the lubricant with engine oil having the most amount of additives.

Adhesive Wear occurs when a peak or asperity from one surface comes in contact with the other surface's peak or asperity. There may be instantaneous micro-welding caused by the friction or heat involved due to the lubricant's loss of film.

Advance Discipline refers to the last 2 disciplines on World Class Maintenance that includes the state-of-the-art non-destructive tools and diagnostics instruments and software such as CMMS or EAM for the maintenance requirements.

Age is the measure of a unit, item, or spare's total exposure to stress which can be expressed as the number of operating hours or other stress units since the asset started operating until their retirement or decommissioned.

Aging is the process where certain parts of the equipment deteriorate or wear out because of stress-induced over a given period. It is also termed as wear and tear or deterioration. TPM refers to this as natural deterioration.

Age Exploration Method increases the interval for Preventive Maintenance overhauling and replacement by 10 % if aging or wear-out is not yet evident when it will be overhauled or replaced.

Analytical Ferrography (ASTM D7690) is an oil analysis test where solid debris suspended in a lubricant is separated and systematically deposited onto a glass slide called a ferrogram. The slide is examined under a microscope to distinguish particle size, concentration, composition, morphology, and surface condition of the ferrous and non-ferrous wear particles

Anthropometric Errors are a type of human error that occurs because the person simply cannot fit into the space provided, cannot reach something, or is not strong enough to move or lift something. Industries aiming for multiskilling should also consider anthropometric factors to avoid human errors.

Antifoaming or Defoamer is the process of removing entrapped air or bubbles in the lubricating oil. Almost all lubricating oil systems contain some form of air. Air is found in four phases: free air, dissolved air, entrained air, and foaming. A defoamer or an anti-foaming agent is a chemical additive in oil that reduces and hinders foam formation in industrial process liquids.

Anti-Wear additives (AW) provide the protective layer on moving parts that minimize metal contact effects. Anti-wear additives are used in many lubricating oils to reduce friction, wear, and scuffing under boundary lubrication conditions, where full-film lubrication cannot be fully maintained. Anti-wear additives are also called boundary lubrication additives.

Asperity, in tribology, in a micro-scale are topographical irregularities that are similar to peaks and valleys of a solid surface. This can only be seen if the surface is enhanced through a Scanning Electron Microscope (SEM). Once the two surfaces' asperities come into contact, lubrication is displaced, and there is metal-to-metal contact and fracture later occurs.

Asset Management is defined as executing that set of processes to realize value as the organization defines it from those assets.

Attrition Rate also referred to as a churn rate, is the rate at which people leave the industry. This refers to the number of people who have left the company, divided by the average number of employees over a given period.

Autonomous Maintenance is one of the main pillars of TPM which refers to the activities in which operators perform their daily inspection, lubrication, parts replacement, minor repair and troubleshooting, accuracy checks, and so forth on their own equipment, aiming to keep the equipment in good running condition.

Availability is the proportion of time the equipment is available for use for its intended purpose, whether it is being utilized or not. The trend should be consistent and should not be lower than 98%. Calculation of availability can be reported on a daily, weekly, or monthly basis.

Available Time is given as the total time in a given period. 24 hours for one day, 168 hours for one week, and 720 hours for 30 days, 744 hours for 31 days, or 8760 hours in a year.

Barcode is a machine-readable form of information on a surface that can be scanned by a barcode scanner. They are often known as UPC codes. The barcode is read by using a special scanner that reads the information directly. This information is transmitted into a database where it can be logged and tracked.

Basic Maintenance Discipline is the fundamental activity that should be performed on the equipment before proceeding with any other advanced or specialized disciplines. It also serves as the foundation of any maintenance management strategy. This includes training, KPI, Basic Equipment Condition, Autonomous Maintenance, and Preventive Maintenance.

Bathtub Curve is a conditional probability curve representing the age-reliability relationship of certain items or characterized by early infant mortality region, a region of relatively constant reliability, and an identifiable wear-out age.

Bending Stress is the force an object encounters when subject to a load at a given point which causes the object to bend or warp. This type of stress usually occurs in objects when subjected to a tensile load.

Beta Filtration Rating is derived from the Multi-pass Test Method for Evaluating Filtration Performance of a Fine Filter Element (ISO 4572). The automatic particle counter measures the upstream particles and quantity per unit volume-of-fluid and the particle size and quantity downstream of the filter during the test.

Blending is a refinery operation that blends different component streams into various grades of gasoline. For gasoline, the fuel is blended to achieve a higher octane rating standard, creating different gasoline types. Most gasoline stations offer three octane levels, including regular, about 87, mid-grade, about 89, and premium, with an octane rating from 91 to 94.

Bottleneck is a constraint or congestion in a production operation that occurs when the products' inventory is built up quicker than the equipment can process. In manufacturing, the bottleneck equipment is those assets that frequently fail in the process, or have a lower UPH (units per hour) than other equipment in the process.

Boundary lubrication is affiliated with metal-to-metal contact between two sliding surfaces as the asperities come in direct contact with one another. This mostly happens during start-up and shutdown.

Breakdown Occurrences refer to the frequency of failure or breakdown of the system or component. The trend should be the lower, the better. This can be reported on a weekly or monthly basis. Care should be taken on what to include and not to include as a breakdown.

Breakdown Loss sometimes referred to as equipment failure loss, is a loss when the machine stops since its function completely failed. The Japanese term for breakdown is kosho.

Brittle Fracture is a fracture that involves little or no permanent deformation on metals. Many non-metals lack ductility and are subject to brittle fractures. A brittle fracture occurs when a part is overloaded and breaks with no visible distortion and deformation.

C-level Executives are high-ranking executives of a company or any organization in charge of making company-wide decision-making. C stands for chief. This will include the Chief Executive Officer (CEO), Chief Operating Officer (COO), and Chief Information Officer (CIO).

Calibration is the process in which instrumentation and equipment used in industries are monitored and maintained to ensure they continue to give accurate and reliable results and ensure that deviations are corrected.

Capital Spares are the items within inventory that are purchased as spare parts for depreciable assets (e.g., capital equipment, backup engines, and redundancies). As such, these capital spares within the inventory can depreciate. In most cases, capital spares are not included as inventories since they are part of a company's property, plant, and equipment (PPE).

Carrying or Holding Costs are the accumulated cost on parts held in stock inside the storeroom. The holding or carrying cost begins to accumulate the day the items are put inside the storeroom. Usually, this will be around 10 % to 30% of the overall costs per year.

Causal Factor in root cause investigation is any unplanned, unintended contributor to an incident or failure. It can also be stated as the contributing cause to the incident. It can also be said that the causal factor is just one of the reasons that contributed to the incident.

Chronic Losses are usually made up of a wide variety of causes and frequently occur over time. The word chronic means repeating or recurring, which can be seen in defects and failures.

Circadian Rhythm which means around the day. It comes from the Latin words circa which means about and dies which means a day. These circadian variations are governed by a biological clock located in the human brain.

Classification is used in FMEA/FMECA to highlight those failure modes with a severity rating of 9 to 10 or failures with the highest impact or consequences on the equipment or system being analyzed in the FMEA analysis.

Cleaning Materials Inventories are items and materials needed to sustain and maintain the plant's cleanliness and facilities, comfort rooms, toilets, and office spaces, used by the janitorial services of the plant.

CMMS or Computerized Maintenance Management Software is a maintenance software used for the maintenance department to streamline and automate the maintenance process.

CMMS should support these functions by capturing and automating administrative maintenance tasks and gathering relevant information to perform this process.

Complex Item refers to an item or asset whose functional failure can result from numerous failure modes.

Compression is when we applied a downward force on a vertical cylinder, the object will have an equal upward force, which is equivalent to the normal force. Once we increase the downward force, then we also increase the upward force.

Conditional Probability of Failure refers to the probability that an item will fail during a particular age interval, given that it survives to enter that interval.

Condition-Based Maintenance or Predictive Maintenance checks the equipment's actual condition using sophisticated measuring non-destructive instruments with precision accuracy. Predictive Maintenance instruments are just a higher form of the human senses. Performing inspection through the use of human senses is also included in On-Condition tasks for RCM.

Consequences of Failure result from a given functional failure at the equipment level and for the operating organization classified in the RCM analysis: which may be a hidden failure consequence, safety consequences, environmental consequences, operational consequences, and non-operational consequences.

Consistency depends on the type and amount of thickener used and the viscosity of the base oil. Consistency is also known as the resistance to deform caused by an outside force. The measure of consistency is called penetration.

Containment is a short-term action being initiated to keep the defects or failures from further damage. Containment is just a temporary solution and will not fix the problem.

Contamination in oil is anything that should not be present in the oil besides its base and additives. Contaminants may be solids, liquids, or even gases in bubbles that cause foaming.

Consumable spares are regularly consumed and used parts and items such as fasteners, seals, belts, oil filters, grease, lubricants, gloves, WD-40, rags, face masks, cleaning materials, chemicals, and so on. These items are stored in the storeroom. They are used by maintenance regularly. They are considered as fast-moving items in the storeroom.

Corrective actions are actions taken to address the probable cause of identified non-conformances or incidents, manage their consequences, and prevent or reduce the likelihood of recurrence of failure. Identification and execution of corrective measures must be done for both the short term and the long term.

Corrective Maintenance is a maintenance task that has different meanings. In the majority of industries, this refers to repairing a failure after it happened. Other industries also refer to this as Predictive Maintenance tasks which are the activities that are done when the P-F interval is nearing its functional failure. For those industries implementing TPM or Total Productive Maintenance, corrective maintenance means performing an improvement on the equipment.

Corrosion is a wear process of materials caused by chemical, electrochemical, or other reactions. The main difference between corrosion and rust is that corrosion occurs due to the chemical reaction on metal surfaces, while rust only affects iron metals exposed to air or moisture.

Corrosion inhibitor is a chemical compound substance that decreases the metal or alloy's corrosion rate when it comes in direct contact with the oil when added to the lubricating oil. Corrosion inhibitors are additives to the fluids that surround the metal or related object. The corrosive inhibitor's nature depends on the material being protected, which are most commonly metal objects, and on the corrosive agent that needs to be neutralized.

Corrosive wear is the deterioration of metal due to chemical or electrochemical reactions with its environment. Corrosion can occur when the substance is exposed to air or some chemicals while rusting mainly occurs when a metal is exposed to air or moisture.

Cortisol is a stress hormone that triggers your heart to work harder. Cortisol works with certain parts of our brain to control our mood, motivation, feelings, anxiety, and fear.

Cost avoidance focuses on taking actions that avoid incurring costs. For industries, these are measures to reduce their expenses. Any actions to avoid an increase in expenses are considered cost avoidance.

Cost-cutting refers to measures implemented by a company to reduce its expenses and improve profitability. Cost-cutting measures are typically implemented during times of financial distress for a company or during economic downturns.

Cost savings is the reduction from last year's spending on the same item. They will appear on the company's budget and in financial statements as a decrease in spending.

Cracking Process is used to maximize the effectiveness of heavier oils. Heavier oil contains large strings of hydrogen and carbon molecules. Using a catalyst, these long strings of molecules are broken into small chains that transform heavier oil into lighter fluids such as gasoline and diesel fuels.

Criticality is a measure of how important an asset is to your process. The more critical the asset, the more impact it will have when failure happens. Criticality is based on what could happen if the failure occurs.

Critical Failure involves the loss of function or damage that could directly affect safety and environmental consequences. It can also have operational consequences with devastating impacts or aftermath.

Culture means how people perceived and do these things around their plants. It refers to their common values and beliefs, while others refer to them as shared thoughts and feelings. According to Schein, culture is the pattern of basic assumptions that a given group had invented, discovered, or developed in learning to cope with its current problems of external adaptation and internal integration that had worked well enough to be considered valid. They have taught new members the correct way to perceive, think, and act concerning their day-to-day activities.

Dedicated Machines refer to manufacturing machines that produce only one type of product and there is no set-up or conversion involved.

Defect and Rework Loss is a type of equipment loss caused when defects on products are found in which the product has to be reworked or scrapped. Defects on products occur for a variety of reasons such as machine, method, raw materials, human error, or environment.

Detection in FMEA/FMECA is an assessment of the design and machinery controls' ability to detect or capture that a failure mode is occurring independently. This is also termed as confidence.

Detergents are oil additives that keep the hot metal components free from deposits and neutralize acids, forming sludge and residues. The detergents used in lubricating oil can be organic soaps and salts of alkaline earth metals such as barium, calcium, and magnesium.

Design Speed Loss is a type of equipment loss on production caused by the difference between the design or theoretical speed and its actual operating speed. This is usually given in the PM manual document provided by the OEM.

Die, Tool and Jig Losses are losses that include the cost of the physical consumption of spare parts or refurbishments of items that are used on the line. Examples of this include the cost of spares, cost of replacement, and maintenance to tooling, dies, and jigs.

Diffusional Interception is a filtration mechanism where the particles that strike through random moving gas molecules impact the medium and are held by adsorptive forces. The probability of removal is further increased by the effect of Direct Interception, Inertial Impaction, and Diffusional Interception combined together.

Direct Interception is where the contaminants are trapped in the pores between the medium fibers. It can also catch contaminants that are smaller than the pore size through bridging. In direct interception, the contaminants come in physical contact and are attached to the filter media.

Dispersants are added to the lubricant to prevent the accumulation or particle attraction of sludge and dirt in the oil. The dispersants' function is to suspend the contaminant in the oil rather than for the contaminant to settle at the bottom of the crankcase and form deposits so that it can be removed by the oil filter easily when the oil flows and circulates inside the engine system.

Dissolved water is when the water is dispersed in the oil in a homogeneous molecular solution. The gas or moisture cannot be removed from a conventional' nominal filtration; however, several absolute oil filtration on the market can remove a certain amount of moisture in the oil.

Distillation Process is a refinery process where crude oil will undergo a process called distilling or the distillation process. The oil is heated in an atmospheric distillation vessel until the crude oil reaches its boiling point, which will turn oil into vapor.

Distribution Loss is a manpower loss that includes losses that occur due to incorrect or inefficient delivery of raw materials, packaging, or products to and from the factory or the production line. Example: Incorrect delivery of materials from supplier to store, late deliveries, excessive handling of deliveries (double handling).

Downtime is the period when equipment or asset is not operating. This is the time that the equipment is unavailable for use. The equipment's unavailability may result from equipment failure, malfunction, or non-machine related such as PM shutdown. Downtime is classified as machine-related (unplanned downtime) and non-machine-related (planned downtime). It refers to the amount of time spent on doing corrective maintenance on the equipment and assets.

Dropping Point is a test on grease that determines the temperature at which the grease drips off the testing unit in a non-decomposed condition. It indicates the grease's heat resistance or up to what temperature the grease can remain firm.

Economic Consequences refer to the consequences of the failure mode, which is evident and does not affect safety and the environment and has no operational consequences. The only consequence, in this case, is the direct cost of the repair.

Economic Order Quantity (EOQ) is one technique that can be used to optimize inventory levels by ordering the right quantity at a specific time interval to minimize inventory cost but still meet the users' demands. EOQ will answer how many orders need to be placed per year and how many items to place per order.

Electronic Data Interchange (EDI) can be best used by Purchasers in automating their transactions with their vendors and suppliers. It provides a direct link between the vendor and the buyer by automating the Purchase Request and Purchase Order system directly linked with an internal system.

Empowerment means power, control, authority, or dominion. The prefix "em" means to put on to or to cover with. Empowering is the passing of authority and responsibility to an individual. Empowerment occurs when the power goes to the operators, who then experience a sense of ownership and control over their jobs.

Emulsified Water can be in the form of microscopic droplets of water distributed in the form of an emulsion. Emulsified water is when the amount of dissolved water is greater than the saturation point. When the water is in an emulsified state in lubricating oil, its color turned into a white or milk state in layman's terms.

Energy Losses are losses in the input energy which cannot be used effectively for processing. Examples can include Start-up losses, insufficient compressed air for pneumatic machines. Usually occurs on the facilities/utilities in industries.

Enterprise Asset Management (EAM) software provides a holistic view of an organization's physical assets and infrastructure throughout their entire life cycle, starting from the design, procurement, installation, commission, operations, and finally, its retirement or disposal.

Environmental Consequences mean that a failure mode has environmental consequences if it causes a loss of function or other damage, leading to the breach of any known environmental standards or regulations. This means that the consequences of the failure go beyond the industry and affect society, people, or any government laws. This is considered the worse consequence of all.

Equipment failure refers to an event in which the equipment cannot accomplish its intended function and objective. It may also mean that the equipment stopped working, is not performing as desired, or is not meeting its target expectations.

Ergonomics originated from the Greek word ergon, meaning work, and nomoi, meaning natural laws. It is the science of refining the design of products to optimize them for human use. It is also sometimes known as human factors engineering. Ergonomics is a science that deals with designing and arranging things so that people can use them easily and safely.

Erosive Wear is a type of wear due to mechanical interaction between the surface and a fluid, a multi-component fluid, impinging liquid, or solid particles. Erosive wear occurs by the impact of particles upon a surface with a resultant material loss due to fracture.

Evidence is a piece of information that supports a conclusion. Evidence in its broadest sense includes anything used to determine or demonstrate the truth of an assertion. It is the lifeblood of any Root Cause Failure Analysis investigation. The three types of evidence include people, paper, and physical evidence.

Evident failures are failures that will become evident to both the maintenance and operators when they occur independently.

External set-up includes the activities, which can be performed while the machine is still running or is running the last production lot before the actual conversion will take place.

Extreme Pressure additives (EP) is an oil additive that is usually used in gearboxes to provide a layer of film on the metal's asperities so that when the two opposite asperities meet, instead of breaking up, they will simply slip or slide each other. It is not recommended to use EP additives on yellow metals.

Fact-Finding Group is a group of people that composed the Principal Investigator and Evidence Gathering Team. Their mission is to find out the root cause of the failure, learn from it, and do something to prevent, mitigate or eliminate its occurrence.

Fail-Safe System is a system whose function is replicated o duplicated so that the function will still be available to the equipment after the failure of one of its sources, making failure tolerable and allowed to happen.

Failure is the inability of the equipment to perform its required function. The failure of a component is viewed as terminating its life. In general, failure refers to the state or condition of not meeting a desirable or intended objective. In life, failure may be viewed as the opposite of success.

Failure Finding Tasks are maintenance tasks designed for hidden or unrevealed failures such as protective devices and redundant functions. These tasks are being performed to reduce or mitigate the risks of multiple failures wherein both the protective device and protected function are both in a failed state. It is also termed as functionality inspection or Detective Maintenance.

Failure Effects describe, most likely, what happens when each failure mode occurs on its own. Writing the failure effect should compose of 20 to 60 words.

Failure Modes are the most likely, or probable cause of failure. These are not the root cause but are considered the possible or most likely cause of why the failure happened.

Failure Mode and Effects Analysis (FMEA) discipline were developed in the United States Military. The procedure for MIL-P-1629, titled Procedures for Performing a Failure Mode, Effects and Criticality Analysis, was written on November 9, 1949. It was used as a reliability evaluation technique to determine the effects of system and equipment failures.

Failure Rate is the expected rate of failure or the number of failures in a specified period. It is expressed in failures per million or billions of hours. It is also the probability of a failure in a stated unit of time and is the reciprocal of MTBF.

False Brinelling is caused by the continuous vibration of the duty equipment, thereby causing fretting damage on the bearing for standby equipment. The best way to address false brinelling is to rotate the shaft manually around 360 to 720 degrees + 60 degrees so that the ball's position changes from its position on the outer raceway of the bearing.

Fatigue Wear is a type of wear that results from a continuous cyclic slip under repetitive load applications for many thousands or millions of load cycles. Fatigue is the phenomenon leading to fracture under repeated or fluctuating stress having a maximum value less than the material's tensile strength. Fatigue fractures are progressive, which can start as micro cracks and propagate into larger ones that cause complete destruction of the components.

Fault-Tree Analysis was developed by Boeing Aerospace in the 1950s for use in the development stages of the design process. This is a mathematical tool that yields probabilities. Its primary intent is to predict the probability of a specific failure.

Filtration can be defined as the process of removing contaminants from a fluid, liquid, or gas stream through some means of a porous medium. An oil filter's function is to remove contamination from a fluid, liquid, or gas medium through some porous medium to achieve a required fluid cleanliness level.

Finished Goods Inventory refers to the finished products of the plant available for customer purchase. These products will either be shipped to their clients, picked up, or ready to sell.

Fire Point is defined as the lowest temperatures at which vapor of the material will catch fire while continue burning even after the ignition source is removed. The fire point is usually higher than the flashpoint, typically plus 10 to 24 °C or 50 to 75 °F beyond the flashpoint. The vapors produced at the flashpoint are not sufficient to ignite the fuel.

Flashpoint is the temperature at which the oil gives off vapors that can be ignited with a flame over the oil. The lower the flashpoint, the greater the oil's tendency to suffer vaporization loss at high temperatures and burn.

Flow rate is how much volume of a liquid can pass a certain point in a given time, measured in gallons per minute or liters per second.

Foaming is the formation of air and bubbles in a petroleum product, lubricant, or fuel oil that can reduce the product's effectiveness. Foaming can cause sluggish operation, air binding of oil pumps, and overflow of tanks or sumps. It can also result in excessive turbulence, improper fluid levels, air leaks, cavitation, or contamination with water or other foreign material.

Four-ball EP (Extreme Pressure) is a grease test called the 4-Ball EP, Four-Ball Extreme Pressure, 4-Ball Weld, or Load Wear Index. The Four-Ball EP test (ASTM D2596) measures the grease's ability to prevent wear during sliding contact under extreme pressure caused by heavy loads.

Fracture refers to the separation of a solid body into two or more pieces imposed upon by stress, which is usually static and at temperatures relative to the material's melting temperature. The applied stress can be considered as tensile, compressive, shear, or torsional.

Free Water can be any water or gas which is not dissolved in the lubricating oil. Free water is when the water and the oil separates. Since the water is denser and heavier than the oil, it will settle at the bottom which can easily be removed from the drain plug.

Friction is a force that is created whenever two surfaces move relatively across each other. As long as there is friction, heat is generated, and wear takes place. The area of contact will always be the point of wear. The main function of lubrication is to reduce friction.

Friction Modifier additives affect the frictional properties between two rubbing surfaces. These additives prevent scoring, reduce wear, noise, and prevent micro-pitting in industrial gear lubricants. Friction modifiers are commonly used in gasoline engine oils.

Function is the normal or characteristic actions of an item, sometimes defined in terms of performance capabilities. This is also what the users want the equipment or asset to do.

Function Loss Failure or failure of the primary function is when a failure occurs where the machine or equipment will totally stop and will definitely halt or stop operations completely.

Functional Failure is defined as the inability of an item to meet its specific performance standard. This was when the asset had failed completely. This can also be called a function-loss breakdown.

Grease is defined as a solid to a semi-fluid product of dispersion of a thickening agent in a lubricant. It comes from the Latin word Crassus, which means fat. Grease is a thick, oily lubricant consisting of inedible lard, the fat of waste animal parts, mineral or petroleum-derived or synthetic oil, plus a thickening agent.

Grease Consistency depends on the type and amount of thickener used and the viscosity of the base oil. Consistency is also known as the resistance to deform caused by an outside force. The measure of consistency is called penetration.

Hard skills are technical knowledge gained from training needed to develop the skills to perform the people's job correctly.

Hershey Number is the dynamic viscosity (η) multiplied by the speed (N) divided by the normal load (P) per length of the tribological contact. The load also means the average pressure.

Heterodyning is a process in ultrasonic analysis that translates these frequencies into the audible range. Heterodyning is the mixing of two waves, which produces both the sum and difference of their original waves, which allows the shifting of a high-frequency sound to the audible or sonic range. It converts the ultrasonic range to an audible or sonic range through this process which is audible to the human ear.

Hidden Failures are failures that will not become evident to the operator or user of the equipment when the failure occurs independently. To consider a hidden failure, a secondary failure should occur. Hidden failures will only be applicable for protective devices and redundant components.

Hidden Failures Consequences refer to the risks of multiple failures due to an undetected earlier failure of a hidden function item. This mostly refers to the aftermath when protective devices and redundancies fail.

Horizontal Replication, also called fan-out, is the process of repeating or doing the proposed tasks derived from the RCM analysis to similar equipment with the same operating context or condition. This may also apply to modification, redesign, or improvement as long as the equipment replicated also possessed the same problems.

Human Cause is the second level of Root Cause Failure Analysis, which refers to human errors, omissions, or commissions resulting from the physical roots. Either someone did something wrong, or the person did the wrong thing.

Human Nature is a concept that denotes the characteristics of human beings such as the way they think, their feelings, the way they feel and react naturally. Humans react based on their instinct.

Human Sensory Perception means humans are gifted with five senses which include smell, touch, hear, feel, and taste. For industries, forget the sense of taste. Some failure modes will give some sort of warning or symptoms that they are on the verge of failing, such as smell, excessive vibration, noise, heat, or anything that can be detected by our human senses.

Hydraulics is the transmission and control of forces and motions through the medium of fluids. It is also the science of transmitting force or motion through the medium of a confined fluid. In a hydraulic system, the power is transmitted by compressing a confined fluid. This transfer of energy takes place because a quantity of liquid is subject to pressure.

Hydrocracking is a flexible catalytic refining process that can upgrade a large variety of petroleum products. Hydrocracking is commonly applied to upgrade the heavier fractions obtained from crude oils' distillation, including their excess.

Hydrodynamic Lubrication (HL) is also termed or called full-film or fluid film lubrication. Hydrodynamic lubrication is where the film is thick enough to separate interacting surfaces to minimize friction.

Hypothesis is a term used in root cause failure analysis which means the probable or most likely causes that need to be verified. In RCM, they are termed failure modes, while in a crime scene, they are referred to as the suspects.

Industrial Internet of Things (IIoT) uses wireless smart sensors and actuators to enhance manufacturing and industrial processes by hooking them on the equipment and providing data directly to your computer.

Inertia Impaction is a process that helps remove contaminants smaller than the pore size of the filter medium. The contaminants are retained mechanically or through adsorption. This type of removing contaminants is more effective in gases than in fluids. This process is efficient for particles greater than 0.5 to 1 micron.

Infant Mortality Failures are failures, which occur at the beginning of life. Others refer to this as commissioning failures, start-up failures, or debugging failures that occur after conducting major Preventive Maintenance activities which include overhauls and replacements.

Infrared is a type of light in the electromagnetic spectrum that is not visible to our naked eyes. Our eyes can only see a small portion in the electromagnetic spectrum, the visible light spectrum, or the rainbow's colors in layman's terms. Infrared radiation lies between the visible and microwave portions of the entire electromagnetic spectrum.

Infrared Thermography is the science of actually allowing us to see the heat. An infrared inspection can detect heat that would normally be invisible to the naked eyes and represent them as an image, which we can see.

Inherent Reliability Level is the level of reliability of an item, equipment, or asset that is attainable by utilizing all the available maintenance tasks on RCM.

Initial Cleaning in Autonomous Maintenance Step 1 removes any form of unwanted abnormalities from the equipment. It consists of removing dirt, contaminants, grime, excess oil, grease, and other foreign objects that affect equipment parts and components that have accumulated over time in the equipment.

Initial task intervals also called proposed tasks, are maintenance task intervals assigned before the service maintenance program, subject to adjustments based on the actual operating experiences.

Inspection Tasks refer to scheduled tasks requiring testing, measurement, visual inspection, or human senses for detecting failure evidence done by both operators and maintenance. For inspecting hidden failures, this will default to Failure Finding tasks, while for evident failure inspection, it will default to On-Condition Tasks.

Instinct can be defined as an inborn impulse or motivation to action typically performed in response to specific external stimuli. It can be considered as an intuition, feeling, impulse, or gut feeling. This is how we react to certain situations.

Insurance Spares is a spare part that will replace a failed part in a piece of equipment whose penalty cost for downtime is very high. These parts do not become obsolete until the equipment is retired from service. In most cases, these parts are big and classified as non-moving parts.

Intangible Measurements cannot be quantified, measured, or contribute to the plant's bottom-line results. They can also be termed as qualitative measurements.

Intermediate Discipline is a discipline on world-class maintenance that refers to the different reliability and maintenance strategies that can be adopted once basic equipment conditions have been established in the asset. This includes Lubrication Management, MRO Spare Parts Management, Life Cycle Management, Root Cause Failure Analysis, Reliability, and Maintenance Strategies.

Internal Set-up are those activities that can be performed during conversion only when the machine had been totally shut down for conversion since a different product will be required to operate on the equipment.

Inventory Turnover is a ratio that measures the number of times an inventory is withdrawn or consumed in a given time by the end-user. Inventory turnover is sometimes called stock turns or stock turnover.

ISO 55000 is a standard described as the parent document of ISO 55001. It provides a general overview of asset management as a discipline and contains definitions of terms used in the ISO 55000 series of standards. This ISO standard aims to establish a clear and consistent understanding of the principles and requirements applied when developing and implementing an Asset Management System.

Karl Fisher Titration Test is an oil analysis instrument that measures the amount of dissolved moisture content in the oil. In this method, water reacts quantitatively with a Karl Fischer reagent. This reagent is a mixture of iodine, sulfur dioxide, pyridine, and methanol.

Kauro Ishikawa developed the Ishikawa or Fishbone Diagram in 1969. A fishbone is constructed by assigning the 4M's and 1E. 4M refers to man, machine, method, and materials, while 1E refers to the environment. The most common probable causes that relate to the problem are listed and grouped accordingly. The team brainstorms and focused on the most likely or probable causes of the failure.

Key Performance Indicators (KPIs) are measurements that are tracked and monitored regularly to indicate if an organization is on the right track to hit its goals and objectives.

Kinematic Viscosity is a viscosity test, which takes the time for the oil to travel through a glass orifice of a capillary tube under the force of gravity.

Kosho is the Japanese term for breakdown. It is lost time due to equipment failure that may or may not cause downtime.

Knowledge-Based Mistakes are types of mistakes that occur when someone is, confronted with a situation that has not yet occurred before and had not been anticipated. In other words, there are no rules or procedures to follow. In situations like this, the person has to decide quickly about an appropriate course of action, and a mistake occurs due to the wrong decision.

Lagging Indicators are indicators and measurements that indicate the bottom-line results. Examples of this include productivity, repair, revenue, and maintenance costs.

Lapse occurs when someone misses out on a key step in a sequence of events or activities. For example, a mechanic leaves a tool behind after working on a machine or simply forgets to fit a key component while reassembling it. If a person will do step 1, step 2, step 3, step 4, and step 5, and the person missed out on step 4 in the process, then a lapse occurred. As humans age, our memory will not be as sharp compared to when we were young, just like in our high school days. Lapse is when people tend to forget things as they age.

Latent Cause is the last level on Root Cause Failure Analysis, which refers to hidden or concealed causes that need to be exposed so industries can learn from the things that go wrong. This is about looking in the mirror and admitting that each of us is also part of the problem.

Leading Indicators are the process of how the results were achieved. These indicators influence the final outcome of the results. These are also the indicators that lead to the lagging results.

Life Data Analysis is when a product, part, or component has operated successfully, or the time it operated before it failed, measured in hours, miles, cycles, strokes, minutes, or other measures.

Life Cycle Cost is the sum of the initial costs and the running costs of the equipment. It is the sum of the overall cost of equipment throughout its entire lifespan, including the costs incurred initially during the design stage up to the stage the equipment will be put out of service, disposed of, or decommissioned.

Line Organization Loss is a loss that results from a shortage of operators on the production floor where the operator needs to operate other machines that were originally planned. This often results in a shorter break time for the operators.

Lubricating oil, also called a lubricant or lube oil, is a class of oil used in equipment, engines, and machinery types to reduce the friction, heat, and wear between mechanical components to avoid metal-to-metal contact.

Lubrication Tasks refer to the scheduled tasks to assure the existence of completeness of lubrication films on the equipment. These tasks can either be scheduled routine tasks or condition-based maintenance tasks.

Machine-Related Downtime, also known as unplanned downtime, is when the equipment is not operating due to machine-related downtimes such as breakdowns, set-up, and conversion, cutting tool change, minor stoppages, and quality defects.

Maintenance backlog is a time indicator that consists of delays in performing the scheduled maintenance works. This consists of pending scheduled planned activities allotted on the maintenance that has already passed its due date.

Maintenance cost is a universal and common measurement for all types of industries, unlike other KPIs. By definition, maintenance costs are any expenses incurred on the equipment, machines, people, or asset by the maintenance function to sustain and preserve it.

Maintenance Induced Failures are direct failures caused by maintenance itself. Examples of maintenance-induced failures are infant mortality failure or doing intrusive maintenance, incorrect overhauling practices, and inducing early failures right after endorsing the equipment back to operators. In this case, the maintenance reassembled the equipment incorrectly.

Maintenance Management is the art or science of managing maintenance resources. It is also the manner of managing to keep our physical assets in an existing state or condition.

Maintenance Prevention according to JIPM, Maintenance Prevention is defined as the use of the latest maintenance data and technology when planning or building new equipment to promote greater reliability, maintainability, economy, operability, and safety while minimizing maintenance cost and deterioration.

Management Loss includes losses caused by the waiting time that is lost due to management problems and delays such as no available MRO spare parts, lack of manpower resources, lack of insufficient utilities, and work instructions.

Mastery is the last level of skill development where the person is capable of transferring both his skill and knowledge to another person.

Marking is the process in a semiconductor industry of identifying, traceability, and distinguishing marks on the integrated circuit package at the end of line process.

Material Hardness is the measure of the material's resistance to plastic deformation. The most common method to determine the hardness of the material is the Rockwell Hardness Test, which can test the hardness of metals and alloys. Usually, a compressive force will be applied.

Maxi-Event or large-scale RCFA is performed on maxi events or failures with a large impact or consequences in the industry. It is recommended that an outsider do this event or an independent third party person act as the principal investigator to avoid bias. There will be three persons assigned as the evidence-gathering team for the maxi event, which will require a stakeholder meeting.

Mean Time between Assists (MTBA) is the average time the equipment performs its intended function between assists. It is also the productive or the operating time divided by the number of assists, errors, or minor stoppages.

Mean Time Between Failure (MTBF) is defined as the average time between failures and therefore is a measure of the trouble-free time. The common unit used is in hours. MTBF is derived from the US MIL-STD 217 for testing electronic parts. It is a reliability engineering term that means the average amount of operating time between the occurrences of breakdowns that requires repair divided by the frequency of failures. The frequency of reporting MTBF should be either on a weekly or monthly basis. MTBF trend should be the higher, the better.

Mean Time to Fail (MTTF) is the expected time to fail of a system, part, or component. It is a basic measure of reliability for non-repairable systems. It is the meantime expected until the first failure of a piece of equipment. This is also equal to the MTBF minus MTTR.

Mean Time to Repair (MTTR) is the average time required to repair a component. Other terms used are Mean Time to Restore, Mean Time to Recover, or Mean Time to React. It is also the average time required to perform corrective maintenance or repair on all removable items in equipment, product, or system.

Mean Units Before Assists (MUBA) indicates how many units were produced before an assist or minor stoppage is encountered on the equipment.

Measurement and Adjustment Loss are losses caused by frequent measurement and adjustment to prevent the recurrence of problems. Example: Excessive inspection integrated into the process as a result of poor quality and failure to find the root cause. Adjustment loss is experienced when adjusting equipment back to the standard after routine cleaning and periodic consumable changes.

Mercaptan was introduced by William Christopher Zeise in 1832. The Latin word mercurium captans means capturing mercury because the thiolate group bonds strongly with mercury compounds. These mercaptans are used as an odorant in sour gasoline into disulfides.

Micron is also known as a micrometer and is exhibited by the Greek symbol Mu or Mμ. A unit of one-micron length is equivalent to 39 millionths of an inch or 0.000039 or 0.0009906 millimeters. The human eye can see around 40 microns and beyond.

Midi-Event or Medium Scale RCFA is performed on midi or medium-scale events. A Principal Investigator leads the investigation with one or two persons assigned as the evidence-gathering group.

Mini-Event or Small Scale RCFA event is usually performed on small scale failures where the Principal Investigator will be the ones collecting the evidence. This type of RCFA will not require a stakeholder meeting.

Molding is the process of modern molding used to encapsulate plastic material components to protect IC or passive devices at the end of line process in semiconductor industries.

Moisture Vapor Transfer Rate (MVTR), sometimes called Water Vapor Transmission Rate (WVTR), is a measure of water vapor passage through a substance. It is a measure of the permeability of vapor barriers. There are many industries where moisture control is critical such as semiconductor industries.

Moment represents the effectiveness of either a rotating, bending or twisting force upon an object. All materials have their respective hardness. The hardness of the material is the measure of the material's resistance to plastic deformation.

Motor oil, also called engine oil, is a lubricant used in internal combustion engines, power cars, motorcycles, diesel engines, engine-generators, mobile, mining equipment, etc. Inside the main engine are metal parts, which move relatively against each other, and the friction between these moving parts consumes more power by converting kinetic energy into heat.

MRO Spare Part is defined as a part of a machine ready to replace an identical part if it becomes faulty. It is also defined as those parts of the machine which are kept on standby to be substituted when a part of equipment fails, a repair is required, or the part simply becomes worn out and needs to be replaced

MRO Wear and Tear Spares are spare parts that must be replaced every time the equipment undergoes a Preventive Maintenance or Shutdown, and the equipment is disassembled and re-assembled for overhauls and parts replacement.

MSG-1 Document is a working paper prepared by the Boeing 747 Maintenance Steering Group published in July 1968 under the title: Handbook Maintenance Evaluation and Program

Development (MSG1), which includes the first use of the decision diagram techniques to develop an initially scheduled maintenance program.

MSG-2 Document is a refinement of the decision diagram procedures in MSG1 published in March 1970 under the title: MSG-2 Airline/Manufacturer Maintenance Program Planning Document, which is the immediate precursor of the RCM document. This document was used to develop a scheduled maintenance program for Lockheed 1011 and Douglas DC10, military aircrafts such as Lockheed S3 and P3, McDonnell F4J.

MSG-3 Document refers to the Operator/Manufacturer Scheduled Maintenance Development developed by the Airlines for America (A4A), formerly the Air Transport Association or ATA. This document was developed in 1980, revised in 1988 and 1993, and to this day process is used for maintaining all types of civil aircraft.

Multipass Test is a controlled laboratory test where the effluent fluid is recirculated through the filter element while a new contaminant is continuously added. This test is used to determine the efficiency and the beta rating of the oil filter.

Multiple Failure is where the protected function had miserably failed because the protective device is said to be in a failed state.

Multi-purpose grease can be defined as grease combining the properties of two or more specialized greases that can be applied in more than one application. This permits the use of a single type of grease for a wide variety of applications.

Near Miss can be considered as minor accidents or close calls that have the potential for an injury, accident, property loss, or even death. A person was standing in front of a Wrecking Ball Crane. After a few minutes, the ball fell which nearly smashed the head of the person just a few feet away.

Net Operating Time is used to identify whether the equipment is operated at the stabilized speed within the unit time.

Noack Volatility test was named after Kurt Noack. It is measured by the principal European test called NOACK. It is the amount of oil lost (light molecules) over time at a given temperature and pressure. It directly impacts high engine temperature and oil effectiveness, especially on viscosity, emissions, and oil consumption. Today's oil has a NOACK volatility limit of 15 %. Oil's Volatility rate should be less than 15%.

Nominal Filtration refers to the average particle size of contaminants that will remain in the fluid after filtration. The efficiency of nominal filtration is 50%, and the beta rating is 2.

Non-Dedicated Machines are those machines in manufacturing that are designed to manufacture different products using the same machine. For example, in manufacturing, some machines are used to manufacture different products, hence, one measurement of importance will be the set-up or conversion time.

Non-Machine Related Downtime, also known as planned downtime, is when the equipment is not operating due to non-machine-related factors such as PM activities, operators break time, meetings, or any other downtime which is not caused by the machine.

Non-Maintenance Induced Failures are failures on the equipment, which are not caused by the maintenance function, but other factors such as how the equipment was operated, design errors, commissioning errors, or management cutting costs.

Non-Operational Consequences is one of the consequences of a failure mode, which does not directly affect safety, environmental, or operational consequences. The only consequences of this failure mode are the direct cost of repair and the chances of secondary damages to the equipment.

No-Scheduled Maintenance is a default task on RCM which means that there is no feasible scheduled maintenance. Failure is allowed to occur since the consequences are minimal and acceptable. Also similar to the terms run to fail, breakdown maintenance, reactive maintenance, or corrective maintenance. When the failure occurs, it will be subject to repair.

Occurrence in FMEA/FMECA is a rating according to the likelihood that a particular failure mode will occur within a specific period. Occurrence is an assessment of the likelihood that a particular failure mode will occur. This is also termed as the probability of failure to occur.

Octane Rating measures the fuel's ability to resist engine knocking caused by the air-fuel mixture. The higher the octane, the greater resistance the fuel has to knock during the combustion process. The octane ratings are a measure of the fuel's stability.

Office Supplies Inventory are inventory supplies used by the different plant offices ranging from printer ink, pencil, ball pen, bond paper, folders, envelopes, scissors, paper clips, memo pads, and so on in industries.

Oil Analysis is a maintenance management tool that allows users to monitor the oil condition for maximum equipment life and maximum lubricant drain interval. It tells us the actual condition of the oil in our equipment.

Oligomerization is a polymerization process in which a few, usually three to ten, of the basic building block of molecules are combined to form the finished product. Therefore, the product is formed with varying molecular weights and viscosities to meet a broad range of requirements.

On-Condition Tasks are maintenance tasks that entail checking the equipment's actual condition or using human senses, pressure, temperature inspections, gauges, and SPC charts.

One-Point Lesson is a learning tool for communicating standards, problems, and improvements in work processes and equipment. Workers and supervisors use one-point lessons to provide key information about everyday work and improvement opportunities.

Operational Consequences are a type of consequence where operations will be affected. The primary function of most equipment in any industry is connected to the need to earn revenue or support revenue-earning activities. A failure mode has operational consequences if it has a direct adverse effect on operational capability.

Operator Motion Loss includes losses generated due to unnecessary or excessive movement by the operator, as a result of poor layout, and work organization due to excessive walking, wasted motion, unnecessary reaching or equipment are far apart where the operator will take a longer time to transport the product to the next process in the manufacturing.

Overall Equipment Effectiveness (OEE) is the primary measurement used for plants and industries initiating Total Productive Maintenance. It is calculated by multiplying the equipment availability by its performance rate and quality rate expressed in percentage.

Oxidation is the breakdown of oil due to the extreme heat in the engine or equipment. This is the reaction between the oil and oxygen, causing acidic gases, sludge, and varnishes to form in the crankcase.

Oxidation inhibitors are added to the oil to extend its operating life. These additives are said to be sacrificial (perhaps like a suicide bomber), consumed while performing their duty of delaying oil oxidation, thus protecting the base oil. They are present in almost every lubricating oil and grease. This inhibitor aims to prevent oxygen from reacting with the oil, thus slowing its aging rate.

P-F interval is the interval between the emanation of potential failure and its decomposition to its final stage, which is a functional failure. At this stage, the equipment has already failed. The P-F interval is actually a warning period or the lead-time to failure better known as the failure development period

P-M Analysis is a problem-solving tool designed to analyze chronic losses according to the inherent principles and natural laws that govern them. It was developed by Mitsugu Kaneda, Shirose Kunio, and Yoshifumi Kimura. P stands for Phenomena and Physical. A phenomenon is a deviation from a normal to an abnormal state. M stands for Mechanism and the 4Ms.

Pareto's 80/20 Rule. Dr. Joseph Juran, a Quality Management pioneer, developed the use of the Pareto Principle for problem-solving. He worked in the US from 1930 to 1940. According to Vilfredo Pareto's observation, 80% of the land in Italy was owned by 20% of its population, where 20 percent of something is always responsible for 80 percent of the results. He recognized a universal principle he called the vital few and trivial many and placed it into writing.

Partial Functional Failure is when the asset is still functioning, but it is below the performance standards that the user wants the asset to function.

Penetration is the measure of the grease's consistency and depends on whether the consistency had been altered by handling. ASTM D217, and D1403 test, measures the penetration of unworked and worked greases.

Performance Rate is sometimes referred to as the throughput or equipment Performance Efficiency. It reflects whether the equipment is running at its full capacity or speed for individual products. Performance Efficiency measures Speed Losses and Idling and Minor Stoppages.

Physical Cause is the first level of Root Cause Failure Analysis that refers to the physical cause of why the part or component failed. This is the technical explanation of why things broke or failed. This is said to be the metallurgical aspect as to why the failure occurred. It is also referred to as the Failure Analysis. The analysis will end on the component or part level.

Physiological Factors are a type of human error that refers to environmental stress affecting human performance. These stresses can include high or low temperatures, loud or irritating noise, excessive humidity, excessive or high vibration, exposure to toxic chemicals, radiation, or working too long without adequate break time, not to mention the day-to-day pressure from the boss.

Planned Maintenance is one of the TPM pillars, aiming to improve the current maintenance activities, whose goal is to zero out all unplanned breakdowns of the equipment or asset. They are also the mentors of Autonomous Maintenance.

Planning is defined as the total process set up to ensure that the right resources and materials arrive at the right place, at the right time, and doing the right job in the right way. Planning also has a macro meaning when reference is made to an array of scheduled shutdowns for the year, the business plan, the marketing plan, budget, and others.

PLCC (Plastic Leaded Chip Carrier) is a plastic, square, surface mount chip IC package that contains leads on all four sides. The lead pins extend down and back under and into tiny indentations in the housing. PLCC is a type of integrated circuit package that can be used to enable us to be mounted on a printed circuit board directly soldered either to the board or within a socket.

Potential Failure is defined as an identifiable physical condition that indicates that a functional failure is about to occur or is in the process of occurring.

Poka-Yoke also referred to as Mistake Proofing, is mostly used by manufacturing industries, especially to help an equipment operator avoid errors and mistakes during operations. Its purpose is to eliminate product defects by preventing, correcting, or drawing attention to human errors which were developed by Shigeo Shingo, as part of the Toyota Production System, developed. It was originally termed as Baka-Yoke, which was called foolproofing or idiot-proofing. Later on, Shigeo Shingo changed it to a milder name Poka-Yoke.

Pour Point is an oil analysis test that refers to the temperature at which oil will solidify due to cold temperature making the oil no longer capable of flowing. This oil analysis test only applies to countries with winter or extremely cold seasons. This refers to the W rating for engine oils. Note that W stands for winter.

Pour point Depressants are critical substances added to the additive compound to prevent the wax formation in the base oil from forming large crystal networks that can inhibit the flow of lubricating oil at cold temperatures, usually during the winter season.

Precision Maintenance involves performing maintenance work in a consistent, precise, and industry-accepted way. If properly implemented, maintenance should yield exactly the same results no matter who is performing the tasks whether the person is the most experienced or least experienced of the craft.

Predictive means to declare or indicate something in advance, especially foretell based on observation, experience, or scientific reason. Predictive comes from the Latin word "pre," meaning before, and "diction," also from the Latin word "dicare," meaning to proclaim.

Predictive Maintenance is a type of maintenance performed on the equipment based on the equipment's actual condition with specialized diagnostic monitoring tools or Predictive Maintenance instruments to determine potential failures. Predictive Maintenance is a maintenance activity geared to indicating where a piece of equipment is on the critical wear curve and predicting its remaining useful life.

Preventive Maintenance is a type of maintenance performed on a fixed schedule or time-based interval. Preventive Maintenance is used when the parts have a wear rate and are directly related to their age.

Primary function explains why the asset was purchased. It has something to do with the volume, output, speed, and quality

Proactive-Corrective-Maintenance, which I can define as any activity done on the equipment such as overhauling, replacement, or other means resulting from Predictive Maintenance tasks before the failure is about to happen.

Proactive Maintenance is about understanding the root cause of why a part keeps on failing and performing corrective measures to eliminate the problem completely and avoiding the recurrence of failure. In Proactive Maintenance, maintenance is ahead of failure.

Protective Device refers to those devices placed in the equipment and assets to protect something. Samples of protective devices include led, beacon, sensors, alarms, emergency stops, lighting arresters, and so on. The protective device's function is to indicate that the protected function is still working and is functional.

Psychological Factors can be grouped according to those, which are unintended, and those errors that are intended. Unintended errors can be grouped into slips and lapses, while intended errors can be grouped according to mistakes and violations.

Pumpability is the ability of the grease to be pumped or pushed through a system, which is similar to the viscosity of the oil. This indicates how easy or hard the grease can flow through lines, nozzles, and grease dispensing units.

Purchasing Cost is also called the Ordering Cost. This is the sum of the fixed cost that is incurred each time an item is ordered. This includes the physical activities required to process an order. Usually, the purchasing cost can include the purchaser's salary, telephone costs, receiving clerk salary, account payable costs, internet costs, electricity, and other costs.

Quality Rate measures the loss incurred due to rejected, reworked, and return products. Quality Rate is the product of the following two factors, the number of items produced and the number of products scrapped, reworked, or rejected.

Radio Frequency Identification (RFID) refers to a technology where the digital data encoded in RFID tags or smart labels are captured by a reader through radio waves. RFID systems consist of three components: an RFID tag, or smart label, an RFID reader, and an antenna. For industries, this is usually used for MRO Spare parts.

Random Failures are failures that can occur at any given period. This means that the probability that an item will fail in any one period is the same as it is in any other period. This means that the conditional probability of failure is not constant. This is also referred to as chance failures and wear out is not identifiable which means that the item can fail at any given time or period.

Raw Materials Inventory: These are inventory parts and items used by industries to produce their finished products. If you work in the automotive industry, this will refer to the different parts needed to assemble a car.

RCFA Logic Tree diagram is a tool that uses deductive logic to guide the thought process used to draw correct conclusions. Therefore, a logic tree is a disciplined methodology that prompts the user to answer questions that will eventually identify the root cause of a failed event.

RCM Decision Diagram is an algorithm that allows the users to make a decision-making process to identify the most appropriate and feasible maintenance tasks to address a particular failure mode

RCM Default Tasks are RCM tasks available when it is not feasible to perform a Preventive or an On-Condition task to address a particular failure mode. Default tasks on RCM include Failure Finding Tasks for hidden failures, Run to Fail, Switching intervals for redundant functions, and redesign or modification.

Reactive-Corrective-Maintenance is the activity of repairing or troubleshooting the equipment after a failure occurs. This will be done after the failure happens, similar to the original definition of corrective maintenance.

Reactive Maintenance is a type of maintenance strategy that simply means fixing it only when it fails. As the saying goes, when it ain't broke, don't fix it; when it fails, then we come and fix it. Other terms used to designate Reactive Maintenance are run to fail, run to destruction, corrective maintenance, fire-fighting mode, and stop the bleeding syndrome. RCM termed this as no-scheduled maintenance.

Recurrence Prevention includes a list of things and actions that need to be done to indicate that the failure has not repeated itself. This will usually be done in the case of the physical, human, and system cause only.

Redesign or Modification means that if the rest of the maintenance tasks are not feasible and worth addressing the failure mode, then the last option for maintenance will be to resort to redesign or modification, especially when the failure mode will have safety or environmental consequences. The goal here is to eliminate or reduce the consequences of failure from happening.

Redundancy or Standby means duplicating the system or component, which is not affected if failures and breakdowns occur. Failures are allowed or being tolerated through some forms of redundancy, standby, or when the asset has some form of duplicated function.

Reforming is a process designed to increase the amount of gasoline that can be produced from crude oil. Hydrocarbons in the naphtha stream have almost the same number of carbon atoms as gasoline, but their structure is generally more complex. Reforming rearranges naphtha hydrocarbons into gasoline molecules. The other products of reforming are light gases and a high-octane gasoline blending component called reformate.

Reliability is the probability that an item will operate without failure throughout a specified interval and that the item will perform its intended function under specified operational and environmental conditions under a given and specified time.

Reliability-Centered Maintenance is a process used to determine the most feasible maintenance requirements in its present operating context or state that it is being operated. This strategy is used to define the correct maintenance tasks for equipment or asset with the aid of an RCM decision diagram or algorithm.

Repair are the activities that are done after an asset or equipment failed. These are the activities to be done to restore the equipment back to Its operating state.

Replacement Asset Value (RAV) is also called the Estimated Asset Value (EAV). It is the cost of maintaining the asset which is measured against the value of the asset. It can be said that as the percentage of the cost to replace the asset. The lower the value of RAV, the more effective we are in maintaining and preserving the asset.

Return on Investment (ROI) tries to directly measure the amount of return for a given investment, relative to the investment's cost. In calculating the Return on Investment (ROI), the benefit or return gained from an investment is divided by the cost of the investment. The result is expressed as a percentage or a ratio.

Risk is something or someone that can cause a problem, harm, injury, damage, danger, hazard, or loss to the industry. Risk is the potential for uncontrolled loss that is of definite value to an organization. It is an intentional interaction with uncertainty.

Risks Priority Number or RPN in FMEA is the sum of severity multiplied by its detection and its occurrence. This is also the sum of the consequences multiplied by its criticality, confidence, and probability.

RNM Cost refers to the total repair and maintenance cost incurred on the equipment or system for a given period. The trend we want should be the lower the RNM cost, the better. RNM costs should be tracked and reported either weekly or every month.

Roll Stability of Grease: ASTM D1831 is a grease test that assesses how stable the grease is when it is subject to operating conditions. This test allows us to determine if the grease will handle the intended load and for how long before it begins to fail.

Root Cause Failure Analysis is a basic investigative tool that allows us to understand why things went wrong by identifying the problem's basic source or origin. Root Cause Failure Analysis is the investigation process or probing on a problem, which is entirely based on the evidence unfolded on equipment-related failures. The level of root causes will include the physical, human, system, and latent cause of the problem.

Rooticians are people who are knowledgeable in conducting a Root Cause Analysis or Root Cause Failure Analysis investigation. They are the fact-finding group deployed to find the cause of a problem or incident.

Rotable Spare Parts are reusable spares or components that can be reconditioned and reused, such as motors and engines. These can be a component or inventory item that can be repeatedly restored or refurbished to be used once again on the equipment into a fully serviceable condition after it failed. The value of rotable spare parts depends on the remaining useful life of the production equipment it supports.

Ruled-Based Mistakes usually occur when people believe that they are following the correct course of action when doing specific tasks based on a specific rule or procedure, but the course of action is inappropriate.

RULER (Remaining Useful Life Evaluation Routine ASTM D-6971-04 3 and ASTM D-6810-02) is an oil analysis test that measures the remaining useful life of the lubricant. This test can be used to determine whether the oil needs to be changed or can be extended. The RULER oil analysis test will measure the remaining level of antioxidant additive in the lubricating oil.

Run to fail is a maintenance strategy, which tells us that it is time to repair the failure when a machine fails. The failure will happen first, and then maintenance reacts by repairing it. This is the very essence of reactive maintenance. Maintenance is done at a point when there is repair or actual breakdown of the equipment. This occurs when repair action is taken on a problem only when the problem results in a machine failure or breakdown.

SAE JA1011 is also known as the Evaluation Criteria for Reliability-Centered Maintenance (RCM) Processes. This document describes the criteria to indicate that any process or analysis to derive the tasks is compliant with the RCM process requirements. This also separates the classical RCM from the Streamlined RCM.

Safety Consequences: A failure consequence where a failure mode can cause injury or kill someone else. Failure modes that can cause accidents or even near-misses, as a result, will be included in this category.

Safety Supplies Inventory refers to items regularly supplied to employees by the EHS, such as gloves, hard hats, goggles, safety shoes, ear protectors, overhaul outfits used as part of the plant's uniform in conducting their regular work routine provided to operators and maintenance people working in the plant.

Saponification is the process used to develop the grease thickeners where fatty acids are used to react with an alkali to form a chemical soap.

Schadenfreude means the pleasure we gained from another person's distress, especially if they perceived the person deserved it because they engaged themselves in a bad deed.

Scheduled Discard Tasks refer to removing an item or spare to be replaced at a specified time interval before the part wears out completely. This is included as a Preventive Maintenance task. Also termed as scheduled replacement of parts. The frequency of scheduled discard tasks is governed by the age at which the item or component shows a rapid increase in the conditional probability of failure.

Scheduled Restoration Tasks are Preventive Maintenance tasks that entail re-manufacturing a single component or overhauling an entire assembly on or before a specified age limit, regardless of its condition at the time, also termed as Scheduled Rework. The frequency of a scheduled restoration task is governed by the age at which the item or component shows a rapid increase in the conditional probability of failure.

Scheduled Maintenance Tasks, also called Preventive Maintenance tasks, are scheduled tasks that need to be done on the equipment or asset at scheduled intervals. In RCM, these refer to scheduled discard and scheduled restoration tasks.

Seal Swell is one of the functions of the oil in which the oil must be compatible with the seals and must not cause the seal to crack, shrink, or degrade; rather, it must cause the seals to expand slightly to ensure proper sealing.

Secondary function refers to the other functions of the asset besides the primary function. Most assets are expected to fulfill more than one function besides their primary functions

Secondary Damage refers to the damages on the equipment that is caused by the primary failure mode. For example, a bearing seizes that it caused a fracture on the shaft or other parts due to excessive vibration.

Set-Up Loss, Conversion, or Changeover in several manufacturing industries, equipment is not dedicated, producing different product types with the same equipment. Hence, when it is time to change one product to another, it will require the time required to remove dies, jigs for one product, clean-up, prepare dies and jigs for the next product, reassemble the equipment, adjust the equipment, perform trial runs and make further adjustments until the product of acceptable quality is now obtained from the equipment.

Severity in FMEA/FMECA refers to the impact of the failure mode and its effects. Severity considers the worst scenario that can happen after a failure or breakdown takes place. It is a rating corresponding to the seriousness of the effects of a potential failure mode. This is termed as the consequence of the failure mode.

Shear force is the sum of the effect of shear stress over a surface that results in a shear strain. The important thing to consider is that the force acting on an object is parallel.

Shear Stress is also called tangential stress. Unlike both tensile and compressive stress, the force applied is not perpendicular but parallel to the area. Once a force is applied in parallel in opposite directions, it can cause the object to deform. Shear stress is a force acting parallel to a surface or to a planar cross-section of an object.

Shutdown Loss is the time when equipment is shut down for scheduled maintenance. However, shut down related work generally affects the operating time of the equipment. Shutdown-related work must be regarded as a planned downtime loss and reduction of shutdown work time which must be reduced.

Single Minute Exchange Of Dies (SMED) is a process developed by Shigeo Shingo to reduce set-up time and changeover for manufacturing. The success of this system was illustrated in 1982 at Toyota when the die punch set up time in the cold-forging process was reduced over three months from one hour and forty minutes to three minutes.

Situation is a set of things happening and the conditions that exist at a particular time and place. Merriam Webster dictionary states that situation is how something is placed concerning its surroundings. The outcome is something that follows as a result or consequence of the situation. It is a result or effect of an action, situation, or event.

Skill is the product of personal motivation and thorough training. The end result is mastery, which is used to perform one's job correctly and efficiently, developed over time.

Soft skills are training that improves our personal well-being that shapes how we work and interact with others. An example of these training includes Supervisory Skills, Leadership, Team-Building, Work Ethics, Effective Communication, Teamwork, Time-Management, and the like.

Soot is a by-product of the combustion process and can escape the piston rings through blow-by. This may be caused either by low compression, too much idling due to stop and start, such as traffic jams or air-fuel mixture. Soot can range from 3 microns and above,

Sour Crude Oil is a crude oil that contains a high amount of sulfur, which is one of the main impurities in the crude oil. It is very common to find crude oil containing sulfur impurities. When the crude oil's total sulfur content is more than 0.5%, the oil is called sour.

Spare Parts Management is the science of managing parts inside the storeroom when operations and maintenance required them in case of a breakdown that seeks replacement or Preventive Maintenance routine and replacement activities.

Speed Operating Rate is the ratio of the Theoretical Cycle Time divided by the Actual Cycle Time.

Squirrel Stores are maintenance secret hiding places for their own spare parts. Squirrel Stores are unofficial stores in which maintenance keeps their own spare parts for their own need and benefit.

Squirrelling is the act where maintenance keeps the spare parts themselves. Why does maintenance do this? It is because there are a lot of stock-outs, which means that the part reflects in the system, but physically it is out of stock.

Slip occurs when somebody does something incorrectly or does something in the wrong sequence. If a person will perform something in sequences such as step 1, step 2, step 3, step 4, and step 5, and what the person did was step 1, step 2, step 4, step 3, and step 5, the person committed a slip. This is when a human action takes place that is not intended. The occurrence of a lack of one's trains of thought is derived mostly from unconscious behavior.

Snake Oil is a quack remedy or panacea. Snake oil is a euphemism for deceptive marketing that promises its customers a silver bullet solution to all their problems. As the word implies, these are considered too good to be true products. Examples of these are lubricant additives commercialized on Home TV shopping performing tests on oil beyond your wildest imagination but only to be recognized by the Federal Trade Commission as a fraud.

Spalling is a fracture of the running surface and subsequent small, discrete material particles removal. It is the pitting or flaking away of bearing material. Spalling can occur in the inner ring, the outer ring of the bearing's rolling elements.

Spectroanalysis (ASTM D5185 and ASTM D4951) is the analysis of metal content and additive package. This test will check 19 elements and reports them in ppm or percentage. This oil analysis tests limit its test from 10 microns and below, and its disadvantage is detecting particles larger than 10 microns.

Sporadic Failures are failures that often indicate sudden large and abrupt deviations from the norm. This refers to catastrophic failures and is the opposite of chronic failures. Usually, their cause is easy to detect since there are just one or a couple of causes.

Stakeholders are a group of people in Root Cause Failure Analysis that will include any of the following; the person being accused, the person whose behavior needs to change, the person who has to spend money, the person accusing, or anyone that the Principal Investigator and evidence gathering team think is involved in the incident.

Start-Up Loss is a type of loss that occurs during starting up the equipment. Start-up loss means that the material loss is caused at the initial stage of product launching, namely the loss caused during the period from a start-up of production to the stabilized production stage.

Strategy is a careful plan or method for achieving a particular goal, usually over a long period. It also includes the carrying out of plans to achieve the goal. It is particularly a long-term plan for success. The strategy also means to maintain and build a competitive advantage over their competition.

Static friction is the force that keeps a motionless object from being pushed or pulled across a surface. The only force that is acting on this block will be the weight of the object.

Strain is the change of shape of the object after compressive or tensile stress is applied, which can also be equal to the Modulus of Elasticity (E), which is equal to the stress divided by the strain of the object. For tensile and compressive stress, the force applied is perpendicular or at 90 degrees to the area. It is also the quantity that describes the amount of deformation that can occur within a material body whenever a load is applied.

Strength in the metallurgical sense is the metal part's property to resist the part's stress imposed upon the part.

Stress is defined as the force per unit area, often considered as the force acting through a small area within a plane.

Stock-Out is when maintenance requests a part in the storeroom where there is an actual inventory in the system, but the actual or physical inventory is zero.

Storeroom's space utilization is equal to the actual space consumed by the storage and spares divided by the total space of the storeroom.

Sweetening is replacing a certain amount of oil so that new additives can be added to the existing used oil inside the equipment. Unlike changing oil, where the oil's total volume is replaced, around 25 to 30% of the used oil will be replaced with new fresh oil during the sweetening process so that new additives will be mixed with the existing oil.

Synthetic oil will fall on Group IV and V Base Stocks for lubricating oil. Synthetic oil is the result of a chemical reaction called synthesis, in which the end result is a uniformly shaped molecules that are more resistant to heat and impossible to achieve through the crude oil refining process compared to mineral-based oil, whose molecules are uneven in size.

TAN / TBN Crossover is where the value of both TAN and TBN becomes equal. When fresh oil enters service, the TAN will be low, while TBN will be high. As the engine runs, TAN will increase, and TBN will fall. When the TAN value becomes higher than TBN, the TAN/TBN crossover had been reached which means that it is time to change the oil.

Tangible Measurements or quantitative measurements are those measurements that are easy to quantify and have a direct impact on the bottom line performance of the plant.

Tensile Stress is when you have an object in a vertical position such as a cylinder, and a downward force is applied to an object, there will always be an opposite force which will be equal to the downward force. Both the upward and downward forces will cause the object to be under tensile stress.

Theoretical Cycle Time also called the Ideal or Design Cycle Time is the Ideal production rate of the machine and is sometimes referred to as the Design Speed. The reciprocal of Theoretical cycle time is the units per hour or UPH.

Tool Cutting Blade Change is an equipment loss in the equipment or machine incurred on swapping or changing any consumable tooling item when it has become worn-out, ineffective, or severely damaged.

Torsional Stress is the twisting of an object caused by force acting on the object's longitudinal axis. The twisting effect is known as the torsion. Torsional stress can lead to deformation. The twisting is called torque which is also called the twisting moment.

Total Acid Number (TAN) is a measure of the total acid concentration present in a lubricant. Occasionally, the decrease of an additive package may cause an initial increase in the TAN of the fresh oil.

Total Base Number (TBN) is a measure of alkaline concentration present in a lubricant. Engine oil is formulated with alkaline additives to combat the build-up of acids in a lubricant as it breaks down. Once alkaline additives are depleted, the lubricant no longer performs its function, and the engine is at risk of oxidation, corrosion, sludge, varnish, and other problems.

Total Functional Failure occurs when there is a total loss of function on the equipment. This can happen both on the primary and secondary functions of the equipment.

Total Productive Maintenance is a plant improvement methodology system that enables continuous and rapid improvement of the manufacturing process through employee involvement, employee empowerment, and closed-loop measurement of results. TPM is a production-driven improvement methodology designed to optimize equipment reliability and ensure efficient management of its assets. TPM aims to build up a corporate culture that thoroughly pursues production system efficiency, improvement, and Overall Equipment Effectiveness. This methodology originated in Japan.

Training gap is defined as the difference between the skills required to complete the job and the existing skillset of any particular team member. Focusing all our training based on the needs may or may not still contribute to the bottom line results.

Training Needs Analysis (TNA) is the process in which the company identifies the training and development needs of its employees so that they can do their job effectively, depending on the specific needs of that function in the organization.

Treating in oil refineries has several options for their treating processes, but the primary purpose of the majority of them is the elimination of unwanted sulfur compounds. A variety of intermediate and finished products, including gasoline, kerosene, jet fuel, and gases, are dried and sweetened. Sweetening is a major refinery treatment of gasoline that treats sulfur compounds to improve color, odor, and oxidation stability.

Tribology was coined by Peter Jost in 1966, derived from the Greek word tribos, meaning rubbing, and translating the word literally on the rubbing science. The word tribology was not widely used until the Oxford English Dictionary defines tribology as the branch of science and technology concerned with interacting surfaces in relative motion and associated matters such as friction, wear, lubrication, and design bearings.

Trim is the cutting of the dambar that shortens the leads together. The form is the forming and bending of the leads into the correct shape and position. Singulation is the cutting of the tie bars attached in the individual units to the lead frame resulting in the individual separation of each unit from the lead frame at the end of line process in the semiconductor industry.

Uptime is the time in which our equipment and asset are operating. It is also the time during which our equipment and asset are working without failure. Other terms to designate uptime includes operating time, productive time, running time, or time the equipment is utilized.

Utilization is the proportion of time that the equipment is utilized for its intended function and purpose. It is also the time the equipment is operating and running.

Varnish is an oil contaminant made of oxidized or carbonaceous adhesive material covering the engine's internal surfaces. As the oil is oxidized, these compounds form a consistent, hard, and shiny substance. Varnish occurs because of many factors, such as oxidization, friction, and heat. These are the products of the evaporation of the oil.

Velocity is the average speed at which the fluid can move from a given point. The velocity is measured in either mile per hour, feet per second, or in SI or International System of Units. This is an important factor in sizing the hydraulic lines.

Vibration can be defined by these common terms, such as swinging back and forth, oscillating, unbalance, and shaking. The vibration occurs when a machine or machine component moves from its neutral or normal position to a lower and upper extreme limit of travel. This happens because the force is applied to the rotating equipment and components. Vibration can also be a cyclic or pulsating motion of a machine or machine components from its point of rest.

Vibration analysis is one of the most common types of Predictive Maintenance techniques that analyzes these plotted vibration signals to diagnose abnormal vibration. This is the most used technique for analyzing the condition of rotating machinery.

Violation occurs when someone knowingly and deliberately commits an error. Violations fall under three categories: routine violation, exceptional violation, and acts of sabotage.

Viscosity is the measurement of a fluid's resistance to flow. Viscosity indicates the oil's capacity to lubricate by providing a thin film to separate metal-to-metal contact. An increase or decrease in viscosity can indicate many things in the oil and the equipment. Viscosity is not a perfect Oil Analysis test as we still need to perform other tests to diagnose the exact problem of why the viscosity of the oil changes.

Viscosity Index (VI) of a lubricant describes how the oil's viscosity will change depending on the temperature change. This means that if the temperatures increase, the viscosity will decrease, and vice versa. The viscosity index is a characteristic used to indicate variations in the viscosity of the lubricating oil concerning temperature changes.

Visual controls are any means or devices used to provide ease in inspection and expose early problems and deviations to the operators. It easily alerts operators of problems in the equipment. It is a business management technique that originated in Japan, where information is communicated using visual signals instead of texts or other written instructions. The design is deliberate in allowing quick recognition of the information being communicated to increase efficiency and clarity. This is also used to spot the anomaly or abnormality on the equipment more easily. Also termed as visual management.

Volatility is the property that describes the degree and rate at which the oil will vaporize under given pressure and temperature. This means that the higher the volatility rate, the higher the oil lost due to evaporation.

Waddington Effect means the less invasive the Preventive Maintenance is, the better the maintenance's outcome can be as more maintenance can lead to early failures. This can be said as the original concept of infant mortality failures.

Water Vapor Transmission Rate (WVTR) is a measure of water vapor passage through a substance. It is a measure of the permeability of vapor barriers.

Water Washout is a test on grease that refers to ASTM D 1264, which covers evaluating the resistance of lubricating grease to water washout from a bearing.

Wear is defined as damage to a solid surface caused by the removal or displacement of materials by any means of mechanical action of a contacting solid, liquid, or gas. It may cause significant surface damage, and the process is usually gradual. The damage is usually thought of as gradual deterioration, correlated with the equipment's operating age.

Wear debris can be defined as metal particles produced from the breakdown of surfaces within a machine. These particles can range from a submicron size to chunks of metal as large as those seen by the naked eye.

Wear Metal Debris Analysis Tests is an oil analysis test that indicates the number of metal contaminants present in the oil. By knowing this information, maintenance can understand what parts inside the equipment are on the verge of wearing out. Each of these elements will have its own specific limits, as specified in the oil analysis tables.

Wear-Out Failures or Age-Related Failures are parts that will eventually survive and reach a specific age before they fail. Age specified may be in running hours, time, number of strokes, calendar days, number of revolutions, number of stress applied, or any other form.

Weibull Distribution is often used to describe the lifetime of parts. It is used to analyze and predict failure rates and describe the failure of parts and equipment.

Wire-bond is a semiconductor machine responsible for adhering or welding a thin wire to a bare semiconductor die-chip pad and the other end of the wire to a conductive pad on a substrate. The wire is usually made of 99% pure gold and is usually very thin, often 0.001 to 0.0013 of an inch.

Work-in-process (WIP) Inventory comprises of all the materials, components, parts, assemblies, and subassemblies that are being processed in production or are still waiting to be processed within the system.

World Class is the ability to compete anywhere globally to meet and beat any competitor anywhere in the world in terms of product, price, quality, and on-time delivery. (Extracted from Terry Wireman's book on World Class Maintenance)

World Class Maintenance Management is the art and science of managing maintenance resources performed by the best-in-class industries worldwide.

Wrench Time is an indicator that is used to measure how much time the maintenance people actually spend on doing actual maintenance work on the equipment. Others refer to this as the tool time or when maintenance is actually holding a tool which of course is not only the wrench.

Yield Losses can be defined as the total loss between the input of raw material and the output of finished goods. This may also include the losses incurred due to reject products or those that had been scrapped.

RSA Maintenance Books Collection in Series

This is not only about a technical book on maintenance; it is a book that makes everyone in maintenance feel proud that they belong to the maintenance function. If you have been living through the day-to-day pressures of doing maintenance, then this is your story. *Rolly Angeles*

I have written these books inspired by the millions of maintenance mankind from different industries who want only the best equipment and assets to function the way we want. These books are written in series, explaining each discipline's maintenance in detail based on its original concept on World Class Maintenance, The 12 Disciplines.

• Volume 1: World Class Maintenance Management – The 12 Disciplines
• Volume 2: Maintenance – Roadmap to Reliability
• Volume 3: Reliability – A Shared Responsibility for Both Operators and Maintenance
• Volume 4: Cutting – Edge Maintenance Management Strategies
• Volume 5: Problems and Solutions on MRO Spare Parts and Storeroom
• Volume 6: Lubrication Tactics for Industries Made Simple
• Volume 7: Decoding Reliability-Centered Maintenance Process for Manufacturing Industries
• Volume 8: RSA Reliability and Maintenance Newsletter Vault Collection, Subscribers Edition
• Volume 9: Investigating Equipment Failures through Root Cause Failure Analysis
• Volume 10: Maintenance Indices – Meaningful Measures of Equipment Performance

Figure F: Rolly Angeles Reliability and Maintenance Books

Available at https://www.amazon.com/Rolly-Angeles/e/B07B3T1TXC/ref=ntt_dp_epwbk_0

RSA Reliability Website: https://rsaonlinebookstore.company.site/

Draf2Digital: https://books2read.com/ap/RWJydN/Rolly-Angeles

Index

www.ingramcontent.com/pod-product-compliance
Lightning Source LLC
Chambersburg PA
CBHW081950130726
48009CB00011B/230